Statistics and Econometrics

Second Edition

Dominick Salvatore, Ph.D.

Professor and Chairperson,
Department of Economics,
Fordham University

Derrick Reagle, Ph.D.

Assistant Professor of Economics,
Fordham University

Schaum's Outline Series

New York Chicago San Francisco Lisbon London Madrid
Mexico City Milan New Delhi San Juan Seoul
Singapore Sydney Toronto

DOMINICK SALVATORE is professor and chairperson of the Department of Economics at Fordham University in New York City. He is the author of the textbook *International Economics*, 7th ed. (2001) and *Managerial Economics in a Global Economy*, 4th ed. (2001). He also authored the Schaum's Outline Series in *International Economics*, 4th ed. (1996) and *Microeconomic Theory*, 3d ed. (1992) and coauthored, with Professor Diulio, the Schaum's Outline in *Principles of Economics*, 2d ed. (1996). His research has been published in numerous leading scholarly journals and presented at many national and international conferences.

DERRICK REAGLE is an assistant professor of economics at Fordham University and a faculty member of Fordham's Graduate Program in International Political Economy and Development. He received his Ph.D. from Vanderbilt University in 1998. Since then, he has contributed significant scholarly research in the areas of statistical hypothesis testing, financial markets, and country risk analysis. Dr. Reagle has presented statistical and applied work at numerous conferences, published in academic journals, and advised the media on economic trends. Dr. Reagle has extensive experience teaching statistics, econometrics, and finance on both the undergraduate and graduate levels. Derrick Reagle currently resides in New York with his wife, Elizabeth, and two sons, Maxwell and Jackson.

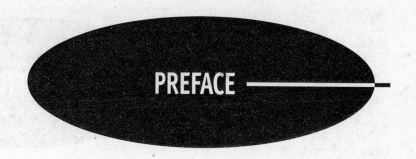

PREFACE

This book presents a clear and concise introduction to statistics and econometrics. A course in statistics or econometrics is often one of the most useful but also one of the most difficult of the required courses in colleges and universities. The purpose of this book is to help overcome this difficulty by using a problem-solving approach.

Each chapter begins with a statement of theory, principles, or background information, fully illustrated with examples. This is followed by numerous theoretical and practical problems with detailed, step-by-step solutions. While primarily intended as a supplement to all current standard textbooks of statistics and/or econometrics, the book can also be used as an independent text, as well as to supplement class lectures.

The book is aimed at college students in economics, business administration, and the social sciences taking a one-semester or a one-year course in statistics and/or econometrics. It also provides a very useful source of reference for M.A. and M.B.A. students and for all those who use (or would like to use) statistics and econometrics in their work. No prior statistical background is assumed.

The book is completely self-contained in that it covers the statistics (Chaps. 1 to 5) required for econometrics (Chaps. 6 to 11). It is applied in nature, and all proofs appear in the problems section rather than in the text itself. Real-world socioeconomic and business data are used, whenever possible, to demonstrate the more advanced econometric techniques and models. Several sources of online data are used, and Web addresses are given for the student's and researcher's further use (App. 12). Topics frequently encountered in econometrics, such as multicollinearity and autocorrelation, are clearly and concisely discussed as to the problems they create, the methods to test for their presence, and possible correction techniques. In this second edition, we have expanded the computer applications to provide a general introduction to data handling, and specific programming instruction to perform all estimations in this book by computer (Chap. 12) using Microsoft Excel, Eviews, or SAS statistical packages. We have also added sections on nonparametric testing, matrix notation, binary choice models, and an entire chapter on time series analysis (Chap. 11), a field of econometrics which has expanded as of late. A sample statistics and econometrics examination is also included.

The methodology of this book and much of its content has been tested in undergraduate and graduate classes in statistics and econometrics at Fordham University. Students found the approach and content of the book extremely useful and made many valuable suggestions for improvement. We have also received very useful advice from Professors Mary Beth Combs, Edward Dowling, and Damodar Gujarati. The following students carefully read through the entire manuscript and made many useful comments: Luca Bonardi, Kevin Coughlin, Sean Hennessy, and James Santangelo. To all of them we are deeply grateful. We owe a great intellectual debt to our former professors of statistics and econometrics: J. S. Butler, Jack Johnston, Lawrence Klein, and Bernard Okun.

We are indebted to the Literary Executor of the late Sir Ronald A. Fisher, F. R. S., to Dr. Frank Yates, F. R. S., and the Longman Group Ltd., London, for permission to adapt and reprint Tables III and IV from their book, *Statistical Tables for Biological, Agricultural and Medical Research*.

In addition to Statistics and Econometrics, the Schaum's Outline Series in Economics includes *Microeconomic Theory, Macroeconomic Theory, International Economics, Mathematics for Economists*, and *Principles of Economics*.

DOMINICK SALVATORE
DERRICK REAGLE

New York, 2001

CONTENTS

CHAPTER 1

Introduction

1.1 THE NATURE OF STATISTICS

Statistics refers to the collection, presentation, analysis, and utilization of numerical data to make inferences and reach decisions in the face of uncertainty in economics, business, and other social and physical sciences.

Statistics is subdivided into descriptive and inferential. *Descriptive statistics* is concerned with summarizing and describing a body of data. *Inferential statistics* is the process of reaching generalizations about the whole (called the *population*) by examining a portion (called the *sample*). In order for this to be valid, the sample must be *representative* of the population and the *probability* of error also must be specified.

Descriptive statistics is discussed in detail in Chap. 2. This is followed by (the more crucial) statistical inference; Chap. 3 deals with probability, Chap. 4 with estimation, and Chap. 5 with hypothesis testing.

EXAMPLE 1. Suppose that we have data on the incomes of 1000 U.S. families. This body of data can be summarized by finding the average family income and the spread of these family incomes above and below the average. The data also can be described by constructing a table, chart, or graph of the number or proportion of families in each income class. This is descriptive statistics. If these 1000 families are representative of all U.S. families, we can then estimate and test hypotheses about the average family income in the United States as a whole. Since these conclusions are subject to error, we also would have to indicate the probability of error. This is statistical inference.

1.2 STATISTICS AND ECONOMETRICS

Econometrics refers to the application of economic theory, mathematics, and statistical techniques for the purpose of testing hypotheses and estimating and forecasting economic phenomena. Econometrics has become strongly identified with *regression analysis*. This relates a dependent variable to one or more independent or explanatory variables. Since relationships among economic variables are generally inexact, a disturbance or error term (with well-defined probabilistic properties) must be included (see Prob. 1.8).

Chapters 6 and 7 deal with regression analysis; Chap. 8 extends the basic regression model; Chap. 9 deals with methods of testing and correcting for violations in the assumptions of the basic regression model; and Chaps. 10 and 11 deal with two specific areas of econometrics, specifically simultaneous-equations and time-series methods. Thus Chaps. 1 to 5 deal with the statistics required for econometrics (Chaps. 6 to 11). Chapter 12 is concerned with using the computer to aid in the calculations involved in the previous chapters.

1

EXAMPLE 2. Consumption theory tells us that, in general, people increase their consumption expenditure C as their disposable (after-tax) income Y_d increases, but not by as much as the increase in their disposable income. This can be stated in explicit linear equation form as

$$C = b_0 + b_1 Y_d \qquad\qquad (1.1)$$

where b_0 and b_1 are unknown constants called *parameters*. The parameter b_1 is the slope coefficient representing the marginal propensity to consume (MPC). Since even people with identical disposable income are likely to have somewhat different consumption expenditures, the theoretically exact and deterministic relationship represented by Eq. (1.1) must be modified to include a random disturbance or error term, u, making it stochastic:

$$C = b_0 + b_1 Y_d + u \qquad\qquad (1.2)$$

1.3   THE METHODOLOGY OF ECONOMETRICS

Econometric research, in general, involves the following three stages:

1. Specification of the model or maintained hypothesis in explicit stochastic equation form, together with the a priori theoretical expectations about the sign and size of the parameters of the function.

2. Collection of data on the variables of the model and estimation of the coefficients of the function with appropriate econometric techniques (presented in Chaps. 6 to 8).

3. Evaluation of the estimated coefficients of the function on the basis of economic, statistical, and econometric criteria.

EXAMPLE 3. The *first stage* in econometric research on consumption theory is to state the theory in explicit stochastic equation form, as in Eq. (1.1), with the expectation that $b_0 > 0$ (i.e., at $Y_d = 0$, $C > 0$ as people dissave and/or borrow) and $0 < b_1 < 1$. The *second stage* involves the collection of data on consumption expenditure and disposable income and estimation of Eq. (1.1). The *third stage* in econometric research involves (1) checking to see if the estimated value of $b_0 > 0$ and if $0 < b_1 < 1$; (2) determining if a "satisfactory" proportion of the variation in C is "explained" by changes in Y_d and if b_0 and b_1 are "statistically significant at acceptable levels" [see Prob. 1.13(c) and Sec. 5.2]; and (3) testing to see if the assumptions of the basic regression model are satisfied or, if not, how to correct for violations. If the estimated relationship does not pass these tests, the hypothesized relationship must be modified and reestimated until a satisfactory estimated consumption relationship is achieved.

Solved Problems

THE NATURE OF STATISTICS

1.1   What is the purpose and function of   (a) The field of study of statistics?   (b) Descriptive statistics?   (c) Inferential statistics?

(a) Statistics is the body of procedures and techniques used to collect, present, and analyze data on which to base decisions in the face of uncertainty or incomplete information. Statistical analysis is used today in practically every profession. The economist uses it to test the efficiency of alternative production techniques; the businessperson may use it to test the product design or package that maximizes sales; the sociologist to analyze the result of a drug rehabilitation program; the industrial psychologist to examine workers' responses to plant environment; the political scientist to forecast voting patterns; the physician to test the effectiveness of a new drug; the chemist to produce cheaper fertilizers; and so on.

(b) Descriptive statistics summarizes a body of data with one or two pieces of information that characterize the whole data. It also refers to the presentation of a body of data in the form of tables, charts, graphs, and other forms of graphic display.

(c) Inferential statistics (both estimation and hypothesis testing) refers to the drawing of generalizations about the properties of the whole (called a *population*) from the specific or a sample drawn from the population. Inferential statistics thus involves inductive reasoning. (This is to be contrasted with deductive reasoning, which ascribes properties to the specific starting with the whole.)

1.2 (a) Are descriptive or inferential statistics more important today? (b) What is the importance of a representative sample in statistical inference? (c) Why is probability theory required?

(a) Statistics started as a purely descriptive science, but it grew into a powerful tool of decision making as its inferential branch was developed. Modern statistical analysis refers primarily to inferential or inductive statistics. However, deductive and inductive statistics are complementary. We must study how to generate samples from populations before we can learn to generalize from samples to populations.

(b) In order for statistical inference to be valid, it must be based on a sample that fully reflects the characteristics and properties of the population from which it is drawn. A representative sample is ensured by random sampling, whereby each element of the population has an equal chance of being included in the sample (see Sec. 4.1).

(c) Since the possibility of error exists in statistical inference, estimates or tests of a population property or characteristic are given together with the chance or probability of being wrong. Thus probability theory is an essential element in statistical inference.

1.3 How can the manager of a firm producing lightbulbs summarize and describe to a board meeting the results of testing the life of a sample of 100 lightbulbs produced by the firm?

Providing the (raw) data on the life of each in the sample of 100 lightbulbs produced by the firm would be very inconvenient and time-consuming for the board members to evaluate. Instead, the manager might summarize the data by indicating that the average life of the bulbs tested is 360 h and that 95% of the bulbs tested lasted between 320 and 400 h. By doing this, the manager is providing two pieces of information (the average life and the spread in the average life) that characterize the life of the 100 bulbs tested. The manager also might want to describe the data with a table or chart indicating the number or proportion of bulbs tested that lasted within each 10-h classification. Such a tubular or graphic representation of the data is also very useful for gaining a quick overview of the data. In summarizing and describing the data in the ways indicated, the manager is engaging in descriptive statistics. It should be noted that descriptive statistics can be used to summarize and describe any body of data, whether it is a sample (as above) or a population (when all the elements of the population are known and its characteristics can be calculated).

1.4 (a) Why may the manager in Prob. 1.3 want to engage in statistical inference? (b) What would this involve and require?

(a) Quality control requires that the manager have a fairly good idea about the average life and the spread in the life of the lightbulbs produced by the firm. However, testing all the lightbulbs produced would destroy the entire output of the firm. Even when testing does not destroy the product, testing the entire output is usually prohibitively expensive and time-consuming. The usual procedure is to take a sample of the output and infer the properties and characteristics of the entire output (population) from the corresponding characteristics of a sample drawn from the population.

(b) Statistical inference requires first of all that the sample be representative of the population being sampled. If the firm produces lightbulbs in different plants, with more than one workshift, and with raw materials from different suppliers, these must be represented in the sample in the proportion in which they contribute to the total output of the firm. From the average life and spread in the life of the bulbs in the sample, the firm manager might estimate, with 95% probability of being correct and 5% probability of being wrong, the average life of all the lightbulbs produced by the firm to be between 320 and 400 h (see Sec. 4.3). Instead, the manager may use the sample information to test, with 95% probability of being correct and 5% probability of being wrong, that the average life of the population of all the bulbs produced by the firm is greater than 320 h (see Sec. 5.2). In estimating or testing the average for a population from sample information, the manager is engaging in statistical inference.

STATISTICS AND ECONOMETRICS

1.5   What is meant by (*a*) Econometrics? (*b*) Regression analysis? (*c*) Disturbance or error term? (*d*) Simultaneous-equations models?

(*a*)   Econometrics is the integration of economic theory, mathematics, and statistical techniques for the purpose of testing hypotheses about economic phenomena, estimating coefficients of economic relationships, and forecasting or predicting future values of economic variables or phenomena. Econometrics is subdivided into theoretical and applied econometrics. *Theoretical econometrics* refers to the methods for measurement of economic relationships in general. *Applied econometrics* examines the problems encountered and the findings in particular fields of economics, such as demand theory, production, investment, consumption, and other fields of applied economic research. In any case, econometrics is partly an art and partly a science, because often the intuition and good judgment of the econometrician plays a crucial role.

(*b*)   Regression analysis studies the causal relationship between one economic variable to be explained (the dependent variable) and one or more independent or explanatory variables. When there is only one independent or explanatory variable, we have *simple regression*. In the more usual case of more than one independent or explanatory variable, we have *multiple regression*.

(*c*)   A (random) disturbance or error must be included in the exact relationships postulated by economic theory and mathematical economics in order to make them stochastic (i.e., in order to reflect the fact that in the real world, economic relationships among economic variables are inexact and somewhat erratic).

(*d*)   Simultaneous-equations models refer to relationships among economic variables expressed with more than one equation and such that the economic variables in the various equations interact. Simultaneous-equations models are the most complex aspect of econometrics and are discussed in Chap. 10.

1.6   (*a*) What are the functions of econometrics? (*b*) What aspects of econometrics (and other social sciences) make it basically different from most physical sciences?

(*a*)   Econometrics has basically three closely interrelated functions. The first is to test economic theories or hypotheses. For example, is consumption directly related to income? Is the quantity demanded of a commodity inversely related to its price? The second function of econometrics is to provide numerical estimates of the coefficients of economic relationships. These are essential in decision making. For example, a government policymaker needs to have an accurate estimate of the coefficient of the relationship between consumption and income in order to determine the stimulating (i.e., the multiplier) effect of a proposed tax reduction. A manager needs to know if a price reduction increases or reduces the total sales revenues of the firm and, if so, by how much. The third function of econometrics is the forecasting of events. This, too, is necessary in order for policymakers to take appropriate corrective action if the rate of unemployment or inflation is predicted to rise in the future.

(*b*)   There are two basic differences between econometrics (and other social sciences) on one hand, and most physical sciences (such as physics) on the other. One is that (as pointed out earlier) relationships among economic variables are inexact and somewhat erratic. The second is that most economic phenomena occur contemporaneously, so that laboratory experiments cannot be conducted. These differences require special methods of analysis (such as the inclusion of a disturbance or error term with the exact relationships postulated by economic theory) and multivariate analysis (such as multiple regression analysis). The latter isolates the effect of each independent or explanatory variable on the dependent variable in the face of contemporaneous change in all explanatory variables.

1.7   In what way and for what purpose are (*a*) economic theory, (*b*) mathematics, and (*c*) statistical analysis combined to form the field of study of econometrics?

(*a*)   Econometrics presupposes the existence of a body of economic theories or hypotheses requiring testing. If the variables suggested by economic theory do not provide a satisfactory explanation, the researcher may experiment with alternative formulations and variables suggested by previous tests or opposing theories. In this way, econometric research can lead to the acceptance, rejection, and reformulation of economic theories.

(*b*) Mathematics is used to express the verbal statements of economic theories in mathematical form, expressing an exact or deterministic functional relationship between the dependent and one or more independent or explanatory variables.

(*c*) Statistical analysis applies appropriate techniques to estimate the inexact and nonexperimental relationships among economic variables by utilizing relevant economic data and evaluating the results.

1.8 What justifies the inclusion of a disturbance or error term in regression analysis?

The inclusion of a (random) disturbance or error term (with well-defined probabilistic properties) is required in regression analysis for three important reasons. First, since the purpose of theory is to generalize and simplify, economic relationships usually include only the most important forces at work. This means that numerous other variables with slight and irregular effects are not included. The error term can be viewed as representing the net effect of this large number of small and irregular forces at work. Second, the inclusion of the error term can be justified in order to take into consideration the net effect of possible errors in measuring the dependent variable, or variable being explained. Finally, since human behavior usually differs in a random way under identical circumstances, the disturbance or error term can be used to capture this inherently random human behavior. This error term thus allows for individual random deviations from the exact and deterministic relationships postulated by economic theory and mathematical economics.

1.9 Consumer demand theory states that the quantity demanded of a commodity D_X is a function of, or depends on, its price P_X, consumer's income Y, and the price of other (related) commodities, say, commodity Z (i.e., P_Z). Assuming that consumers' tastes remain constant during the period of analysis, state the preceding theory in (*a*) specific or explicit linear form or equation and (*b*) in stochastic form. (*c*) Which are the coefficients to be estimated? What are they called?

(*a*) $$D_X = b_0 + b_1 P_X + b_2 Y + b_3 P_Z \qquad (1.3)$$

(*b*) $$D_X = b_0 + b_1 P_X + b_2 Y + b_3 P_Z + u \qquad (1.4)$$

(*c*) The coefficients to be estimated are b_0, b_1, b_2, and b_3. They are called *parameters*.

THE METHODOLOGY OF ECONOMETRICS

1.10 With reference to the consumer demand theory in Prob. 1.9, indicate (*a*) what the first step is in econometric research and (*b*) what the a priori theoretical expectations are of the sign and possible size of the parameters of the demand function given by Eq. (1.4).

(*a*) The first step in econometric analysis is to express the theory of consumer demand in stochastic equation form, as in Eq. (1.4), and indicate the a priori theoretical expectations about the sign and possibly the size of the parameters of the function.

(*b*) Consumer demand theory postulates that in Eq. (1.4), $b_1 < 0$ (indicating that price and quantity are inversely related), $b_2 > 0$ if the commodity is a normal good (indicating that consumers purchase more of the commodity at higher incomes), $b_3 > 0$ if X and Z are substitutes, and $b_3 < 0$ if X and Z are complements.

1.11 Indicate the second stage in econometric research (*a*) in general and (*b*) with reference to the demand function specified by Eq. (1.4).

(*a*) The second stage in econometric research involves the collection of data on the dependent variable and on each of the independent or explanatory variables of the model and utilizing these data for the empirical estimation of the parameters of the model. This is usually done with multiple regression analysis (discussed in Chap. 7).

(*b*) In order to estimate the demand function given by Eq. (1.4), data must be collected on (1) the quantity demanded of commodity X by consumers, (2) the price of X, (3) consumer's incomes, and (4) the price of commodity Z per unit of time (i.e., per day, month, or year) and over a number of days, months, or years. Data on P_X, Y, and P_Z are then regressed against data on D_X and estimates of parameters b_0, b_1, b_2, and b_3 obtained.

1.12 How does the type of data required to estimate the demand function specified by Eq. (1.4) differ from the type of data that would be required to estimate the consumption function *for a group of families at one point in time*?

In order to estimate the demand function given by Eq. (1.4), numerical values of the variables are required over a period of time. For example, if we want to estimate the demand function for coffee, we need the numerical value of the quantity of coffee demanded, say, per year, over a number of years, say, from 1960 to 1980. Similarly, we need data on the average price of coffee, consumers' income, and the price, of say, tea (a substitute for coffee) per year from 1960 to 1980. Data that give numerical values for the variables of a function from period to period are called *time-series data*. However, to estimate the consumption function for a group of families at one point in time, we need *cross-sectional data* (i.e., numerical values for the consumption expenditures and disposable incomes of each family in the group at a particular point in time, say, in 1982).

1.13 What is meant by (*a*) The third stage in econometric analysis? (*b*) A priori theoretical criteria? (*c*) Statistical criteria? (*d*) Econometric criteria? (*e*) The forecasting ability of the model?

(*a*) The *third stage in econometric research* involves the evaluation of the estimated model on the basis of the a priori criteria, statistical and econometric criteria, and the forecasting ability of the model.

(*b*) The *a priori economic criteria* refer to the sign and size of the parameters of the model postulated by economic theory. If the estimated coefficients do not conform to those postulated, the model must be revised or rejected.

(*c*) The *statistical criteria* refer to (1) the proportion of variation in the dependent variable "explained" by changes in the independent or explanatory variables and (2) verification that the dispersion or spread of each estimated coefficient around the true parameter is sufficiently narrow to give us "confidence" in the estimates.

(*d*) The *econometric criteria* refer to tests that the assumptions of the basic regression model, and particularly those about the disturbance or error term, are satisfied.

(*e*) The *forecasting ability* of the model refers to the ability of the model to accurately predict future values of the dependent variable based on known or expected future value(s) of the independent or explanatory variable(s).

1.14 How can the *estimated* demand function given by Eq. (1.4) be evaluated in terms of (*a*) The a priori criteria? (*b*) The statistical criteria? (*c*) The econometric criteria? (*d*) The forecasting ability of the model?

(*a*) The estimated demand function given by Eq. (1.4) can be evaluated in terms of the a priori theoretical criteria by checking that the estimated coefficients conform to the theoretical expectations with regard to sign and possible size, as postulated in Prob. 1.10(*b*). The demand theory given by Eq. (1.4) is confirmed only if $b_1 < 0$, if $b_2 > 0$ (if X is a normal good), and if $b_3 > 0$ (if Z is a substitute for X), as postulated by demand theory.

(*b*) The statistical criteria are satisfied only if a "high" proportion of the variation in D_X over time is "explained" by changes in P_X, Y, and P_Z, and if the dispersion of estimated b_1, b_2, and b_3 around the true parameters are "sufficiently narrow." There is no generally accepted answer as to what is a "high" proportion of the variation in D_X "explained" by P_X, Y, and P_Z. However, because of common trends in time-series data, we would expect more than 50 to 70% of the variation in the dependent variable to be explained by the independent or explanatory variables for the model to be judged satisfactory. Similarly, in order for each estimated coefficient to be "statistically significant," we would expect the dispersion of each estimated coefficient about the true parameter (measured by its standard deviation; see Sec. 2.3) to be generally less than half the estimated value of the coefficient.

(c) The econometric criteria are used to determine if the assumptions of the econometric methods used are satisfied in the estimation of the demand function of Eq. (1.4). Only if these assumptions are satisfied will the estimated coefficients have the desirable properties of unbiasedness, consistency, efficiency, and so forth (see Sec. 6.4).

(d) One way to test the forecasting ability of the demand model given by Eq. (1.4) is to use the estimated function to predict the value of D_X for a period not included in the sample and checking that this predicted value is "sufficiently close" to the actual observed value of D_X for that period.

1.15 Present in schematic form the various stages of econometric research.

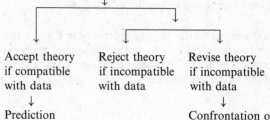

Stage 1:Economic theory
↓
Mathematical model
↓
Econometric (stochastic) model

Stage 2:Collection of appropriate data
↓
Estimation of the parameters of the model

Stage 3:Evaluation of the model on the basis of economic, statistical, and econometric criteria

Accept theory
if compatible
with data
↓
Prediction

Reject theory
if incompatible
with data

Revise theory
if incompatible
with data
↓
Confrontation of
revised theory
with new data

Supplementary Problems

THE NATURE OF STATISTICS

1.16 (a) To which field of study is statistical analysis important? (b) What are the most important functions of descriptive statistics? (c) What is the most important function of inferential statistics?
Ans. (a) To economics, business, and other social and physical sciences (b) Summarizing and describing a body of data (c) Drawing inferences about the characteristics of a population from the corresponding characteristics of a sample drawn from the population.

1.17 (a) Is statistical inference associated with deductive or inductive reasoning? (b) What are the conditions required in order for statistical inference to be valid?
Ans. (a) Inductive reasoning (b) A representative sample and probability theory

STATISTICS AND ECONOMETRICS

1.18 Express in the form of an explicit linear equation the statement that the level of investment spending I is inversely related to rate of interest R.
Ans. $I = b_0 + b_1 R$ with b_1 postulated to be negative(1.5)

1.19   What is the answer to Prob. 1.18 an example of?
Ans.   An economic theory expressed in (exact or deterministic) mathematical form

1.20   Express Eq. (1.5) in stochastic form.
Ans.
$$I = b_0 + b_1 R + u \qquad\qquad (1.6)$$

1.21   Why is a stochastic form required in econometric analysis?
Ans.   Because the relationships among economic variables are inexact and somewhat erratic as opposed to the exact and deterministic relationships postulated by economic theory and mathematical economics

THE METHODOLOGY OF ECONOMETRICS

1.22   What are stages   (*a*) one,   (*b*) two, and   (*c*) three in econometric research?
Ans.   (*a*) Specification of the theory in stochastic equation form and indication of the expected signs and possible sizes of estimated parameters   (*b*) Collection of data on the variables of the model and estimation of the coefficients of the function   (*c*) Economic, statistical, and econometric evaluation of the estimated parameters

1.23   What is the first stage of econometric analysis for the investment theory in Prob. 1.18?
Ans.   Stating the theory in the form of Eq. (1.6) and predicting $b_1 < 0$

1.24   What is the second stage in econometric analysis for the investment theory in Prob. 1.18?
Ans.   Collection of time-series data on I and R and estimation of Eq. (1.6)

1.25   What is the third stage of econometric analysis for the investment theory in Prob. 1.18?
Ans.   Determination that the estimated coefficient of $b_1 < 0$, that an "adequate" proportion of the variation in I over time is "explained" by changes in R, that b_1 is "statistically significant at customary levels," and that the econometric assumptions of the model are satisfied

CHAPTER 2

Descriptive Statistics

2.1 FREQUENCY DISTRIBUTIONS

It is often useful to organize or arrange a body of data into a *frequency distribution*. This breaks up the data into groups or classes and shows the number of observations in each class. The number of classes is usually between 5 and 15. A *relative frequency distribution* is obtained by dividing the number of observations in each class by the total number of observations in the data as a whole. The sum of the relative frequencies equals 1. A *histogram* is a bar graph of a frequency distribution, where classes are measured along the horizontal axis and frequencies along the vertical axis. A *frequency polygon* is a line graph of a frequency distribution resulting from joining the frequency of each class plotted at the class midpoint. A *cumulative frequency distribution* shows, for each class, the total number of observations in all classes up to and including that class. When plotted, this gives a *distribution curve*, or *ogive*.

EXAMPLE 1. A student received the following grades (measured from 0 to 10) on the 10 quizzes he took during a semester: 6, 7, 6, 8, 5, 7, 6, 9, 10, and 6. These grades can be arranged into frequency distributions as in Table 2.1 and shown graphically as in Fig. 2-1.

Table 2.1 Frequency Distributions of Grades

Grades	Absolute Frequency	Relative Frequency
5	1	0.1
6	4	0.4
7	2	0.2
8	1	0.1
9	1	0.1
10	1	0.1
	10	1.0

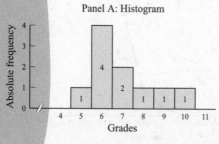
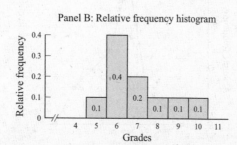

Fig. 2-1

9

EXAMPLE 2.   The cans in a sample of 20 cans of fruit contain net weights of fruit ranging from 19.3 to 20.9 oz, as given in Table 2.2. If we want to group these data into 6 classes, we get *class intervals* of 0.3 oz [(21.0 − 19.2)/6 = 0.3 oz].   The weights given in Table 2.2 can be arranged into the frequency distributions given in Table 2.3 and shown graphically in Fig. 2-2.

Table 2.2   Net Weight in Ounces of Fruit

19.7	19.9	20.2	19.9	20.0	20.6	19.3	20.4	19.9	20.3
20.1	19.5	20.9	20.3	20.8	19.9	20.0	20.6	19.9	19.8

Table 2.3   Frequency Distribution of Weights

Weight, oz	Class Midpoint	Absolute Frequency	Relative Frequency	Cumulative Frequency
19.2–19.4	19.3	1	0.05	1
19.5–19.7	19.6	2	0.10	3
19.8–20.0	19.9	8	0.40	11
20.1–20.3	20.2	4	0.20	15
20.4–20.6	20.5	3	0.15	18
20.7–20.9	20.8	2	0.10	20
		20	1.00	

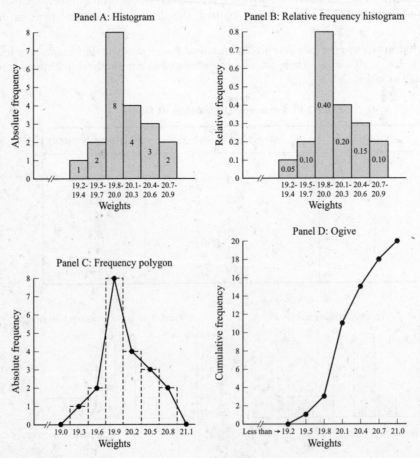

Fig. 2-2

2.2 MEASURES OF CENTRAL TENDENCY

Central tendency refers to the location of a distribution. The most important measures of central tendency are (1) the *mean*, (2) the *median*, and (3) the *mode*. We will be measuring these for populations (i.e., the collection of all the elements that we are describing) and for samples drawn from populations, as well as for grouped and ungrouped data.

1. The *arithmetic mean* or *average*, of a population is represented by μ (the Greek letter mu); and for a sample, by $\overline{X}$ (read "X bar"). For *ungrouped* data, μ and $\overline{X}$ are calculated by the following formulas:

$$\mu = \frac{\sum X}{N} \quad \text{and} \quad \overline{X} = \frac{\sum X}{n} \qquad (2.1a, b)$$

where $\sum X$ refers to the sum of all the observations, while N and n refer to the number of observations in the population and sample, respectively. For *grouped* data, μ and $\overline{X}$ are calculated by

$$\mu = \frac{\sum fX}{N} \quad \text{and} \quad \overline{X} = \frac{\sum fX}{n} \qquad (2.2a, b)$$

where $\sum fX$ refers to the sum of the frequency of each class f times the class midpoint X.

2. The *median* for *ungrouped* data is the value of the middle item when all the items are arranged in either ascending or descending order in terms of values:

$$\text{Median} = \text{the } \left(\frac{N+1}{2}\right)\text{th item in the data array} \qquad (2.3)$$

where N refers to the number of items in the population (n for a sample). The median for *grouped* data is given by the formula

$$\text{Median} = L + \frac{n/2 - F}{f_m} c \qquad (2.4)$$

where L = lower limit of the median class (i.e., the class that contains the middle item of the distribution

n = the number of observations in the data set

F = sum of the frequencies up to but not including the median class

f_m = frequency of the median class

c = width of the class interval

3. The *mode* is the value that occurs most frequently in the data set. For *grouped* data, we obtain

$$\text{Mode} = L + \frac{d_1}{d_1 + d_2} c \qquad (2.5)$$

where L = lower limit of the modal class (i.e., the class with the greatest frequency)

d_1 = frequency of the modal class minus the frequency of the previous class

d_2 = frequency of the modal class minus the frequency of the following class

c = width of the class interval

The mean is the most commonly used measure of central tendency. The mean, however, is affected by extreme values in the data set, while the median and the mode are not. Other measures of central tendency are the *weighted mean*, the *geometric mean*, and the *harmonic mean* (see Probs. 2.7 to 2.9).

EXAMPLE 3. The mean grade for the population on the 10 quizzes given in Example 1, using the formula for ungrouped data, is

$$\mu = \frac{\sum X}{N} = \frac{6+7+6+8+5+7+6+9+10+6}{10} = \frac{70}{10} = 7 \text{ points}$$

To find the median for the ungrouped data, we first arrange the 10 grades in ascending order: 5, 6, 6, 6, 6, 7, 7, 8, 9, 10. Then we find the grade of the $(N+1)/2$ or $(10+1)/2 = 5.5$th item. Thus the median is the average of the 5th and 6th item in the array, or $(6+7)/2 = 6.5$. The mode for the ungrouped data is 6 (the value that occurs most frequently in the data set).

EXAMPLE 4. We can *estimate the mean* for the *grouped* data given in Table 2.3 with the aid of Table 2.4:

$$\overline{X} = \frac{\sum fX}{n} = \frac{401.6}{20} = 20.08 \text{ oz}$$

This calculation could be simplified by coding (see Prob. 2.6).

Table 2.4   Calculation of the Sample Mean for the Data in Table 2.3

Weight, oz	Class Midpoint X	Frequency f	fX
19.2–19.4	19.3	1	19.3
19.5–19.7	19.6	2	39.2
19.8–20.0	19.9	8	159.2
20.1–20.3	20.2	4	80.8
20.4–20.6	20.5	3	61.5
20.7–20.9	20.8	2	41.6
		$\sum f = n = 20$	$\sum fX = 401.6$

We can estimate the *median* (med) for the same grouped data as follows:

$$\text{Med} = L + \frac{n/2 - F}{f_m} c = 19.8 + \frac{20/2 - 3}{8} 0.3 = 19.8 + \frac{7}{8} 0.3$$
$$= 19.8 + 0.2625 \cong 20.06 \text{ oz}$$

where $L = 19.8 =$ lower limit of the median class (i.e., the 19.8–20.0 class which contains the 10th and 11th observations)

$n = 20 =$ number of observations or items

$F = 3 =$ sum of frequencies up to but not including the median class

$f_m = 8 =$ frequency of the median class

$c = 0.3 =$ width of class interval

Similarly

$$\text{Mode} = L + \frac{d_1}{d_1 + d_2} c = 19.8 + \frac{6}{6 + 4} 0.3 = 19.8 + \frac{1.8}{10} = 19.8 + 0.18 = 19.98 \text{ oz}$$

As noted in Prob. 2.4, the mean, median, and mode for grouped data are estimates used when only the grouped data are available or to reduce calculations with a large ungrouped data set.

2.3 MEASURES OF DISPERSION

Dispersion refers to the variability or spread in the data. The most important measures of dispersion are (1) the *average deviation*, (2) the *variance*, and (3) the *standard deviation*. We will measure these for populations and samples, as well as for grouped and ungrouped data.

1. *Average deviation.* The *average deviation* (AD), also called the *mean absolute deviation* (MAD), is given by

$$\text{AD} = \frac{\sum |X - \mu|}{N} \qquad \text{for populations} \tag{2.6a}$$

and

$$\text{AD} = \frac{\sum |X - \overline{X}|}{n} \qquad \text{for samples} \tag{2.6b}$$

where the two vertical bars indicate the absolute value, or the values omitting the sign, with the other symbols having the same meaning as in Sec. 2.2. For grouped data

$$\text{AD} = \frac{\sum f |X - \mu|}{N} \qquad \text{for populations} \tag{2.7a}$$

and

$$\text{AD} = \frac{\sum f |X - \overline{X}|}{n} \qquad \text{for samples} \tag{2.7b}$$

where f refers to the frequency of each class and X to the class midpoints.

2. *Variance.* The population variance σ^2 (the Greek letter sigma squared) and the sample variance s^2 for ungrouped data are given by

$$\sigma^2 = \frac{\sum (x - \mu)^2}{N} \qquad \text{and} \qquad s^2 = \frac{\sum (X - \overline{X})^2}{n - 1} \tag{2.8a,b}$$

For grouped data

$$\sigma^2 = \frac{\sum f (X - \mu)^2}{N} \qquad \text{and} \qquad s^2 = \frac{\sum f (X - \overline{X})^2}{n - 1} \tag{2.9a,b}$$

3. *Standard deviation.* The population standard deviation σ and sample standard deviation s are the positive square roots of their respective variances. For ungrouped data

$$\sigma = \sqrt{\frac{\sum (X - \mu)^2}{N}} \qquad \text{and} \qquad s = \sqrt{\frac{\sum (X - \overline{X})^2}{n - 1}} \tag{2.10a,b}$$

For grouped data

$$\sigma = \sqrt{\frac{\sum f (X - \mu)^2}{N}} \qquad \text{and} \qquad s = \sqrt{\frac{\sum f (X - \overline{X})^2}{n - 1}} \tag{2.11a,b}$$

The most widely used measure of (absolute) dispersion is the standard deviation. Other measures (besides the variance and average deviation) are the *range*, the *interquartile range*, and the *quartile deviation* (see Probs. 2.11 and 2.12).

4. The *coefficient of variation* (V) measures *relative* dispersion:

$$V = \frac{\sigma}{\mu} \qquad \text{for populations} \tag{2.12a}$$

and

$$V = \frac{s}{\overline{X}} \qquad \text{for samples} \tag{2.12b}$$

EXAMPLE 5. The average deviation, variance, standard deviation, and coefficient of variation for the ungrouped data given in Example 1 can be found with the aid of Table 2.5 ($\mu = 7$; see Example 3):

$$AD = \frac{\sum |X - \mu|}{N} = \frac{12}{10} = 1.2 \text{ points}$$

$$\sigma^2 = \frac{\sum (X - \mu)^2}{N} = \frac{22}{10} = 2.2 \text{ points squared}$$

$$\sigma = \sqrt{\frac{\sum (X - \mu)^2}{N}} = \sqrt{\frac{22}{10}} = \sqrt{2.2} \cong 1.48 \text{ points}$$

$$V = \frac{\sigma}{\mu} \cong \frac{1.48}{7} \cong 0.21, \quad \text{or} \quad 21\%$$

Table 2.5   Calculations on the Data in Example 1

Grade X	μ	$X - \mu$	$|X - \mu|$	$(X - \mu)^2$
6	7	−1	1	1
7	7	0	0	0
6	7	−1	1	1
8	7	1	1	1
5	7	−2	2	4
7	7	0	0	0
6	7	−1	1	1
9	7	2	2	4
10	7	3	3	9
6	7	−1	1	1
		$\sum(X - \mu) = 0$	$\sum |X - \mu| = 12$	$\sum(X - \mu)^2 = 22$

EXAMPLE 6.   The average deviation, variance, standard deviation, and coefficient of variation for the frequency distribution of weights (grouped data) given in Table 2.3 can be found with the aid of Table 2.6 ($\overline{X} = 20.08$ oz; see Example 4):

$$AD = \frac{\sum f |X - \overline{X}|}{n} = \frac{6.36}{20} = 0.318 \text{ oz}$$

$$s^2 = \frac{\sum f(X - \overline{X})^2}{n - 1} = \frac{2.9520}{19} \cong 0.1554 \text{ oz squared}$$

$$s = \sqrt{\frac{\sum f(X - \overline{X})^2}{n - 1}} = \sqrt{\frac{2.9520}{19}} = \sqrt{0.1544} \cong 0.3942 \text{ oz}$$

$$V = \frac{s}{\overline{X}} \cong \frac{0.3942 \text{ oz}}{20.08 \text{ oz}} \cong 0.0196, \quad \text{or} \quad 1.96\%$$

Note that in the formula for s^2 and s, $n - 1$ rather than n is used in the denominator (see Prob. 2.16 for the reason). From the formulas for σ^2, σ, s^2, and s given in this section, others may be derived that will simplify the calculations for a large body of data (see Probs. 2.17 to 2.19 for their derivation and application).

Table 2.6   Calculations on the Data in Table 2.4

Weight, oz	Class Midpoint X	Frequency f	Mean $\overline{X}$	$X - \overline{X}$	$|X - \overline{X}|$	$\sum f |X - \overline{X}|$	$(X - \overline{X})^2$	$f(X - \overline{X})^2$
19.20–19.40	19.30	1	20.08	−0.78	0.78	0.78	0.6084	0.6084
19.50–19.70	19.60	2	20.08	−0.48	0.48	0.96	0.2304	0.4608
19.80–20.00	19.90	8	20.08	−0.18	0.18	1.44	0.0324	0.2592
20.10–20.30	20.20	4	20.08	0.12	0.12	0.48	0.0144	0.0576
20.40–20.60	20.50	3	20.08	0.42	0.42	1.26	0.1764	0.5292
20.70–20.90	20.80	2	20.08	0.72	0.72	1.44	0.5184	1.0368
		$\sum f = n = 20$				$\sum f |X - \overline{X}| = 6.36$		$\sum f(X - \overline{X})^2 = 2.9520$

2.4 SHAPE OF FREQUENCY DISTRIBUTIONS

The shape of a distribution refers to   (1) its symmetry or lack of it (*skewness*) and   (2) its peakedness (*kurtosis*).

1. *Skewness.*   A distribution has zero skewness if it is symmetrical about its mean.   For a symmetrical (unimodal) distribution, the mean, median, and mode are equal.   A distribution is *positively skewed* if the right tail is longer.   Then, mean > median > mode.   A distribution is *negatively skewed* if the left tail is longer.   Then, mode > median > mean (see Fig. 2-3).

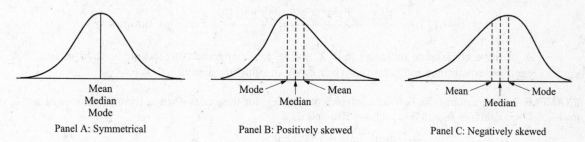

	Mean		Mode		Mean	Mean		Mode
	Median			Median			Median	
	Mode							

Panel A: Symmetrical   Panel B: Positively skewed   Panel C: Negatively skewed

Fig. 2-3

Skewness can be measured by the *Pearson coefficient of skewness*:

$$Sk = \frac{3(\mu - \text{med})}{\sigma} \qquad \text{for populations} \tag{2.13a}$$

and

$$Sk = \frac{3(\overline{X} - \text{med})}{s} \qquad \text{for samples} \tag{2.13b}$$

Mean and variance are the first and second moments of a distribution, respectively.   Skewness can also be measured by the third moment [the numerator of Eq. (2.14*a,b*)] divided by the cube of the standard deviation:

$$Sk = \frac{\sum f(X - \mu)^3}{\sigma^3} \qquad \text{for populations} \tag{2.14a}$$

and

$$Sk = \frac{\sum f(X - \overline{X})^3}{s^3} \qquad \text{for samples} \tag{2.14b}$$

For symmetric distributions, Sk = 0.

2. *Kurtosis.*   A peaked curve is called *leptokurtic*, as opposed to a flat one (*platykurtic*), relative to one that is *mesokurtic* (see Fig. 2-4).   Kurtosis can be measured by the *fourth moment* [the numerator of Eq. (2.15*a,b*)] divided by the standard deviation raised to the fourth power.   The kurtosis for a mesokurtic curve is 3.

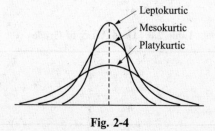

Fig. 2-4

$$\text{Kurtosis} = \frac{\sum f(X - \mu)^4}{\sigma^4} \qquad \text{for populations} \qquad (2.15a)$$

and

$$\text{Kurtosis} = \frac{\sum f(X - \overline{X})^4}{s^4} \qquad \text{for samples} \qquad (2.15b)$$

3. *Joint moment.* The comovement of two separate distributions can be measured by *covariance*:

$$\text{cov}(X, Y) = \frac{\Sigma(X - \overline{X})(Y - \overline{Y})}{N} = \frac{\Sigma(XY)}{N} - \overline{X}\overline{Y} \qquad \text{for populations}$$

$$\text{cov}(X, Y) = \frac{\Sigma(X - \overline{X})(Y - \overline{Y})}{n} = \frac{\Sigma(XY)}{n} - \overline{X}\overline{Y} \qquad \text{for samples}$$

A positive covariance indicates that X and Y move together in relation to their means. A negative covariance indicates that they move in opposite directions.

EXAMPLE 7. We can find the Pearson coefficient of skewness for the grades given in Example 1 by using $\mu = 7$, med $= 6.5$ (see Example 3), and $\sigma = 1.48$ (see Example 5):

$$\text{Sk} = \frac{3(\mu - \text{med})}{\sigma} \cong \frac{3(7 - 6.5)}{1.48} \cong \frac{3(0.5)}{1.48} \cong 1.01 \qquad \text{(see Fig. 2-1)}$$

Similarly, by using $\overline{X} = 20.08$ oz, med $= 20.06$ oz (see Example 4), and $s = 0.39$ oz (see Example 6), we can find the Pearson coefficient of skewness for the frequency distribution of weights in Table 2.3 as follows:

$$\text{Sk} = \frac{3(\overline{X} - \text{med})}{s} \cong \frac{3(20.08 - 20.06)}{0.39} \cong 0.15 \qquad \text{(see Fig. 2-2c)}$$

For kurtosis, see Prob. 2.23.

Solved Problems

FREQUENCY DISTRIBUTIONS

2.1 Table 2.7 gives the grades on a quiz for a class of 40 students. (*a*) Arrange these grades (raw data set) into an array from the lowest grade to the highest grade. (*b*) Construct a table showing class intervals and class midpoints and the absolute, relative, and cumulative frequencies for each grade. (*c*) Present the data in the form of a histogram, relative-frequency histogram, frequency polygon, and ogive.

Table 2.7 Grades on a Quiz for a Class of 40 Students

7	5	6	2	8	7	6	7	3	9
10	4	5	5	4	6	7	4	8	2
3	5	6	7	9	8	2	4	7	9
4	6	7	8	3	6	7	9	10	5

(*a*) See Table 2.8.

Table 2.8 Data Array of Grades

2	2	2	3	3	3	4	4	4	4
4	5	5	5	5	5	6	6	6	6
6	6	7	7	7	7	7	7	7	7
8	8	8	8	9	9	9	9	10	10

(b) See Table 2.9. Note that since we are dealing here with discrete data (i.e., data expressed in whole numbers), we used the actual grades as the class midpoints.

Table 2.9   Frequency Distribution of Grades

Grade	Class Midpoint	Absolute Frequency	Relative Frequency	Cumulative Frequency
1.5–2.4	2	3	0.075	3
2.5–3.4	3	3	0.075	6
3.5–4.4	4	5	0.125	11
4.5–5.4	5	5	0.125	16
5.5–6.4	6	6	0.150	22
6.5–7.4	7	8	0.200	30
7.5–8.4	8	4	0.100	34
8.5–9.4	9	4	0.100	38
9.5–10.4	10	2	0.050	40
		40	1.000	

(c) See Fig. 2-5.

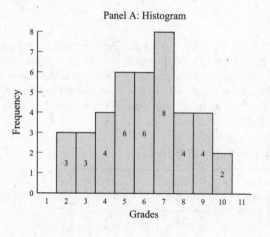

Panel A: Histogram

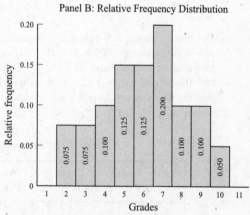

Panel B: Relative Frequency Distribution

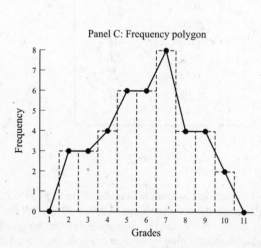

Panel C: Frequency polygon

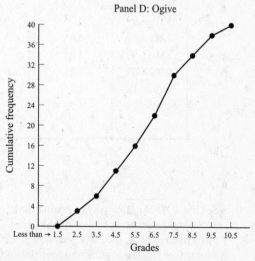

Panel D: Ogive

Fig. 2-5

2.2 A sample of 25 workers in a plant receive the hourly wages given in Table 2.10. (*a*) Arrange these raw data into an array from the lowest to the highest wage. (*b*) Group the data into classes. (*c*) Present the data in the form of a histogram, relative-frequency histogram, frequency polygon, and ogive.

Table 2.10   Hourly Wages in Dollars

3.65	3.78	3.85	3.95	4.00	4.10	4.25	3.55	3.85	3.96
3.60	3.90	4.26	3.75	3.95	4.05	4.08	4.15	3.80	4.05
3.88	3.95	4.06	4.18	4.05					

(*a*) See Table 2.11.

Table 2.11   Data Array of Wages in Dollars

3.55	3.60	3.65	3.75	3.78	3.80	3.85	3.85	3.88	3.90
3.95	3.95	3.95	3.96	4.00	4.05	4.05	4.05	4.06	4.08
4.10	4.15	4.18	4.25	4.26					

(*b*) The hourly wages in Table 2.10 range from \$3.55 to \$4.26. This can be conveniently subdivided into 8 equal classes of \$0.10 each. That is, $(\$4.30 - \$3.50)/8 = \$0.80/8 = \$0.10$. Note that the range was extended from \$3.50 to \$4.30 so that the lowest wage, \$3.55, falls *within* the lowest class and the largest wage, \$4.26, falls *within* the largest class. It is also convenient (and needed for plotting the frequency polygon) to find the class mark or midpoint of each class. These are shown in Table 2.12.

Table 2.12   Frequency Distribution of Wages

Hourly Wage, $	Class Midpoint, $	Absolute Frequency	Relative Frequency	Cumulative Frequency
3.50–3.59	3.55	1	0.04	1
3.60–3.69	3.65	2	0.08	3
3.70–3.79	3.75	2	0.08	5
3.80–3.89	3.85	4	0.16	9
3.90–3.99	3.95	5	0.20	14
4.00–4.09	4.05	6	0.24	20
4.10–4.19	4.15	3	0.12	23
4.20–4.29	4.25	2	0.08	25
		25	1.00	

(*c*) See Fig. 2-6. Another way of getting the ogive is to plot the cumulative frequencies up to \$3.595, 3.695, 3.795, and so on (so as to include the upper limit of each class). The values \$3.595, 3.695, 3.795, etc. are often referred to as the *class boundaries* or *exact limits*. Note that the class midpoints are obtained by adding together the lower and upper class boundaries and dividing by 2. For example, the second class midpoint is given by $(3.595 + 3.695)/2 = 7.290/2 = 3.65$ (see Table 2.12).

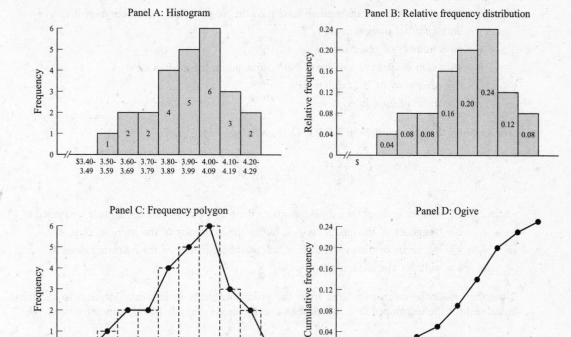

Fig. 2-6

MEASURES OF CENTRAL TENDENCY

2.3 Find the mean, median, and mode (*a*) for the grades on the quiz for the class of 40 students given in Table 2.7 (the ungrouped data) and (*b*) for the grouped data of these grades given in Table 2.9.

(*a*) Since we are dealing with *all* grades, we want the *population mean:*

$$\mu = \frac{\sum X}{N} = \frac{7 + 5 + 6 + \cdots + 5}{40} = \frac{240}{40} = 6 \text{ points}$$

That is, μ is obtained by adding together all the 40 grades given in Table 2.7 and dividing by 40 [the three centered dots (ellipses) were put in to avoid repeating the 40 values in Table 2.7]. The *median* is given by the values of the $[(N + 1)/2]$th item in the data array in Table 2.8. Therefore, the median is the value of the $(40 + 1)/2$ or 20.5th, or the average of the 20th and 21st item. Since they are both equal to 6, the median is 6. The *mode* is 7 (the value that occurs most frequently in the data set).

(*b*) We can find the *population mean* for the grouped data in Table 2.9 with the aid of Table 2.13:

$$\mu = \frac{\sum fX}{N} = \frac{240}{40} = 6$$

This is the same mean we found for the ungrouped data. Note that the sum of the frequencies, $\sum f$, equals the number of observations in the population, N, and $\Sigma X = \sum fX$. The *median* for the grouped data of Table 2.13 is given by

$$\text{Med} = L + \frac{N/2 - F}{f_m} c = 5.5 + \frac{40/2 - 16}{6} 1 = 5.5 + 0.67 = 6.17$$

where $L = 5.5 =$ lower limit of the median class (i.e., the 5.5–6.4 class, which contains the 20th and 21st observations)

$N = 40 =$ number of observations

$F = 16 =$ sum of observations up to but not including the median class

$f_m = 6 =$ frequency of the median class

$c = 1 =$ width of class interval

The *mode* for the grouped data in Table 2.13 is given by

$$\text{Mode} = L + \frac{d_1}{d_1 + d_2}c = 6.5 + \frac{2}{2 + 4}1 = 6.5 + 0.33 = 6.83$$

where $L = 6.5 =$ lower limit of the modal class (i.e., the 6.5–7.4 class with the highest frequency of 8)

$d_1 = 2 =$ frequency of the modal class, 8, minus the frequency of the previous class, 6

$d_2 = 4 =$ frequency of the modal class, 8, minus the frequency of the following class, 4

$c = 1 =$ width of the class interval

Note that while the mean calculated from the grouped data is in this case identical to the mean calculated for the ungrouped data, the median and the mode are only (good) approximations.

Table 2.13   Calculation of the Population Mean for the Grouped Data in Table 2.9

Grade	Class Midpoint X	Frequency f	fX
1.5–2.4	2	3	6
2.5–3.4	3	3	9
3.5–4.4	4	5	20
4.5–5.4	5	5	25
5.5–6.4	6	6	36
6.5–7.4	7	8	56
7.5–8.4	8	4	32
8.5–9.4	9	4	36
9.5–10.4	10	2	20
		$\sum f = N = 40$	$\sum fX = 240$

2.4 Find the mean, median, and mode (*a*) for the sample of hourly wages received by the 25 workers recorded in Table 2.10 (the ungrouped data) and (*b*) for the grouped data of these wages given in Table 2.12.

(*a*) $$\overline{X} = \frac{\sum X}{n} = \frac{\$3.65 + \$3.78 + \$3.85 + \cdots + \$4.05}{25} = \frac{\$98.65}{25} = \$3.946 \quad \text{or} \quad \$3.95$$

Median = \$3.95 [the value of the $(n + 1)/2 = (25 + 1) = $ 13th item in the data array in Table 2.11]. Mode = \$3.95 and \$4.05, since there are three of each of these wages. Thus the distribution is *bimodal* (i.e., it has two modes).

(*b*) We can find the sample mean for the grouped data in Table 2.12 with the aid of Table 2.14:

$$\overline{X} = \frac{\sum fX}{n} = \frac{\$98.75}{25} = \$3.95$$

Note that in this case $\sum fX = \$98.75 \neq \sum X = \$98.65$ (found in part *a*) since the average of the observations in each class is not equal to the class midpoint for all classes [as in Prob. 2.3(*b*)].

Thus $\overline{X}$ calculated from the grouped data is only a very good approximation for the true value of $\overline{X}$ calculated for the ungrouped data. In the real world, we often have only the grouped data, or if we have a very large body of ungrouped data, it will save on calculations to *estimate* the mean by first grouping the data.

$$\text{Med} = L + \frac{n/2 - F}{f_m} c = \$3.90 + \frac{25/2 - 9}{5}(0.10) = \$3.90 + \$0.07 = \$3.97$$

as compared with the true median of \$3.95 found from the ungrouped data (see part *a*).

$$\text{Mode} = L + \frac{d_1}{d_1 + d_2} c = \$4.00 + \frac{1}{1 + 3}(0.10) = \$4.00 + \$0.025 = \$4.025 \quad \text{or} \quad \$4.03$$

as compared with the true modes of \$3.95 and \$4.05 found from the ungrouped data (see part *a*). Sometimes the mode is simply given as the midpoint of the modal class.

Table 2.14   Calculation of the Sample Mean for the Grouped Data in Table 2.12

Hourly Wage, \$	Class Midpoint X, \$	Frequency f	fX
3.50–3.59	3.55	1	3.55
3.60–3.69	3.65	2	7.30
3.70–3.79	3.75	2	7.50
3.80–3.89	3.85	4	15.40
3.90–3.99	3.95	5	19.75
4.00–4.09	4.05	6	24.30
4.10–4.19	4.15	3	12.45
4.20–4.29	4.25	2	8.45
		$\sum f = n = 25$	$\sum fX = \$98.75$

2.5   Compare the advantages and disadvantages of (*a*) the mean, (*b*) the median, and (*c*) the mode as measures of central tendency.

(*a*) The advantages of the mean are (1) it is familiar and understood by virtually everyone, (2) all the observations in the data are taken into account, and (3) it is used in performing many other statistical procedures and tests. The disadvantages of the mean are (1) it is affected by extreme values, (2) it is time-consuming to compute for a large body of ungrouped data, and (3) it cannot be calculated when the last class of grouped data is open-ended (i.e., it includes the lower limit of the last class "and over").

(*b*) The advantages of the median are (1) it is not affected by extreme values, (2) it is easily understood (i.e., half the data are smaller than the median and half are greater), and (3) it can be calculated even when the last class is open-ended and when the data are qualitative rather than quantitative. The disadvantages of the mean are (1) it does not use much of the information available, and (2) it requires that observations be arranged into an array, which is time-consuming for a large body of ungrouped data.

(*c*) The advantages of the mode are the same as those for the median. The disadvantages of the mode are (1) as for the median, the mode does not use much of the information available, and (2) sometimes no value of the data is repeated more than once, so that there is no mode, while at other times there may be many modes. In general, the mean is the most frequently used measure of central tendency and the mode is the least used.

2.6 Find the mean for the grouped data in Table 2.12 by *coding* (i.e., by assigning the value of $\mu = 0$ to the 4th or 5th classes and $\mu = -1$, $\mu = -2$, etc. to each lower class and $\mu = 1$, $\mu = 2$, etc. to each larger class and then using the formula

$$\bar{X} = X_0 + \frac{\sum f\mu}{n} c \tag{2.16}$$

where X_0 is the midpoint of the class assigned $\mu = 0$ and c is the width of the class intervals). See Table 2.15.

Table 2.15   Calculation of the Sample Mean by Coding for the Grouped Data in Table 2.12

Hourly Wage, $	Class Midpoint X, $	Code μ	Frequency f	$f\mu$
3.50–3.59	3.55	−3	1	−3
3.60–3.69	3.65	−2	2	−4
3.70–3.79	3.75	−1	2	−2
3.80–3.89	3.85	0	4	0
3.90–3.99	3.95	1	5	5
4.00–4.09	4.05	2	6	12
4.10–4.19	4.15	3	3	9
4.20–4.29	4.25	4	2	8
			$\sum f = n = 25$	$\sum f\mu = 25$

$$\bar{X} = X_0 + \frac{\sum f\mu}{n} c = \$3.85 + \frac{25}{25}(\$0.10) = \$3.85 + \$0.10 = \$3.95$$

$\bar{X}$ for the grouped data formed by coding is identical to that found in Prob. 2.4*b* without coding. Coding eliminates the problem of having to deal with possibly large and inconvenient class midpoints; thus it may simplify the calculations.

2.7 A firm pays a wage of $4 per hour to its 25 unskilled workers, $6 to its 15 semiskilled workers, and $8 to its 10 skilled workers. What is the *weighted average*, or *weighted mean*, wage paid by this firm?

In find the weighted mean, or weighted average, of a population, μ_w, or sample, $\bar{X}_w$, the weights, w, have the same function as the frequency in finding the mean for the grouped data. Thus

$$\bar{X}_w \quad \text{or} \quad \mu_w = \frac{\sum wX}{\sum w} \tag{2.17}$$

For this problem, the weights are the number of workers employed at each wage, and $\sum w$ equals the sum of all the workers:

$$\mu_w = \frac{(\$4)(25) + (\$6)(15) + (\$8)(10)}{25 + 15 + 10} = \frac{\$100 + \$90 + \$80}{50} = \frac{\$270}{50} = \$5.40$$

This weighted average compares with the simple average of $6 [($4 + $6 + $8)/3 = $6]$ and is a better measure of the average wages.

2.8 A nation faces a rate of inflation of 2% in one year, 5% in the second year, and 12.5% in the third year. Find the geometric mean of the inflation rates (the *geometric mean*, μ_G or $\bar{X}_G$, of a set of n positive numbers is the nth root of their product and is used mainly to average rates of change and index numbers):

$$\mu_G \quad \text{or} \quad \bar{X}_G = \sqrt[n]{X_1 \cdot X_2 \cdots X_n} \tag{2.18}$$

where $X_1, X_2, \ldots, X_n$ refer to the n (or N) observations.

$$\mu_G = \sqrt[3]{(2)(5)(12.5)} = \sqrt[3]{125} = 5\%$$

This compares with $\mu = (2 + 5 + 12.5)/3 = 19.5/3 = 6.5\%$. When all the numbers are equal, μ_G equals μ; otherwise μ_G is smaller than μ. In practice, μ_G is calculated by logarithms:

$$\log \mu_G = \frac{\sum \log x}{N} \tag{2.19}$$

The geometric mean is used primarily in the mathematics of finance and financial management.

2.9   A commuter drives 10 mi on the highway at 60 mi/h and 10 mi on local streets at 15 mi/h. Find the harmonic mean. The harmonic mean μ_H is used primarily to average ratios:

$$\mu_H = \frac{N}{\sum(1/X)} \tag{2.20}$$

$$= \frac{2}{(1/60) + (1/15)} = \frac{2}{(1+4)/60}$$

$$= \frac{2}{5/60} = 2\frac{60}{5} = \frac{120}{5} = 24 \text{ mi/h}$$

as compared with $\mu = \sum X/N = (60 + 15)/2 = 75/2 = 37.5 \text{ mi/h}$. Note that if the commuter had averaged 37.5 mi/h, it would have taken her $(20 \text{ mi}/37.5 \text{ mi})60 \text{ min} = 32 \text{ min}$ to drive the 20 mi. Instead she drives 6 min on the highway (10 mi at 60 mi/h) and 40 min on local streets (10 mi at 15 mi/h) for a total of 50 min, and this is the (correct) answer we get by using $\mu_H = 24 \text{ mi/h}$. That is, $(20 \text{ mi}/24 \text{ mi/h}) \times 60 \text{ min} = 50 \text{ min}$.

2.10   (a) For the ungrouped data in Table 2.7, find the first, second, and third quartiles and the third deciles and sixtieth percentiles.   (b) Do the same for the grouped data in Table 2.12. (*Quartiles* divide the data into 4 parts, *deciles* into 10 parts, and *percentiles* into 100 parts.)

(a) Q_1 (first quartile) = 4 (the average of the 10th and 11th values in Table 2.8)

Q_2 (second quartile) = 6 = the value of the 20.5th item = the median

Q_3 (third quartile) = 7.5 = the value of the 30.5th item

D_3 (third decile) = 5 = the value of the 12.5th item

P_{60} (sixtieth percentile) = 7 = the value of the 24.5th item

(b)

$$Q_1 = L + \frac{n/4 - F}{f_1}c$$

$$= \$3.80 + \frac{25/4 - 5}{4}(\$0.10) = \$3.80 + \$0.03125 \cong \$3.83 \tag{2.21}$$

$$Q_2 = L + \frac{n/2 - F}{f_2}c$$

$$= \$3.90 + \frac{25/2 - 9}{5}(\$0.10) = \$3.90 + \$0.07 = \$3.97 = \text{median} \tag{2.22}$$

$$Q_3 = L + \frac{3n/4 - F}{f_3}c$$

$$= \$4.00 + \frac{75/4 - 14}{6}(\$0.10) = \$4.00 + \$0.0792 \cong \$4.08 \tag{2.23}$$

$$D_3 = L + \frac{3n/10 - F}{f_3} c$$

$$= \$3.80 + \frac{75/10 - 5}{4} (\$0.10) = \$3.80 + \$0.0625 = \$3.86 \qquad (2.24)$$

$$P_{60} = L + \frac{60n/100 - F}{f_{60}} c$$

$$= \$4.00 + \frac{1500/100 - 14}{6} (\$0.10) = \$4.00 + \$0.0167 \cong \$4.02 \qquad (2.25)$$

MEASURES OF DISPERSION

2.11 (*a*) Find the range for the ungrouped data in Table 2.7. (*b*) Find the range for the ungrouped data in Table 2.10 and for the grouped data in Table 2.12. (*c*) What are the advantages and disadvantages of the range?

(*a*) The range for ungrouped data is equal to the value of the largest observation minus the value of the smallest observation in the data set. The range for the ungrouped data in Table 2.7 is from 2 to 10, or 8 points.

(*b*) The range for the ungrouped data in Table 2.10 is from \$3.55 to \$4.26, or \$0.71. For grouped data, the range extends from the lower limit of the smallest class to the upper limit of the largest class. For the grouped data in Table 2.12, the range extends from \$3.50 to \$4.29.

(*c*) The advantages of the range are that it is easy to find and understand. Its disadvantages are that it considers only the lowest and highest values of a distribution, it is greatly influenced by extreme values, and it cannot be found for open-ended distributions. Because of these disadvantages, the range is of limited usefulness (except in quality control).

2.12 Find the interquartile range and the quartile deviation (*a*) for the ungrouped data in Table 2.7 and (*b*) for the grouped data in Table 2.12.

(*a*) The interquartile range is equal to the difference between the third and first quartiles:

$$\text{IR} = Q_3 - Q_1 \qquad (2.26)$$

For the ungrouped data in Table 2.7, IR $= 7.5 - 4 = 3.5$ points [utilizing the values of Q_3 and Q_4 found in Prob. 2.10 (*a*)]. Note that the interquartile range is not affected by extreme values because it utilizes only the middle half of the data. It is thus better than the range, but it is not as widely used as the other measures of dispersion. For the quartile deviation,

$$\text{QD} = \frac{Q_3 - Q_1}{2} \qquad (2.27)$$

Therefore, QD $= (7.5 - 4)/2 = 3.5/2 = 1.75$ points. Quartile deviation measures the average range of one-fourth of the data.

(*b*) IR $= Q_3 - Q_1 = \$4.08 - \$3.83 = \$0.25$ [utilizing the values of Q_3 and Q_1 found in Prob. 2.10(*b*)]:

$$\text{QD} = \frac{Q_3 - Q_1}{2} = \frac{\$4.08 - \$3.83}{2} = \$0.125$$

2.13 Find the average deviation for (*a*) the ungrouped data in Table 2.7 and (*b*) for the grouped data in Table 2.9.

(*a*) Since $\mu = 6$ [see Prob. 2.3(*a*)],

$$\sum |X - \mu| = 1 + 1 + 0 + 4 + 2 + 1 + 0 + 1 + 3 + 3 + 4 + 2 + 1 + 1 + 2 + 0 + 1 + 2 + 2 + 4$$
$$+ 3 + 1 + 0 + 1 + 3 + 2 + 4 + 2 + 1 + 3 + 2 + 0 + 1 + 2 + 3 + 0 + 1 + 3 + 4 + 1$$
$$= 72$$

$$\text{AD} = \frac{\sum |X - \mu|}{N} = \frac{72}{40} = 1.8 \text{ points}$$

Note that the average deviation takes every observation into account. It measures the average of the absolute deviation of each observation from the mean. It takes the absolute value (indicated by the two vertical bars) because $\sum(X - \mu) = 0$ (see Example 5).

(b) We can find the average deviation for the same grouped data with the aid of Table 2.16:

$$AD = \frac{\sum f|X - \mu|}{N} = \frac{72}{40} = 1.8 \text{ points}$$

the same as we found for the ungrouped data.

Table 2.16 Calculations for the Average Deviation for the Grouped Data in Table 2.9

Grade	Class Midpoint X	Frequency f	Mean μ	$X - \mu$	$|X - \mu|$	$f|X - \mu|$
1.5–2.4	2	3	6	−4	4	12
2.5–3.4	3	3	6	−3	3	9
3.5–4.4	4	5	6	−2	2	10
4.5–5.4	5	5	6	−1	1	5
5.5–6.4	6	6	6	0	0	0
6.5–7.4	7	8	6	1	1	8
7.5–8.4	8	4	6	2	2	8
8.5–9.4	9	4	6	3	3	12
9.5–10.4	10	2	6	4	4	8
		$\sum f = N = 40$				$\sum|X - \mu| = 72$

2.14 Find the average deviation for the grouped data in Table 2.12.

We can find the average deviation for the grouped data of hourly wages in Table 2.12 with the aid of Table 2.17 $[\overline{X} = \$3.95$; see Prob. 2.4(b)]:

$$AD = \frac{\sum f|X - \overline{X}|}{n} = \frac{\$3.60}{25} = \$0.144$$

Note that the average deviation found for the grouped data is an *estimate* of the "true" average deviation that could be found for the ungrouped data. It usually differs slightly from the true average deviation because we use the estimate of the mean for the grouped data in our calculations [compare the values of $\overline{X}$ found in Prob. 2.4(a) and (b)].

Table 2.17 Calculations for the Average Deviation for the Grouped Data in Table 2.12

Hourly Wage, $	Class Midpoint X, $	Frequency f	Mean $\overline{X}$, $	$X - \overline{X}$, $	$|X - \overline{X}|$, $	$f|X - \overline{X}|$, $
3.50–3.59	3.55	1	3.95	−0.40	0.40	0.40
3.60–3.69	3.65	2	3.95	−0.30	0.30	0.60
3.70–3.79	3.75	2	3.95	−0.20	0.20	0.40
3.80–3.89	3.85	4	3.95	−0.10	0.10	0.40
3.90–3.99	3.95	5	3.95	0.00	0.00	0.00
4.00–4.09	4.05	6	3.95	0.10	0.10	0.60
4.10–4.19	4.15	3	3.95	0.20	0.20	0.60
4.20–4.29	4.25	2	3.95	0.30	0.30	0.60
		$\sum f = n = 25$				$\sum f|X - \overline{X}| = \$3.60$

2.15 Find the variance and the standard deviation for (a) the ungrouped data in Table 2.7 and (b) the grouped data in Table 2.9. (c) What is the advantage of the standard deviation over the variance?

(a)
$$\sigma^2 = \frac{\sum(X-\mu)^2}{N} \quad \text{and} \quad \mu = 6 \quad \text{(see Prob. 2.3a)}$$

$$\sum(X-\mu)^2 = 1+1+0+16+4+1+0+1+9+9+16+4+1+1+4+0+1+4+4+16$$
$$+9+1+0+1+9+4+16+4+1+9+4+0+1+4+9+0+1+9+16+1$$
$$= 192$$

$$\sigma^2 = \frac{\sum(X-\mu)^2}{N} = \frac{192}{40} = 4.8 \text{ points squared}$$

$$\sigma = \sqrt{\frac{\sum(X-\mu)^2}{N}} = \sqrt{\frac{192}{40}} = \sqrt{4.8} \cong 2.19 \text{ points}$$

(b) We can find the variance and the standard deviation for the grouped data of grades with the aid of Table 2.18:

$$\sigma^2 = \frac{\sum f(X-\mu)^2}{N} = \frac{192}{40} = 4.8 \text{ points squared}$$

and
$$\sigma = \sqrt{\sigma^2} = \sqrt{4.8} \cong 2.19 \text{ points}$$

the same as we found for the ungrouped data.

Table 2.18 Calculations for the Variance and Standard Deviation for the Data in Table 2.9

Grade	Class Midpoints X	Frequency f	Mean μ	$X - \mu$	$(X-\mu)^2$	$f(X-\mu)^2$
1.5–2.4	2	3	6	−4	16	48
2.5–3.4	3	3	6	−3	9	27
3.5–4.4	4	5	6	−2	4	20
4.5–5.4	5	5	6	−1	1	5
5.5–6.4	6	6	6	0	0	0
6.5–7.4	7	8	6	1	1	8
7.5–8.4	8	4	6	2	4	16
8.5–9.4	9	4	6	3	9	36
9.5–10.4	10	2	6	4	16	32
		$\sum f = N = 40$				$\sum f(X-\mu)^2 = 192$

(c) The advantage of the standard deviation over the variance is that the standard deviation is expressed in the same units as the data rather than in "the width squared," which is how the variance is expressed. The standard deviation is by far the most widely used measure of (absolute) dispersion.

2.16 Find the variance and the standard deviation for the grouped data in Table 2.10.

We can find the variance and the standard deviation for the grouped data of hourly wages with the aid of Table 2.19 [$\overline{X} = \$3.95$; see Prob. 2.4(b)]:

$$s^2 = \frac{\sum f(X-\overline{X})^2}{n-1} = \frac{0.82}{24} \cong 0.0342 \text{ dollars squared}$$

and
$$s = \sqrt{\frac{\sum f(X-\overline{X})^2}{n-1}} = \sqrt{0.0342} = \$0.18$$

Table 2.19　Calculations for the Variance and Standard Deviation for the Data in Table 2.12

Hourly Wage, $	Class Midpoint X, $	Frequency f	Mean $\overline{X}$, $	$X - \overline{X}$, $	$(X - \overline{X})^2$	$f(X - \overline{X})^2$
3.50–3.59	3.55	1	3.95	−0.40	0.16	0.16
3.60–3.69	3.65	2	3.95	−0.30	0.09	0.18
3.70–3.79	3.75	2	3.95	−0.20	0.04	0.08
3.80–3.89	3.85	4	3.95	−0.10	0.01	0.04
3.90–3.99	3.95	5	3.95	0.00	0.00	0.00
4.00–4.09	4.05	6	3.95	0.10	0.01	0.06
4.10–4.19	4.15	3	3.95	0.20	0.04	0.12
4.20–4.29	4.25	2	3.95	0.30	0.09	0.18
		$\sum f = n = 25$				$\sum f(X - \overline{X})^2 = 0.82$

Note that in the formula for s^2 and s, $n - 1$ rather than n is used in the denominator. The reason for this is that if we take many samples from a population, the average of the sample variances does not tend to equal population variance, σ^2, unless we use $n - 1$ in the denominator of the formula for s^2 (more will be said on this in Chap. 5). Furthermore, s^2 and s for the grouped data are *estimates* for the true s^2 and s that could be found for the ungrouped data because we use the estimate of $\overline{X}$ from the grouped data in our calculations.

2.17　Starting with the formula for σ^2 and s^2 given in Sec. 2.3, prove that

(a)
$$\sigma^2 = \frac{\sum X^2 - N\mu^2}{N} \quad \text{and} \quad s^2 = \frac{\sum X^2 - n\overline{X}^2}{n - 1} \qquad (2.28a, b)$$

(b)
$$\sigma^2 = \frac{\sum fX^2 - N\mu^2}{N} \quad \text{and} \quad s^2 \cong \frac{\sum fX^2 - n\overline{X}^2}{n - 1} \qquad (2.29a, b)$$

(a)
$$\sigma^2 = \frac{\sum(X - \mu)^2}{N} = \frac{\sum(X^2 - 2X\mu + \mu^2)}{N} = \frac{\sum X^2 - 2\mu\sum X + N\mu^2}{N}$$
$$= \frac{\sum X^2}{N} - 2\mu^2 + \mu^2 = \frac{\sum X^2 - N\mu^2}{N}$$

We can get s^2 by simply replacing μ with $\overline{X}$ and N with n in the numerator and N with $n - 1$ in the denominator of the formula for σ^2.

(b)
$$\sigma^2 = \frac{\sum f(X - \mu)^2}{N} = \frac{\sum f(X^2 - 2X\mu + \mu^2)}{N} = \frac{\sum fX^2 - 2\mu\sum fX + N\mu^2}{N}$$
$$= \frac{\sum fX^2}{N} - 2\mu^2 + \mu^2 = \frac{\sum fX^2 - N\mu^2}{N}$$

We can get s^2 in the same way as we did in part a. The preceding formulas will simplify the calculations for σ^2 and s^2 for a large body of data. *Coding* also helps (see Prob. 2.6).

2.18　Find the variance and the standard deviation for　(a) the ungrouped data in Table 2.7 and (b) the grouped data in Table 2.9, *using the simpler computational formulas in Prob. 2.17.*

(a)

$$\sum X^2 = 49 + 25 + 36 + 4 + 64 + 49 + 36 + 49 + 9 + 81 + 100 + 16 + 25 + 25$$
$$+ 16 + 36 + 49 + 18 + 64 + 4 + 9 + 25 + 36 + 49 + 81 + 64 + 4 + 16 + 49$$
$$+ 81 + 16 + 36 + 49 + 64 + 9 + 36 + 49 + 81 + 100 + 25$$
$$= 1,632$$

$$\mu = \frac{\sum X}{N} = \frac{240}{40} = 6$$

$$\sigma^2 = \frac{\sum X^2 - N\mu^2}{N} = \frac{1,632 - (40)(36)}{40} = \frac{1,632 - 1,440}{40} = \frac{192}{40} = 4.8 \text{ points squared}$$

$$\sigma = \sqrt{\sigma^2} = \sqrt{4.8} \cong 2.19 \text{ points}$$

the same as in Prob. 2.15(a).

(b) We can find σ^2 and σ for the grouped data in Table 2.9 with the aid of Table 2.20:

$$\mu = \frac{\sum fX}{N} = \frac{240}{6} = 6$$

$$\sigma^2 = \frac{\sum fX^2 - N\mu^2}{N} = \frac{1,632 - (40)(36)}{40} = \frac{1,632 - 1,440}{40} = \frac{192}{40} = 4.8 \text{ points squared}$$

$$\sigma = \sqrt{\sigma^2} = \sqrt{4.8} \cong 2.19 \text{ points}$$

the same as in part a and Prob. 2.15.

Table 2.20   Calculations for the Variance and Standard Deviation for the Grouped Data in Table 2.9

Grade	Class Midpoint X	Frequency f	fX	X^2	fX^2
1.5–2.4	2	3	6	4	12
2.5–3.4	3	3	9	9	27
3.5–4.4	4	5	20	16	80
4.5–5.4	5	5	25	25	125
5.5–6.4	6	6	36	36	216
6.5–7.4	7	8	56	49	392
7.5–8.4	8	4	32	64	256
8.5–9.4	9	4	36	81	324
9.5–10.4	10	2	20	100	200
		$\sum f = N = 40$	$\sum fX = 240$		$\sum fX^2 = 1,632$

2.19 Find the variance and the standard deviation for the grouped data in Table 2.12 using the simpler computational formula given in Prob. 2.17(b).

We can find s^2 and s for the grouped data in Table 2.12 with the aid of Table 2.21:

$$\overline{X} = \frac{\sum fX}{n} = \frac{98.75}{25} = \$3.95$$

$$s^2 = \frac{\sum fX^2 - n\overline{X}^2}{n-1} = \frac{390.8825 - (25)(15.6025)}{24} = \frac{390.8825 - 390.0625}{24} = \frac{0.82}{24}$$
$$\cong 0.0342 \text{ dollars squared}$$

and

$$s \cong \sqrt{0.0342} \cong \$0.18$$

the same as we found in Prob. 2.16.

Table 2.21   Calculations for the Variance and Standard Deviation for the Grouped Data in Table 2.12

Hourly Wage, $	Class Midpoint X, $	Frequency f	fX, $	X^2	fX^2
3.50–3.59	3.55	1	3.55	12.6025	12.6025
3.60–3.69	3.65	2	7.30	13.3225	26.6450
3.70–3.79	3.75	2	7.50	14.0625	28.1250
3.80–3.89	3.85	4	15.40	14.8225	59.2900
3.90–3.99	3.95	5	19.75	15.6025	78.0125
4.00–4.09	4.05	6	24.30	16.4025	98.4150
4.10–4.19	4.15	3	12.45	17.2225	51.6675
4.20–4.29	4.25	2	8.50	18.0625	36.1250
		$\sum f = n = 25$	$\sum fX = \$98.75$		$\sum fX^2 = 390.8825$

2.20 Find the coefficient of variation V for the data in (a) Table 2.7 and (b) Table 2.12. (c) What is the usefulness of the coefficient of variation?

(a) with $\mu = 6$ and $\sigma \cong 2.19$ (see Prob. 2.19)

$$V = \frac{\sigma}{\mu} \cong \frac{2.19 \text{ points}}{6 \text{ points}} \cong 0.635, \quad \text{or} \quad 6.35\%$$

(b) With $\overline{X} = \$3.95$ and $s \cong \$0.18$ (see Prob. 2.19)

$$V = \frac{s}{\overline{X}} \cong \frac{\$0.18}{\$3.95} \cong 0.046, \quad \text{or} \quad 4.6\%$$

(c) The coefficient of variation measures the *relative* dispersion in the data and is expressed as a pure number without any units.   This is to be contrasted with standard deviation and other measures of *absolute* dispersion, which are expressed in the units of the problem.   Thus the coefficient of variation can be used to compare the relative dispersion of two or more distributions expressed in different units, as well as when the true mean values differ.   For example, we can say that the dispersion of the data in Table 2.7 is greater than that in Table 2.12.   The coefficient of variation also can be used to compare the relative dispersion of the same type of data over different time periods (when μ or $\overline{X}$ and σ or s change).

SHAPE OF FREQUENCY DISTRIBUTIONS

2.21 Find the Pearson coefficient of skewness for the (grouped) data in (a) Table 2.9 and (b) Table 2.12.

(a) With $\mu = 6$, med $= 6.17$ [see Prob. 2.3(b)], and $\sigma \cong 2.19$ [see Prob. 2.15(b)]

$$\text{Sk} = \frac{3(\mu - \text{med})}{\sigma} \cong \frac{3(6 - 6.17)}{2.19} \cong \frac{3(-0.17)}{2.19} \cong -0.23 \quad \text{(a pure number)}$$

Note that median is greater than mean and that the distribution is slightly negatively skewed (see Fig. 2-5c).

(b) With $\overline{X} = \$3.95$, med $= \$3.97$ [see Prob. 2.4(b)], and $s \cong \$0.18$ (see Prob. 2.16)

$$\text{Sk} = \frac{3(\overline{X} - \text{med})}{s} \cong \frac{3(3.95 - 3.97)}{0.18} = \frac{3(-0.02)}{0.18} = -0.33$$

(see Fig. 2-6c).

2.22   Using the formula for skewness based on the third moment, find the coefficient of skewness for the data in   (*a*) Table 2.9 and   (*b*) Table 2.12.

(*a*)   We can find the coefficient of skewness for the data in Table 2.9 using the formula based on the third moment with the aid of Table 2.22:

$$\text{Sk} = \frac{\sum f(X - \mu)^3}{s^3} \cong \frac{-42}{2.19^3} = \frac{-42}{10.50349} \cong -4$$

This indicates that this distribution is negatively skewed, but the *degree* of skewness is measured differently than in Prob. 2.21.

Table 2.22   Calculations for Skewness for the Data in Table 2.9

Grade	Class Midpoint X	Frequency f	Mean μ	$X - \mu$	$(X - \mu)^3$	$f(X - \mu)^3$
1.5–2.4	2	3	6	−4	−64	−192
2.5–3.4	3	3	6	−3	−27	−81
3.5–4.4	4	5	6	−2	−8	−40
4.5–5.4	5	5	6	−1	−1	−5
5.5–6.4	6	6	6	0	0	0
6.5–7.4	7	8	6	1	1	8
7.5–8.4	8	4	6	2	8	32
8.5–9.4	9	4	6	3	27	108
9.5–10.4	10	2	6	4	64	128
		$\sum f = N = 40$				$\sum f(X - \mu)^3 = -42$

(*b*)   See Table 2.23.

$$\text{Sk} = \frac{\sum f(X - \overline{X})^3}{s^3} = \frac{-0.054}{0.18^3} \cong \frac{-0.054}{0.006} \cong -9$$

Note that regardless of the measure of skewness used, the distributions of the data in Tables 2.9 and 2.12 are negatively skewed, with the latter more negatively skewed than the former.

Table 2.23   Calculations for Skewness for the Data in Table 2.12

Hourly Wages, $	Class Midpoint X, $	Frequency f	Mean $\overline{X}$, $	$X - \overline{X}$, $	$(X - \overline{X})^3$	$f(X - \overline{X})^3$
3.50–3.59	3.55	1	3.95	−0.40	−0.064	−0.064
3.60–3.69	3.65	2	3.95	−0.30	−0.027	−0.054
3.70–3.79	3.75	2	3.95	−0.20	−0.008	−0.016
3.80–3.89	3.85	4	3.95	−0.10	−0.001	−0.004
3.90–3.99	3.95	5	3.95	0	0	0
4.00–4.09	4.05	6	3.95	0.10	0.001	0.006
4.10–4.19	4.15	3	3.95	0.20	0.008	0.024
4.20–4.29	4.25	2	3.95	0.30	0.027	0.054
						$\sum f(X - \overline{X})^3 = -0.054$

2.23 Find the coefficient of kurtosis for the data in (a) Table 2.9 and (b) Table 2.12.

(a) We can find the coefficient of kurtosis for the data in Table 2.9 with the aid of Table 2.24:

$$\text{Kurtosis} = \frac{\sum f(X - \mu)^4}{\sigma^4} \cong \frac{2{,}004}{2.19^4} \cong \frac{2{,}004}{23.00} \cong 87.13 \quad \text{(a pure number)}$$

Thus the distribution of grades is very peaked (leptokurtic; see Fig. 2-5c).

Table 2.24 Calculations for Kurtosis for the Data in Table 2.9

Grade	Class Midpoint X	Frequency f	Mean μ	$X - \mu$	$(X - \mu)^4$	$f(X - \mu)^4$
1.5–2.4	2	3	6	−4	256	768
2.5–3.4	3	3	6	−3	81	243
3.5–4.4	4	5	6	−2	16	80
4.5–5.4	5	5	6	−1	1	5
5.5–6.4	6	6	6	0	0	0
6.5–7.4	7	8	6	1	1	8
7.5–8.4	8	4	6	2	16	64
8.5–9.4	9	4	6	3	81	324
9.5–10.4	10	2	6	4	256	512
		$\sum f = N = 40$				$\sum f(X - \mu)^4 = 2{,}004$

(b) Table 2.25 will aid us here:

$$\text{Kurtosis} = \frac{\sum f(X - \overline{X})^4}{s^2} \cong \frac{0.067}{0.001} \cong 67$$

Thus the distribution of wages is also leptokurtic (see Fig. 2-6c), but less than the distribution of grades.

Table 2.25 Calculations for Kurtosis for the Data in Table 2.12

Hourly Wages, $	Class Midpoint X, $	Frequency f	Mean $\overline{X}$, $	$X - \overline{X}$, $	$(X - \overline{X})^4$	$f(X - \overline{X})^4$
3.50–3.59	3.55	1	3.95	−0.40	0.0256	0.0256
3.60–3.69	3.65	2	3.95	−0.30	0.081	0.0162
3.70–3.79	3.75	2	3.95	−0.20	0.0016	0.0032
3.80–3.89	3.85	4	3.95	−0.10	0.0001	0.0004
3.90–3.99	3.95	5	3.95	0	0	0
4.00–4.09	4.05	6	3.95	0.10	0.0001	0.0006
4.10–4.19	4.15	3	3.95	0.20	0.0016	0.0048
4.20–4.29	4.25	2	3.95	0.30	0.0081	0.0162
						$\sum f(X - \overline{X})^4 = 0.0670$

2.24   Find the covariance between hourly wage X and education Y, measured in years of schooling in the data in Table 2.26.

Table 2.26   Employee Hourly Wages and Years of Schooling

Employee Number	Hourly Wage X, $	Years of Schooling Y
1	8.50	12
2	12.00	14
3	9.00	10
4	10.50	12
5	11.00	16
6	15.00	16
7	25.00	18
8	12.00	18
9	6.50	12
10	8.25	10

From the calculations in Table 2.27, $\operatorname{cov}(X, Y) = (103.55/10) = 10.355$.   When X and Y are both above or below their means, covariance is increased.   When X and Y move in opposite directions relative to their means (employee 5), covariance is decreased.   Since in this case $\operatorname{cov}(X, Y) > 0$, X and Y move together relative to their means.

Table 2.27   Calculations for Covariance

Employee Number	Hourly Wage X, $	Years of Schooling Y	$(X - \overline{X})$	$(Y - \overline{Y})$	$(X - \overline{X})(Y - \overline{Y})$
1	8.50	12	−3.275	−1.8	5.895
2	12.00	14	0.225	0.2	0.045
3	9.00	10	−2.775	−3.8	10.545
4	10.50	12	−1.275	−1.8	2.295
5	11.00	16	−0.775	2.2	−1.705
6	15.00	16	3.225	2.2	7.095
7	25.00	18	13.225	4.2	55.545
8	12.00	18	0.225	4.2	0.945
9	6.50	12	−5.275	−1.8	9.495
10	8.25	10	−3.525	−3.8	13.395
	$\overline{X} = 11.775$	$\overline{Y} = 13.8$			$\Sigma(X - \overline{X})(Y - \overline{Y}) = 103.55$

2.25 Compute the covariance from Table 2.26 using the alternate formula.
Computations are given in Table 2.28. $\text{cov}(X, Y) = (1728.5/10) - (11.775)(13.8) = 172.85 - 162.495 = 10.355$.

Table 2.28 Calculations for Covariance with Alternate Formula

Employee Number	Hourly Wage X, $	Years of Schooling Y	XY
1	8.50	12	102
2	12.00	14	168
3	9.00	10	90
4	10.50	12	126
5	11.00	16	176
6	15.00	16	240
7	25.00	18	450
8	12.00	18	216
9	6.50	12	78
10	8.25	10	82.5
	$\overline{X} = 11.775$	$\overline{Y} = 13.8$	$\Sigma XY = 1,728.5$

Supplementary Problems

FREQUENCY DISTRIBUTIONS

2.26 Table 2.29 gives the frequency for gasoline prices at 48 stations in a town. Present the data in the form of a histogram, a relative-frequency histogram, a frequency polygon, and an ogive.

Table 2.29 Frequency Distribution of Gasoline Prices

Price, $	Frequency
1.00–1.04	4
1.05–1.09	6
1.10–1.14	10
1.15–1.19	15
1.20–1.24	8
1.25–1.29	5

2.27 Table 2.30 gives the frequency distribution of family incomes for a sample of 100 families in a city. Graph the data into a histogram, a relative-frequency histogram, a frequency polygon, and an ogive.

Table 2.30 Frequency Distribution of
Family Incomes

Family Income, $	Frequency
10,000–11,999	12
12,000–13,999	14
14,000–15,999	24
16,000–17,999	15
18,000–19,999	13
20,000–21,999	7
22,000–23,999	6
24,000–25,999	4
26,000–27,999	3
28,000–29,999	2
	100

MEASURES OF CENTRAL TENDENCY

2.28 Find (a) the mean, (b) the median, and (c) the mode for the grouped data in Table 2.29.
Ans. (a) $\mu = \$1.15$ (b) Median = $1.16 (c) Mode = $1.17

2.29 Find (a) the mean, (b) the median, and (c) the mode for the frequency distribution of incomes in Table 2.30.
Ans. (a) $\overline{X} = \$17,000$, (b) Median = $16,000 (c) Mode = $15,053

2.30 Find the mean for the grouped data in (a) Table 2.29 and (b) Table 2.30 by coding.
Ans. (a) $\mu = \$1.15$ (b) $\overline{X} = \$17,000$

2.31 A firm pays 5/12 of its labor force an hourly wage of $5, 1/3 of the labor force a wage of $6, and 1/4 a wage of $7. What is the weighted average paid by this firm?
Ans. $\mu_w \cong \$5.83$

2.32 For the same amount of capital invested in each of 3 years, an investor earned a rate of return of 1% during the first year, 4% during the second year, and 16% during the third. (a) Find μ_G. (b) Find μ. (c) Which is appropriate?
Ans. (a) $\mu_G = 4\%$ (b) $\mu = 7\%$ (c) μ_G

2.33 A plane traveled 200 mi at 600 mi/h and 100 mi at 500 mi/h. What was its average speed?
Ans. $\mu_H = 562.5$ mi/h

2.34 A driver purchases $10 worth of gasoline at $0.90 a gallon and $10 at $1.10 a gallon. What is the average price per gallon?
Ans. $\mu_H \cong \$0.99$ per gallon

2.35 For the grouped data of Table 2.29, find (a) the first quartile, (b) the second quartile, (c) the third quartile, (d) the fourth decile, and (e) the seventieth percentile.
Ans. (a) $Q_1 = \$1.11$ (b) $Q_2 \cong \$1.16$ (c) $Q_3 \cong \$1.21$ (d) $D_4 = \$1.146$ (e) $P_{70} \cong \$1.195$

2.36 For the grouped data in Table 2.30, find (a) the first quartile, (b) the third quartile, (c) the third decile, and (d) the sixtieth percentile.
Ans. (a) $Q_1 \cong \$13,857$ (b) $Q_3 \cong \$19,538$ (c) $D_3 \cong \$14,333$ (d) $P_{60} \cong \$17,333$

MEASURES OF DISPERSION

2.37 What is the range of the distribution of (a) gasoline prices in Table 2.29 and (b) family incomes in Table 2.30?
    Ans. (a) $0.29   (b) $10,000 to $29,999, or $20,000

2.38 Find the interquartile range and quartile deviation for the data in (a) Table 2.29 and (b) Table 2.30.
    Ans. (a) IR $\cong$ $0.10 and QD $\cong$ $0.05   (b) IR $\cong$ $476 and QD $\cong$ $238

2.39 Find the average deviation for the data in (a) Table 2.29 and (b) Table 2.30.
    Ans. (a) $0.0575   (b) $3,520

2.40 Find (a) the variance and (b) the standard deviation for the frequency distribution of gasoline prices in Table 2.29.
    Ans. (a) $\sigma^2 \cong 0.0048$ dollars squared   (b) $\sigma \cong$ $0.0693

2.41 Find (a) the variance and (b) the standard deviation for the frequency distribution of family incomes in Table 2.30.
    Ans. (a) $s^2 = 19,760,000$ dollars squared   (b) $s \cong$ $4,445.22

2.42 Using the *easier computational formulas*, find (a) the variance and (b) the standard deviation for the distribution of gasoline prices in Table 2.29.
    Ans. (a) $\sigma^2 \cong 0.0048$ dollars squared   (b) $\sigma \cong$ $0.0693

2.43 Using the *easier computational formulas*, find (a) the variance and (b) the standard deviation for the family incomes in Table 2.30.
    Ans. (a) $s^2 = 19,760,000$ dollars squared   (b) $s \cong$ $4,445.22

2.44 Find the coefficient of variation V for (a) the data in Table 2.29 and (b) the data in Table 2.30. (c) Which data have the greater dispersion?
    Ans. (a) 0.060, or 6%   (b) 0.261, or 26.1%   (c) The data of Table 2.30.

SHAPE OF FREQUENCY DISTRIBUTIONS

2.45 Find the Pearson coefficient of skewness for the data in (a) Table 2.29 and (b) Table 2.30.
    Ans. (a) -0.43 (b) 0.67

2.46 Find the coefficient of skewness using the formula based on the third moment for the data in (a) Table 2.29 and (b) Table 2.30.
    Ans. (a) -1.88 (b) 755

2.47 Find the coefficient of kurtosis for the data in (a) Table 2.29 and (b) Table 2.30.
    Ans. (a) 177   (b) 300

2.48 For covariance, (a) in what range should the covariance for directly related data fall? (b) for inversely related data? (c) for unrelated data?
    Ans. (a) cov > 0 (b) cov < 0 (c) cov $= 0$

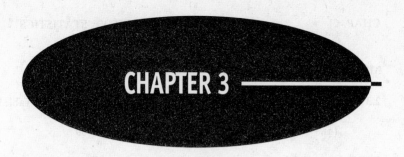

CHAPTER 3

Probability and Probability Distributions

3.1 PROBABILITY OF A SINGLE EVENT

If event A can occur in n_A ways out of a total of N possible and equally likely outcomes, the probability that event A will occur is given by

$$P(A) = \frac{n_A}{N} \qquad (3.1)$$

where $P(A)$ = probability that event A will occur

$\quad n_A$ = number of ways that event A can occur

$\quad N$ = total number of equally possible outcomes

Probability can be visualized with a *Venn diagram*. In Fig. 3-1, the circle represents event A, and the *total* area of the rectangle represents all possible outcomes.

$P(A)$ ranges between 0 and 1:

$$0 \leq P(A) \leq 1 \qquad (3.2)$$

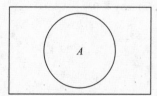

Fig. 3-1

If $P(A) = 0$, event A cannot occur. If $P(A) = 1$, event A will occur with certainty.

36

If $P(A')$ represents the probability of *nonoccurrence* of event A, then

$$P(A) + P(A') = 1 \qquad\qquad (3.3)$$

EXAMPLE 1.   A head (H) and a tail (T) are the two equally possible outcomes in tossing a balanced coin.   Thus

$$P(H) = \frac{n_H}{N} = \frac{1}{2}$$
$$P(T) = \frac{n_T}{N} = \frac{1}{2}$$

and $\qquad\qquad P(H) + P(T) = 1$

EXAMPLE 2.   In rolling a fair die once, there are six possible and equally likely outcomes: 1, 2, 3, 4, 5, and 6.   Thus

$$P(1) = P(2) = P(3) = P(4) = P(5) = P(6) = \frac{1}{6}$$

The probability of not rolling a 1 is

$$P(1') = 1 - P(1) = 1 - \frac{1}{6} = \frac{5}{6}$$

and $\qquad\qquad P(1) + P(1') = \frac{1}{6} + \frac{5}{6} = \frac{6}{6} = 1$

EXAMPLE 3.   A card deck has 52 cards divided into 4 suits (diamonds, hearts, clubs, and spades) with 13 cards in each suit (1, 2, 3, . . . , 10, jack, queen, king).   If the deck is well-shuffled, each of the 52 cards is equally likely to be picked.   Since there are 4 jacks, the probability of picking a jack, J, on a single pick is

$$J = \frac{n_J}{N} = \frac{4}{52} = \frac{1}{13}$$

Since there are 13 diamonds, D

$$P(D') = 1 - P(D) = 1 - \frac{13}{52} = 1 - \frac{1}{4} = \frac{3}{4}$$

and $\qquad\qquad P(D) + P(D') = \frac{1}{4} + \frac{3}{4} = 1$

EXAMPLE 4.   Suppose that in 100 tosses of a balanced coin, we get 53 heads and 47 tails.   The relative frequency of heads is 53/100, or 0.53.   This is the *relative frequency* or *empirical probability*, which is to be distinguished from the *a priori* or *classical probability* of $P(H) = 0.5$.   As the number of tosses increases and approaches infinity in the limit, the relative frequency or empirical probability approaches the a priori or classical probability.   For example, the relative frequency or empirical probability might be 0.517 or 1000 tosses, 0.508 for 10,000 tosses, and so on.

3.2   PROBABILITY OF MULTIPLE EVENTS

1. *Rule of addition for nonmutually exclusive events.*   Two events, A and B, are *not mutually exclusive* if the occurrence of A does not preclude the occurrence of B, or vice versa.   Then

$$P(A \text{ or } B) = P(A) + P(B) - P(A \text{ and } B) \qquad\qquad (3.4)$$

   $P(A \text{ and } B)$ is subtracted to avoid double counting.   This can be seen with the Venn diagram in Fig. 3.2.

2. *Rule of addition for mutually exclusive events.*   Two events, A and B, are *mutually exclusive* if the occurrence of A precludes the occurrence of B, or vice versa $[P(A \text{ and } B) = 0]$.   Then

$$P(A \text{ and } B) = P(A) + P(B) \qquad\qquad (3.5)$$

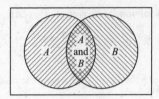

Fig. 3-2

3. *Rule of multiplication for dependent events.* Two events are *dependent* if the occurrence of one is connected in some way with the occurrence of the other. Then the *joint probability* of A and B is

$$P(A \text{ and } B) = P(A) \cdot P(B/A) \tag{3.6}$$

This reads: "The probability that *both* events A and B will take place equals the probability of event A times the probability of event B, given that event A has already occurred."

$$P(B/A) = \text{conditional probability of } B, \text{ given that } A \text{ has already occurred} \tag{3.7}$$

and

$$P(A \text{ and } B) = P(B \text{ and } A) \tag{3.8}$$

See Prob. 3.15(c) and (d).

4. *Rule of multiplication for independent events.* Two events, A and B, are *independent* if the occurrence of A is not connected in any way to the occurrence of B. $[P(B/A) = P(B)]$. Then

$$P(A \text{ and } B) = P(A) \cdot P(B) \tag{3.9}$$

EXAMPLE 5. On a single toss of a die, we can get only one of six possible outcomes: 1, 2, 3, 4, 5, or 6. These are *mutually exclusive* events. If the die is fair, $P(1) = P(2) = P(3) = P(4) = P(5) = P(6) = 1/6$. The probability of getting a 2 *or* a 3 on a single toss of the die is

$$P(2 \text{ or } 3) = P(2) + P(3) = \frac{1}{6} + \frac{1}{6} = \frac{2}{6} = \frac{1}{3}$$

Similarly

$$P(2 \text{ or } 3 \text{ or } 4) = P(2) + P(3) + P(4) = \frac{1}{6} + \frac{1}{6} + \frac{1}{6} = \frac{3}{6} = \frac{1}{2}$$

EXAMPLE 6. Picking at random a spade or a king on a single pick from a well-shuffled card deck does *not* constitute two *mutually exclusive* events because we could pick the king of spades. Thus

$$P(\text{S or K}) = P(\text{S}) + P(\text{K}) - P(\text{S and K}) = \frac{13}{52} + \frac{4}{52} - \frac{1}{52} = \frac{16}{52} = \frac{4}{13}$$

Using *set theory*, the preceding statement can be rewritten in an equivalent way as

$$P(\text{S} \cup \text{K}) = P(\text{S}) + P(\text{K}) - P(\text{S} \cap \text{K}) = \frac{13}{52} + \frac{4}{52} - \frac{1}{52} = \frac{16}{52} = \frac{4}{13}$$

where the symbol $\cup$ (read "union") replaces *or* and $\cap$ (read "intersection") replaces *and*.

EXAMPLE 7. The outcomes of two successive tosses of a balanced coin are *independent* events. The outcome of the first toss in no way affects the outcome on the second toss. Thus

$$P(\text{H and H}) = P(\text{H} \cap \text{H}) = P(\text{H}) \cdot P(\text{H}) = \frac{1}{2} \cdot \frac{1}{2} = \frac{1}{4}, \text{ or } 0.25$$

Similarly,

$$P(\text{H and H and H}) = P(\text{H} \cap \text{H} \cap \text{H}) = P(\text{H}) \cdot P(\text{H}) \cdot P(\text{H}) = \frac{1}{2} \cdot \frac{1}{2} \cdot \frac{1}{2} = \frac{1}{8}, \text{ or } 0.125$$

EXAMPLE 8. The probability that on the first pick from a deck we get the king of diamonds is

$$P(\text{K}_\text{D}) = \frac{1}{52}$$

If the first card picked was indeed the king of diamonds and if the first card was not replaced, the probability of getting another king on the second pick is *dependent* on the first pick because there are now only 3 kings and 51 cards left in the deck. The *conditional probability* of picking another king, given that the king of dimaonds was already picked and not replaced, is

$$P(K/K_D) = \frac{3}{51}$$

Thus the probability of picking the king of diamonds on the first pick and, without replacement, picking another king on the second pick is

$$P(K_D \text{ and } K) = P(K_D) \cdot P(K/K_D) = \frac{1}{52} \cdot \frac{3}{51} = \frac{3}{2652}$$

or about 1 in 1000. Related to conditional probability is Bayes' theorem (see Prob. 3.17). Problem 3.18 reviews combinations and permutations, or "counting techniques."

3.3 DISCRETE PROBABILITY DISTRIBUTIONS: THE BINOMIAL DISTRIBUTION

A *random variable* is a variable whose values are associated with some probability of being observed. A *discrete* (as opposed to *continuous*) random variable is one that can assume only finite and distinct values. The set of all possible values of a random variable and its associated probabilities is called a *probability distribution*. The sum of all probabilities equals 1 (see Example 9).

One discrete probability distribution is the *binomial distribution*. This is used to find the probability of X number of occurrences or successes of an event, $P(X)$, in n trials of the same experiment when (1) there are only *two* possible and mutually *exclusive outcomes*, (2) the n trials are *independent*, and (3) the probability of occurrence or success, p, remains *constant* in each trial. Then

$$P(X) = \frac{n!}{X!(n-X)!} p^X (1-p)^{n-X} \tag{3.10}$$

where $n!$ (read "n factorial") $= n \cdot (n-1) \cdot (n-2) \cdots 3 \cdot 2 \cdot 1$, and $0! = 1$ by definition (see Prob. 3.18).

The mean of the binomial distribution is

$$\mu = np \tag{3.11}$$

The standard deviation is

$$\sigma = \sqrt{np(1-p)} \tag{3.12}$$

If $p = 1 - p = 0.5$, the binomial distribution is symmetrical; if $p < 0.5$, it is skewed to the right; and if $p > 0.5$, it is skewed to the left.

EXAMPLE 9. The possible outcomes in 2 tosses of a balanced coin are TT, TH, HT, and HH. Thus

$$P(0\text{H}) = \frac{1}{4} \qquad P(1\text{H}) = \frac{1}{2} \quad \text{and} \quad P(2\text{H}) = \frac{1}{4}$$

The number of heads is therefore a discrete random variable, and the set of all possible outcomes with their associated probabilities is a discrete probability distribution (see Table 3.1 and Fig. 3-3).

Table 3.1   Probability Distribution of Heads in Two Tosses of a Balanced Coin

Number of Heads	Possible Outcomes	Probability
0	TT	0.25
1	TH, HT	0.50
2	HH	0.25
		1.00

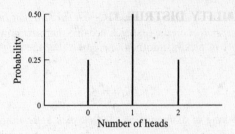

**Fig. 3-3   Probability Distribution of Heads in
Two Tosses of a Balanced Coin**

EXAMPLE 10.   Using the binomial distribution, we can find the probability of 4 heads in 6 flips of a balanced coin as follows:

$$P(4) = \frac{6!}{4!(6-4)!}(1/2)^4(1/2)^2 = \frac{6 \cdot 5 \cdot 4 \cdot 3 \cdot 2 \cdot 1}{4 \cdot 3 \cdot 2 \cdot 1 \cdot 2 \cdot 1}(1/16)(1/4) = 15(1/64) = \frac{15}{64} \cong 0.23$$

When n and X are large numbers, lengthy calculations to find probabilities can be avoided by using App. 1.   The *expected* number of heads in 6 flips $= \mu = np = (6)(1/2) = 3$ heads.   The standard deviation of the probability distribution of 6 flips is

$$\sigma = \sqrt{np(1-p)} = \sqrt{(6)(1/2)(1/2)} = \sqrt{6/4} = \sqrt{1.5} \cong 1.22 \text{ heads}$$

Because $p = 0.5$, this probability distribution is symmetrical.   If we were not dealing with a coin and the trials were not dependent (as in sampling without replacement), we would have had to use the *hypergeometric distribution* (see Prob. 3.27).

3.4   THE POISSON DISTRIBUTION

The *Poisson distribution* is another discrete probability distribution.   It is used to determine the probability of a designated number of successes *per unit of time*, when the events or successes are independent and the average number of successes per unit of time remains constant.   Then

$$P(X) = \frac{\lambda^X e^{-\lambda}}{X!} \tag{3.13}$$

where $X =$ designated number of successes

$\quad P(X) =$ probability of X number of successes

$\quad\quad \lambda =$ (Greek letter lambda) $=$ average number of successes per unit of time

$\quad\quad e =$ base of the natural logarithmic system, or 2.71828

Given the value of λ (the expected value or mean *and* variance of the Poisson distribution), we can find $e^{-\lambda}$ from App. 2, substitute in Eq. (*3.13*), and find $P(X)$.

EXAMPLE 11.   A police department receives an average of 5 calls per hour.   The probability of receiving 2 calls in a randomly selected hour is

$$P(X) = \frac{\lambda^X e^{-\lambda}}{X!} = \frac{5^2 e^{-5}}{2!} = \frac{(25)(0.00674)}{2} = 0.08425$$

The Poisson distribution can be used as an approximation to the binomial distribution when n is large and p or $1 - p$ is small [say, $n \geq 30$ and $np < 5$ or $n(1 - p) < 5$].   See Prob. 3.30.

3.5 CONTINUOUS PROBABILITY DISTRIBUTIONS: THE NORMAL DISTRIBUTION

A *continuous random variable* X is one that can assume an infinite number of values within any given interval.   The probability that X falls within any interval is given by the area under the probability distribution (or density function) within that interval.   The total area (probability) under the curve is 1 (see Prob. 3.31).

The *normal distribution* is a continuous probability distribution and the most commonly used distribution in statistical analysis (see Prob. 3.32).   The normal curve is bell-shaped and symmetrical about its mean.   It extends indefinitely in both directions, but most of the area (probability) is clustered around the mean (see Fig. 3-4); 68.26% of the area (probability) under the normal curve is included within one standard deviation of the mean (i.e., within $\mu \pm 1\sigma$), 95.44% within $\mu \pm 2\sigma$, and 99.74% within $\mu \pm 3\sigma$.

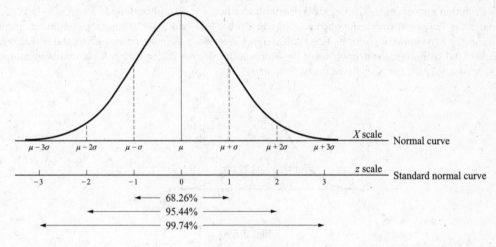

Fig. 3-4

The *standard normal distribution* is a normal distribution with a mean of 0 and a standard deviation of 1 (i.e., $\mu = 0$ and $\sigma = 1$).   Any normal distribution (X scale in Fig. 3-4) can be converted into a standard normal distribution by letting $\mu = 0$ and expressing deviations from μ in standard deviation units (z scale).

To find probabilities (areas) for problems involving the normal distribution, we first convert the X value into its corresponding z value, as follows:

$$z = \frac{X - \mu}{\sigma} \tag{3.14}$$

Then we look up the z value in App. 3.   This gives the proportion of the area (probability) included under the curve between the mean and that z value.

EXAMPLE 12.   The area (probability) under the standard normal curve between $z = 0$ and $z = 1.96$ is obtained by looking up the value of 1.96 in App. 3.   We move down the z column in the table to 1.9 and then across unitl we are below the column headed 0.06.   The value that we get is 0.4750.   This means that 47.50% of the total area (of 1, or 100%) under the curve lies between $z = 0$ and $z = 1.96$ (the shaded area in the figure above the table).   Because of symmetry, the area between $z = 0$ and $z = -1.96$ (not given in the table) is also 0.4750, or 47.50%.

EXAMPLE 13.   Suppose that X is a normally distributed random variable with $\mu = 10$ and $\sigma^2 = 4$ and we want to find the probability of X assuming a value between 8 and 12.   We first calculate the z values corresponding to the X values of 8 and 12 and then look up these z values in App. 3:

$$z_1 = \frac{X_1 - \mu}{\sigma} = \frac{8 - 10}{2} = -1 \quad \text{and} \quad z_2 = \frac{X_2 - \mu}{\sigma} = \frac{12 - 10}{2} = +1$$

For $z = 1$, we get 0.3413 from App. 3. Then, $z = \pm 1$ equals 2(0.3413), or 0.6826. This means that the probability of X assuming a value between 8 and 12, or $P(8 < X < 12)$, is 68.26% (see Fig. 3-4).

EXAMPLE 14. Suppose again that X is a normally distributed random variable with $\mu = 10$ and $\sigma^2 = 4$. The probability that X will assume a value between 7 and 14 can be found as follows:

$$z_1 = \frac{X_1 - \mu}{\sigma} = \frac{7 - 10}{2} = -1.5 \quad \text{and} \quad z_2 = \frac{X_2 - \mu}{\sigma} = \frac{14 - 10}{2} = 2$$

For $z_1 = -1.5$, we look up 1.50 in App. 3 and get 0.4322. For $z_2 = 2$, we get 0.4772. Therefore, $P(7 < X < 14) = 0.4332 + 0.4772 = 0.9104$, or 91.04% (see Fig. 3-5). Therefore, the probability of X assuming a value *smaller than* 7 or *larger than* 14 (the unshaded tail *areas* in Fig. 3-5) is $1 - 0.9104 = 0.0896$, 8.96%. The normal distribution approximates the binomial distribution when $n \geq 30$ and both $np > 5$ and $n(1 - p) > 5$, and it approximates the Poisson distribution when $\lambda \geq 10$ (see Probs. 3.37 and 3.38). Another continuous probability distribution is the *exponential distribution* (see Prob. 3.39). *Chebyshev's theorem*, or *inequality*, states that regardless of the shape of a distribution, the proportion of the observations or area falling within K standard deviations of the mean is at least $1 - 1/K^2$, for $K \geq 1$ (see Probs. 3.40 and 3.72).

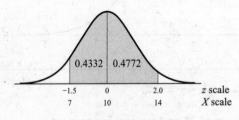

0.4332	0.4772

| -1.5 | 0 | 2.0 | z scale |
| 7 | 10 | 14 | X scale |

Fig. 3-5

Solved Problems

PROBABILITY OF A SINGLE EVENT

3.1 (*a*) Distinguish among classical or a priori probability, relative frequency or empirical probability, and subjective or personalistic probability. (*b*) What is the disadvantage of each? (*c*) Why do we study probability theory?

(*a*) According to *classical probability*, the probability of an event A is given by

$$P(A) = \frac{n_A}{N}$$

where $P(A)$ = probability that event A will occur

n_A = number of ways event A can occur

N = total number of equally possible outcomes

By the classical approach, we can make probability statements about balanced coins, fair dice, and standard card decks a priori, or without tossing a coin, rolling a die, or drawing a card. *Relative frequency* or *empirical probability* is given by the ratio of the number of times an event occurs to the total number of actual outcomes or observations. As the number of experiments or trials (such as the tossing of a coin) increases, the relative frequency or empirical probability approaches the classical or a priori probability. *Subjective* or *personalistic probability* refers to the *degree of belief* of an individual that the event will occur, based on whatever evidence is available to the individual.

(b) The classical or a priori approach to probability can only be applied to games of chance (such as tossing a balanced coin, rolling a fair die, or picking cards from a standard deck of cards) where we can determine a priori, or without experimentation, the probability that an event will occur.   In real-world problems of economics and business, we often cannot assign probabilities a priori and the classical approach cannot be used.   The relative-frequency or empirical approach overcomes the disadvantages of the classical approach by using the relative frequencies of past occurrences as probabilities.   The difficulty with the relative-frequency or empirical approach is that we get different probabilities (relative frequencies) for different numbers of trials or experiments.   These probabilities stabilize, or approach a limit, as the number of trials or experiments increases.   Because this may be expensive and time-consuming, people may end up using it without a "sufficient" number of trials or experiments.   The disadvantage of the subjective or personalistic approach to probability is that different people faced with the same situation may come up with completely different probabilities.

(c) Most of the decisions we face in economics, business, science, and everyday life involve risks and probabilities.   These probabilities are easier to understand and illustrate for games of choice because objective probabilities can easily be assigned to various events.   However, the primary reason for studying probability theory is to help us make intelligent decisions in economics, business, science, and everyday life when risk and uncertainty are involved.

3.2 What is the probability of (a) A head in one toss of a balanced coin? A tail? A head or a tail? (b) A 2 in one rolling of a fair die?   Not a 2?   A 2 or not a 2?

(a)
$$P(\text{H}) = \frac{n_\text{H}}{N} = \frac{1}{2}$$

$$P(\text{T}) = \frac{n_\text{T}}{N} = \frac{1}{2}$$

$$P(\text{H}) + P(\text{T}) = \frac{1}{2} + \frac{1}{2} = 1$$

(b) Since each of the 6 sides of a fair die is equally likely to come up and a 2 is one of the possibilities

$$P(2) = \frac{n_2}{N} = \frac{1}{6}$$

The probability of not rolling a 2 [that is, $P(2')$] is given by

$$P(2') = 1 - P(2) = 1 - \frac{1}{6} = \frac{5}{6}$$

$$P(2) + P(2') = \frac{1}{6} + \frac{5}{6} = \frac{6}{6} = 1, \text{ or certainty}$$

3.3 What is the probability that by picking one card from a well-shuffled deck, the card is (a) a king, (b) a spade, (c) the king of spades, (d) not the king of spades, or (e) the king of spades or not the king of spades?

(a) Since there are 4 kings K in the 52 cards of the standard deck

$$P(\text{K}) = \frac{n_\text{K}}{N} = \frac{4}{52} = \frac{1}{13}$$

(b) Since there are 13 spades S in the 52 cards, $P(\text{S}) = 13/52 = 1/4$

(c) There is only one king of spades in the deck, therefore $P(\text{K}_\text{S}) = 1/52$

(d) The probability of not picking the king of spades is $P(\text{K}_\text{S}') = 1 - 1/52 = 51/52$

(e) $P(\text{K}_\text{S}) + P(\text{K}_\text{S}') = 1/52 + 51/52 = 52/52 = 1$, or certainty

3.4 An urn (vase) contains 10 balls that are exactly alike except that 5 are red, 3 are blue, and 2 are green. What is the probability that, in picking up a single ball, the ball is (*a*) Red? (*b*) Blue? (*c*) Green? (*d*) Nonblue? (*e*) Nongreen? (*f*) Green or nongreen? (*g*) What are the odds of picking a blue ball? (*h*) What are the odds of not picking a blue ball?

(*a*)
$$P(R) = \frac{N_R}{N} = \frac{5}{10} = 0.5$$

(*b*)
$$P(B) = \frac{n_B}{N} = \frac{3}{10} = 0.3$$

(*c*)
$$P(G) = \frac{n_G}{N} = \frac{2}{10} = 0.2$$

(*d*)
$$P(B') = 1 - P(B) = 1 - 0.3 = 0.7$$

(*e*)
$$P(G') = 1 - P(G) = 1 - 0.2 = 0.8$$

(*f*)
$$P(G) + P(G') = 0.2 + 0.8 = 1$$

(*g*) The odds of picking a blue ball are given by the ratio of the number of ways of picking a blue ball to the number of ways of not picking a blue ball. Since there are 3 blue balls and 7 nonblue balls, the odds in favor of picking a blue ball are 3 to 7, or 3 : 7.

(*h*) The odds of not (against) picking a blue ball are 7 to 3, or 7 : 3.

3.5 Suppose that a 3 comes up 106 times in 600 tosses of a die. (*a*) What is the relative frequency of the 3? How does this differ from classical or a priori probability? (*b*) What would you expect to be the relative frequency or empirical probability if you increased the number of times the die is rolled?

(*a*) The relative frequency or empirical probability of the 3 is given by the ratio of the number of times 3 comes up (106) out of the total number of times the die is rolled (600). Thus the relative frequency or empirical probability of the 3 is $106/600 \cong 0.177$ in 600 rolls. According to the classical or a priori approach (and without rolling the die at all), $P(3) = 1/6 \cong 0.167$. If the die is fair, we *expect* the 3 to come up 100 times in 600 rolls of the die as compared with the actual, observed, or empirical 106 times.

(*b*) If the number of times the same die is rolled is increased from 600, we expect the relative frequency or empirical probability to approach (i.e., to become less unequal with) the classical or a priori probability.

3.6 The production process results in 27 defective items for each 1000 items produced. (*a*) What is the relative frequency or empirical probability of a defective item? (*b*) How many defective items do you expect out of the 1600 items produced each day?

(*a*) The relative frequency or empirical probability of a defective item is $27/1000 = 0.027$.

(*b*) By multiplying the number of items produced each day (1600) by the relative frequency or empirical probability of a defective item (0.027), we get the number of defective items we expect out of each day's output. This is $(1600)(0.027) = 43$, to the nearest item.

PROBABILITY OF MULTIPLE EVENTS

3.7 Define and give some examples of events that are (*a*) mutually exclusive, (*b*) not mutually exclusive, (*c*) independent, and (*d*) dependent.

(*a*) Two or more events are *mutually exclusive*, or *disjoint*, if the occurrence of one of them precludes or prevents the occurrence of the other(s). When one event takes place, the other(s) will not. For example, in a single flip of a coin, we get either a head or a tail, but not both. Heads and tails are therefore mutually exclusive events. In a simple toss of a die, we get one and only one of six possible outcomes: 1, 2, 3, 4, 5, or 6. The outcomes are therefore mutually exclusive. A card picked at random can be of only one suit: diamonds, hearts, clubs, or spades. A child is born either a boy or a girl. An item produced on an assembly line is either good or defective.

(b) Two or more events are *not mutually exclusive* if they may occur at the same time. The occurrence of one does not preclude the occurrence of the other(s). For example, a card picked at random from a deck of cards can be both an ace and a club. Therefore, aces and clubs are not mutually exclusive events, because we could pick the ace of clubs. Because we could have inflation and recession at the same time, inflation and recession are not mutually exclusive events.

(c) Two or more events are *independent* if the occurrence of one of them in no way affects the occurrence of the other(s). For example, in two successive flips of a balanced coin, the outcome of the second flip in no way depends on the outcome of the first flip. The same is true for two successive tosses of a pair of dice or picks of two cards from a deck with replacement.

(d) Two or more events are dependent if the occurrence of one of them affects the probability of the occurrence of the other(s). For example, if we pick a card from a deck and do not replace it, the probability of picking the same card on the second pick is 0. All other probabilities also are affected, since there are now only 51 cards in the deck. Similarly, if the proportion of defective items is greater for the evening than for the morning shift, the probability that an item picked at random from the evening output is defective is greater than for the morning output.

3.8 Draw a Venn diagram for (a) mutually exclusive events and (b) not mutually exclusive events. (c) Are mutually exclusive events dependent or independent? Why?

(a) Figure 3-6 illustrates the Venn diagram for events A and B which are mutually exclusive.

(b) Figure 3-7 illustrates the Venn diagram for events A and B which are not mutually exclusive.

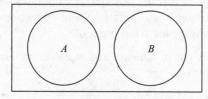

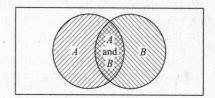

Fig. 3-6   Fig. 3-7

(c) Mutually exclusive events are dependent events. When one event occurs, the probability of the other occurring is 0. Thus the occurrence of the first affects (precludes) the occurrence of the other.

3.9 What is the probability of getting (a) Less than 3 on a single roll of a fair die? (b) Hearts or clubs on a single pick from a well-shuffled standard deck of cards? (c) A red or a blue ball from an urn containing 5 red balls, 3 blue balls, and 2 green balls? (d) More than 3 on a single roll of a fair die?

(a) Getting less than 3 on a single roll of a fair die means getting a 1 or a 2. These are mutually exclusive events. Applying the rule of addition for mutually exclusive events, we get

$$P(1 \text{ or } 2) = P(1) + P(2) = \frac{1}{6} + \frac{1}{6} = \frac{2}{6} = \frac{1}{3}$$

Using set theory, $P(1 \text{ or } 2)$ can be rewritten in an equivalent way as $P(1 \cup 2)$, where $\cup$ is read "union" and stands for *or*.

(b) Getting a heart or a club on a single pick from a well-shuffled deck of cards also constitutes two mutually exclusive events. Applying the rule of addition, we get

$$P(\text{H or C}) = P(\text{H} \cup \text{C}) = \frac{13}{52} + \frac{13}{52} = \frac{26}{52} = \frac{1}{2}$$

(c) $$P(\text{R or B}) = P(\text{R} \cup \text{B}) = \frac{5}{10} + \frac{3}{10} = \frac{8}{10} = \frac{4}{5} = 0.8$$

(d) $$P(4 \text{ or } 5 \text{ or } 6) = P(4 \cup 5 \cup 6) = P(4) + P(5) + P(6) = \frac{1}{6} + \frac{1}{6} + \frac{1}{6} = \frac{3}{6} = \frac{1}{2}$$

3.10 (a) What is the probability of getting an ace or a club on a single pick from a well-shuffled standard deck of cards? (In all remaining problems, it will be implicitly assumed that coins are balanced, die are fair, and decks of cards are standard and well shuffled and cards are picked at random without replacement.)   (b) What is the function of the negative term in the rule of addition for events that are not mutually exclusive?

(a) Getting an ace or a club does not constitute two mutually exclusive events because we could get the ace of clubs.   Applying the rule of addition for events that are not mutually exclusive, we get

$$P(A \text{ or } C) = P(A) + P(C) - P(A \text{ and } C) = \frac{4}{52} + \frac{13}{52} - \frac{1}{52} = \frac{16}{52} = \frac{4}{13}$$

The preceding probability statement can be rewritten in an equivalent way using set theory as

$$P(A \cup C) = P(A) + P(C) - P(A \cap C)$$

where $\cap$ is read "intersection" and stands for *and*.

(b) The function of the negative term in the rule of addition for events that are not mutually exclusive is to avoid double counting.   For example, in calculating $P(A \text{ or } C)$ in part *a*, the ace of clubs is counted twice, once as an ace and once as a club.   Therefore, we subtract the probability of getting the ace of clubs in order to avoid this double counting.   If the events are mutually exclusive, the probability that both events will occur simultaneously is 0, and no double counting is involved.   This is why the rule of addition for mutually exclusive events does not contain a negative term.

3.11 What is the probability of (a) Inflation I or recession R if the probability of inflation is 0.3,   the probability of recession is 0.2, and the probability of inflation and recession is 0.06? (b) Drawing an ace, a club, or a diamond on a single pick from a deck?

(a) Since the probability of inflation *and* recession is not 0, inflation and recession are not mutually exclusive events.   Applying the rule of addition, we get

$$P(I \text{ or } R) = P(I) + P(R) - P(I \text{ and } R)$$

or

$$P(I \cup R) = P(I) + P(R) - P(I \cap R)$$

and

$$P(I \text{ or } R) = P(I \cup R) = 0.3 + 0.2 - 0.06 = 0.44$$

(b) Getting an ace, a club, or a diamond does not constitute mutually exclusive events because we could get the ace of clubs or the ace of diamonds.   Applying the rule of addition for events that are not mutualy exclusive, we get

$$P(A \text{ or } C \text{ or } D) = P(A) + P(C) + P(D) - P(A \text{ and } C) - P(A \text{ and } D)$$

$$P(A \text{ or } C \text{ or } D) = \frac{4}{52} + \frac{13}{52} + \frac{13}{52} - \frac{1}{52} - \frac{1}{52} = \frac{28}{52} = \frac{7}{13}$$

3.12 What is the probability of (a) Two 6s on 2 rolls of a die? (b) A 6 on each die in rolling 2 dice once?   (c) Two blue balls in 2 successive picks with replacement from the urn in Prob. 3.4? (d) Three girls in a family with 3 children?

(a) Getting a 6 on each of 2 rolls of a die constitutes independent events.   Applying the rule of multiplication for independent events, we get

$$P(6 \text{ and } 6) = P(6 \cap 6) = P(6) \cdot P(6) = \frac{1}{6} \cdot \frac{1}{6} = \frac{1}{36}$$

(b) Getting a 6 on each die in rolling 2 dice once also constitutes independent events.   Therefore

$$P(6 \text{ and } 6) = P(6 \cap 6) = P(6) \cdot P(6) = \frac{1}{6} \cdot \frac{1}{6} = \frac{1}{36}$$

(c) Since we replace the first ball picked, the probability of getting a blue ball on the second pick is the same as on the first pick.   The events are independent.   Therefore

$$P(\text{B and B}) = P(\text{B} \cap \text{B}) = P(\text{B}) \cdot P(\text{B}) = \frac{3}{10} \cdot \frac{3}{10} = \frac{9}{100} = 0.09$$

(d) The probability of a girl, G, on each birth constitutes independent events, each with a probability of 0.5. Therefore

$$P(\text{G and G and G}) = P(\text{G} \cap \text{G} \cap \text{G}) = P(\text{G}) \cdot P(\text{G}) \cdot P(\text{G}) = (0.5) \cdot (0.5) \cdot (0.5) = 0.125$$

or 1 chance in 8.

3.13 (a) List all possible outcomes in rolling 2 dice simultaneously. (b) What is the probability of getting a total of 5 in rolling 2 dice simultaneously? (c) What is the probability of getting a total of 4 or less in rolling 2 dice simultaneously? More than 4?

(a) Each die has 6 possible and equally likely outcomes, and the outcome on each die is independent. Since each of the 6 outcomes on the first die can be associated with each of the 6 outcomes on the second die, there are a total of 36 possible outcomes; that is, the *sample space N* is 36. (In Table 3.2, the first number refers to the outcome on the first die, and the second number refers to the second die. The dice can be distinguished by different colors.) The total of the 36 possible outcomes also can be shown by a *tree (or sequential) diagram*, as in Fig. 3-8.

Table 3.2   Outcomes in Rolling Two Dice Simultaneously

1, 1	2, 1	3, 1	4, 1	5, 1	6, 1
1, 2	2, 2	3, 2	4, 2	5, 2	6, 2
1, 3	2, 3	3, 3	4, 3	5, 3	6, 3
1, 4	2, 4	3, 4	4, 4	5, 4	6, 4
1, 5	2, 5	3, 5	4, 5	5, 5	6, 5
1, 6	2, 6	3, 6	4, 6	5, 6	6, 6

(b) Out of the 36 possible and equally likely outcomes, 4 of them give a total of 5. These are 1, 4; 2, 3; 3, 2; and 4, 1. Thus the probability of a total of 5 (event A) in rolling 2 dice simultaneously is given by

$$P(A) = \frac{n_A}{N} = \frac{4}{36} = \frac{1}{9}$$

(c) Rolling a total of 4 or less involves rolling a total of 2, 3, or 4. There are 6 possible and equally likely ways of rolling a total of 4 or less. These are 1, 1; 1, 2; 1, 3; 2, 1; 2, 2; and 3, 1. Thus if event A is defined as rolling a total of 4 or less, $P(A) = 6/36 = 1/6$. The probability of getting a total of more than 4 equals 1 minus the probability of getting a total of 4 or less. This is $1 - 1/6 = 5/6$.

3.14 What is the probability of (a) Picking a second red ball from the urn in Prob. 3.4 when a red ball was already obtained on the first pick and not replaced? (b) A red ball on the second pick when the first ball picked was not red and was not replaced? (c) A red ball on the third pick when a red and a nonred ball were obtained on the first two picks and were not replaced?

(a) Picking a second red ball from the urn when a red ball was already picked on the first pick and was not replaced is a *dependent* event, since there are now only 4 red balls and 5 nonred balls remaining in the urn. The *conditional probability* of picking a second red ball when a red ball was already obtained on the first pick and was not replaced is $P(R/R) = 4/9$.

(b) The conditional probability of obtaining a red ball on the second pick when the first ball picked was not red (R′) and was not replaced in the urn before the second ball is picked is $P(R/R') = 5/9$.

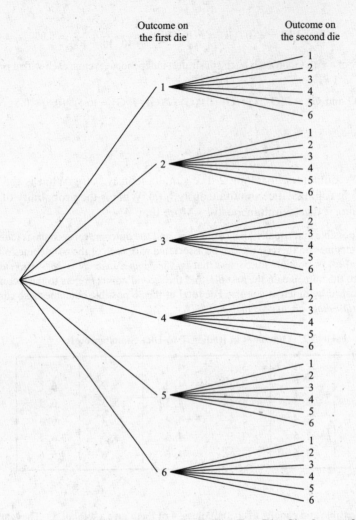

Fig. 3-8   Tree Diagram for Rolling Two Dice Simultaneously

(c)   Since 2 balls, one of which was red, were already picked and not replaced, there remains a total of 8 balls, of which 4 are red, in the urn.   The (conditional) probability of picking another red ball is
$P(R/R \text{ and } R') = P(R/R' \text{ and } R) = 4/8 = 1/2$.

3.15   What is the probability of obtaining   (a) Two red balls from the urn in Prob. 3.4 in 2 picks without replacement?   (b) Two aces from a deck in 2 picks without replacement?   (c) The ace of clubs and a spade in *that order* in 2 picks from a deck without replacement?   (d) A spade and the ace of clubs *in that order* in 2 picks from a deck without replacement?   (e) Three red balls from the urn of Prob. 3.4 in 3 picks without replacement?   (f) Three red balls from the same urn in 3 picks *with replacement*?

(a)   Applying the rule of multiplication for dependent events, we get

$$P(R \text{ and } R) = P(R \cap R) = P(R) \cdot P(R/R) = \frac{5}{10} \cdot \frac{4}{9} = \frac{20}{90} = \frac{2}{9}$$

(b)
$$P(A \text{ and } A) = P(A \cap A) = P(A) \cdot P(A/A) = \frac{4}{52} \cdot \frac{3}{51} = \frac{12}{2652} = \frac{1}{221}$$

(c)
$$P(A_C \text{ and } S) = P(A_C \cap S) = P(A_C) \cdot P(S/A_C) = \frac{1}{52} \cdot \frac{13}{51} = \frac{13}{2652}$$

(d) $$P(S \text{ and } A_C) = P(S \cap A_C) = P(S) \cdot P(A_C/S) = \frac{13}{52} \cdot \frac{1}{51} = \frac{13}{2652} = P(A_C \text{ and } S)$$

(e) $$P(R \text{ and } R \text{ and } R) = P(R \cap R \cap R) = P(R) \cdot P(R/R) \cdot P(R/R \text{ and } R)$$

$$= \frac{5}{10} \cdot \frac{4}{9} \cdot \frac{3}{8} = \frac{60}{720} = \frac{1}{12}$$

(f) With replacement, picking three balls from an urn constitutes three independent events. Therefore

$$P(R \text{ and } R \text{ and } R) = P(R) \cdot P(R) \cdot P(R) = \frac{5}{10} \cdot \frac{5}{10} \cdot \frac{5}{10} = \frac{125}{1000} = \frac{1}{8} = 0.125$$

3.16 Past experience has shown that for every 100,000 items produced in a plant by the morning shift, 200 are defective, and for every 100,000 items produced by the evening shift, 500 are defective. During a 24-h period, 1000 items are produced by the morning shift and 600 by the evening shift. What is the probability that an item picked at random from the total of 1600 items produced during the 24-h period (a) Was produced by the morning shift and is defective? (b) Was produced by the evening shift and is defective? (c) Was produced by the evening shift and is *not* defective? (d) Is defective, whether produced by the morning or the evening shift?

(a) The probabilities of picking an item produced by the morning shift M and evening E are

$$P(M) = \frac{1000}{1600} = 0.625 \quad \text{and} \quad P(E) = \frac{600}{1600} = 0.375$$

The probabilities of picking a defective item D from the morning and evening outputs separately are

$$P(D/M) = \frac{20}{100,000} = 0.002 \quad \text{and} \quad P(D/E) = \frac{500}{100,000} = 0.005$$

The probability that an item picked at random from the total of 1600 items produced during the 24-h period was produced by the morning shift and is defective is

$$P(M \text{ and } D) = P(M) \cdot P(D/M) = (0.625)(0.002) = 0.00125$$

(b) $$P(E \text{ and } D) = P(E) \cdot P(D/E) = (0.375)(0.005) = 0.001875$$

(c) $$P(E \text{ and } D') = P(E) \cdot P(D'/E) = (0.375)\frac{99,500}{100,000} = 0.373125$$

(d) The expected number of defective items from the morning shift is equal to the probability of a defective item from the morning output times the number of items produced by the morning shift; that is, $(0.002)(1000) = 2$. From the evening shift we expect $(0.005)(600) = 3$ defective items. Thus we expect 5 defective items from the 1600 items produced during the 24-h period. If there are indeed 5 defective items, the probability of picking at random any of the 5 defective items out of a total of 1600 items is 5/1600 or 1/320 or 0.003125.

3.17 (a) From the rule of multiplication for dependent events B and A, derive the formula for $P(A/B)$ in terms of $P(B/A)$ and $P(B)$. This is known as *Bayes' theorem* and is used to revise probabilities when additional relevant information becomes available. (b) Using Bayes' theorem, find the probability that a defective item picked at random from the 24-h output of 1600 items in Prob. 3.16 was produced by the morning shift; by the evening shift.

(a) $$P(B \text{ and } A) = P(B) \cdot P(A/B)$$

By dividing both sides by $P(B)$ and rearranging, we get

$$P(A/B) = \frac{P(B \text{ and } A)}{P(B)}$$

However, $P(B \text{ and } A) = P(A \text{ and } B)$; see Prob. 3.15(c) and (d). Therefore

$$P(A/B) = \frac{P(A \text{ and } B)}{P(B)} \quad \text{and} \quad P(A/B) = \frac{P(A) \cdot P(B/A)}{P(B)} \qquad \textit{Bayes' theorem} \qquad (3.15)$$

(b) Applying Bayes' theorem to the statement in Prob. 3.16, letting A signify the morning shift M and B signify defective D, and utilizing the results of Prob. 3.16, we get

$$P(M/D) = \frac{P(M) \cdot P(D/M)}{P(D)} = \frac{(0.625)(0.002)}{0.003125} = \frac{0.00125}{0.003125} = 0.4$$

That is, the probability that a *defective* item picked at random from the total 24-h output of 1600 items was produced by the morning shift is 40%. Similarly

$$P(E/D) = P(E) \cdot P(D/E) = \frac{(0.375)(0.005)}{0.003125} = \frac{0.001875}{0.003125} = 0.6, \text{ or } 60\%$$

Bayes' theorem can be generalized, for example, to find the probability that a defective item B picked at random was produced by any of n plants ($A_i, i = 1, 2, \ldots, n$), as follows:

$$P(A_i/B) = \frac{P(A_i) \cdot P(B/A_i)}{\sum P(A_i) \cdot P(B/A_i)} \qquad (3.16)$$

where $\sum$ refers to the summation over the n plants (the only ones producing the output). Bayes' theorem is applied in business decision theory, but is seldom used in the field of economics. (However, bayesian econometrics is becoming increasingly important.)

3.18 A club has 8 members. (a) How many different committees of 3 members each can be formed from the club? (Two committees are different even when only one member is different.) (b) How many committees of 3 members each can be formed from the club if each committee is to have a president, a treasurer, and a secretary?

(a) We are interested here in finding the number of *combinations* of 8 people taken 3 at a time *without concern for the order*

$$_8C_3 = \frac{8!}{3!(8-3)!} = \frac{8!}{3!5!} = \frac{8 \cdot 7 \cdot 6 \cdot 5 \cdot 4 \cdot 3 \cdot 2 \cdot 1}{3 \cdot 2 \cdot 1 \cdot 5 \cdot 4 \cdot 3 \cdot 2 \cdot 1} = \frac{8 \cdot 7 \cdot 6}{3 \cdot 2 \cdot 1} = \frac{336}{6} = 56$$

In general, the number of arrangements of n things taken X at a time without conern for the order is a *combination* given by

$$_nC_X = \binom{n}{X} = \frac{n!}{X!(n-X)!} \qquad (3.17)$$

where $n!$ (read n factorial) $= n \cdot (n-1) \cdot (n-2) \cdots 3 \cdot 2 \cdot 1$ and $0! = 1$ by definition.

(b) Since each committee of 3 has to have a president, a treasurer, and a secretary, we are now interested in finding the number of *permutations* of 8 people taken 3 at a time, *when the order is important*:

$$_8P_3 = \frac{8!}{(8-3)!} = \frac{8!}{5!} = \frac{8 \cdot 7 \cdot 6 \cdot 5 \cdot 4 \cdot 3 \cdot 2 \cdot 1}{5 \cdot 4 \cdot 3 \cdot 2 \cdot 1} = 8 \cdot 7 \cdot 6 = 336$$

In general, the number of arrangements, *in a definite order*, of n things taken X at a time is a permutation given by

$$_nP_X = \frac{n!}{(n-X)!} \qquad (3.18)$$

Permutations and combinations (often referred to as *counting techniques*) are helpful in counting the number of equally likely ways event A can occur in relation to the total of all possible and equally likely outcomes. Combinations and permutations were not used in previous problems because those problems were simple enough without them.

DISCRETE PROBABILITY DISTRIBUTIONS: THE BINOMIAL DISTRIBUTION

3.19 Define what is meant by and give an example of   (a) a random variable,   (b) a discrete random variable, and   (c) a discrete probability distribution.   (d) What is the distinction between a probability distribution and a relative-frequency distribution?

(a) A *random variable* is a variable whose values are associated with some probability of being observed. For example, on 1 roll of a fair die, we have 6 mutually exclusive outcomes (1, 2, 3, 4, 5, or 6), each associated with a probability occurrence of 1/6. Thus the outcome from the roll of a die is a random variable.

(b) A *discrete random variable* is one that can assume only finite or distinct values. For example, the outcomes from rolling a die constitute discrete random variables because they are limited to the values 1, 2, 3, 4, 5, and 6. This is to be contrasted with *continuous variables*, which can assume an infinite number of values within any given interval [see Prob. 3.31(a)].

(c) A *discrete probability distribution* refers to the set of all possible values of a (discrete) random variable and their associated probabilities. The set of the 6 outcomes in rolling a die and their associated probabilities is an example of a discrete probability distribution. The sum of the probabilities associated with all the values that the discrete random variable can assume always equals 1.

(d) A *probability distribution* refers to the *classical* or *a priori probabilities* associated with all the values that a random variable can assume. Because those probabilities are assigned a priori and without any experimentation, a probability distribution is often referred to as a *theoretical* (relative) frequency distribution. This differs from an *empirical* (relative) frequency distribution, which refers to the ratio of the number of times each outcome actually occurs to the total number of actual trials or observations. For example, in actually rolling a die a number of times, we are not likely to get each outcome exactly 1/6 of the times. However, as the number of rolls increases, the *empirical* (relative) frequency distribution stabilizes at the (uniform) *probability* or *theoretical* relative-frequency distribution of 1/6.

3.20 Derive the formula for   (a) the mean μ or expected value $E(X)$ and   (b) the variance for a *discrete probability distribution*.

(a) The formula for the arithmetic mean for grouped population data [Eq. (2.2a)] is

$$\mu = \frac{\sum fX}{N}$$

where $\sum fX$ is the sum of the frequency of each class f times the class midpoint X and $N = \sum f$, which is the number of all observations or frequencies. In dealing with probability distributions, the mean μ is often referred to as the "expected value" $E(X)$. The formula for μ or $E(X)$ for a discrete probability distribution can be derived by starting with Eq. (2.2a) and letting $f = P(X)$, which is the probability of each of the possible outcomes X. Then, $\sum fX = \sum XP(X)$, which is the sum of the value of each outcome times its probability of occurrence, and $N = \sum f = \sum P(X)$, which is the sum of the probabilities of each outcome, which is 1. Thus

$$E(X) = \mu = \sum XP(X) \qquad (3.19)$$

(b) The formula for the variance of grouped population data [Eq. (2.9a)] is

$$\sigma^2 = \frac{\sum f(X - \mu)^2}{N} \qquad (3.20)$$

Once again, letting $f = P(X) =$ probability of each outcome and $N = \sum f = \sum P(X) = 1$, we can get the formula for the variance of a discrete probability distribution:

$$\text{Var } X = \sigma_X^2 = \sum [X - E(X)]^2 P(X) = \sum X^2 P(X) - [E(X)]^2 = E(X^2) - \mu^2 \qquad (3.21)$$

3.21 Table 3.3 gives the number of job applications processed at a small employment agency during the past 100-day period. Determine the expected number of applications processed and the variance and standard deviation.

Table 3.3 Number of Job Applications Processed during the Past 100-Day Period

Number of Job Applicants	Number of Days Achieved
7	10
8	10
10	20
11	30
12	20
14	10
	100

To the extent that we believe that the experience of the past 100 days is typical, we can find the relative-frequency distribution and equate its probability distribution. This and the other calculations to find $E(X)$ and Var X are shown in Table 3.4:

$$\text{Var } X = \sigma_X^2 = \sum X^2 P(X) - [\Sigma X P(X)]^2 = 116 - (10.6)^2 = 116 - 112.36 = 3.64 \text{ applications squared}$$

$$\text{SD } X = \sigma_X = \sqrt{\sigma_X^2} = \sqrt{3.64} \cong 1.91 \text{ applications}$$

Table 3.4 Calculations to Find the Expected Value and Variance

Number, X	Days, f	$P(X)$	$XP(X)$	X^2	$X^2 P(X)$
7	10	0.1	0.7	49	4.9
8	10	0.1	0.8	64	6.4
10	20	0.2	2.0	100	20.0
11	30	0.3	3.3	121	36.3
12	20	0.2	2.4	144	28.8
14	10	0.1	1.4	196	19.6
$N = \sum f = 100$		$\sum P(X) = 1.0$	$\sum XP(X) = 10.6$		$\sum X^2 P(X) = 116.0$
$E(X) = \mu = \sum XP(X) = 10.6$ applications					

3.22 (*a*) State the conditions required to apply the binomial distribution. (*b*) What is the probability of 3 heads in 5 flips of a balanced coin? (*c*) What is the probability of less than 3 heads in 5 flips of a balanced coin?

(*a*) The binomial distribution is used to find the probability of X number of occurrences or successes of an event, $P(X)$, in n trials of the same experiment when (1) there are only 2 mutually exclusive outcomes, (2) the n trials are independent, and (3) the probability of occurrence or success, p, remains constant in each trial.

(b)
$$P(X) = nC_X p^X (1-p)^{n-X} = \binom{n}{X} p^X (1-p)^{n-X} = \frac{n!}{X!(n-X)!} p^X (1-p)^{n-X}$$

See Eqs. (3.10) and (3.17). In some books, $1-p$ (the probability of failure) is defined as q. Here $n = 5$, $X = 3$, $p = 1/2$, and $1-p = 1/2$. Substituting these values into the preceding equation, we get

$$P(3) = \frac{5!}{3!(5-3)!} (1/2)^3 (1/2)^{5-3} = \frac{5!}{3!2!} (1/2)^3 (1/2)^2 = \frac{5 \cdot 4 \cdot 3 \cdot 2 \cdot 1}{3 \cdot 2 \cdot 1 \cdot 2 \cdot 1} (1/2)^5 = 10(1/32) = 0.3125$$

(c)
$$P(X < 3) = P(0) + P(1) + P(2)$$

$$P(0) = \frac{5!}{0!5!} (1/2)^0 (1/2)^5 = \frac{1}{32} = 0.03125$$

$$P(1) = \frac{5!}{1!(5-1)!} (1/2)^1 (1/2)^4 = \frac{5}{32} = 0.15625$$

$$P(2) = \frac{5!}{2!(5-2)!} (1/2)^2 (1/2)^3 = \frac{10}{32} = 0.3125$$

Thus $$P(X < 3) = P(0) + P(1) + P(2) = 0.03125 + 0.15625 + 0.3125 = 0.5$$

3.23 (a) Suppose that the probability of parents having a child with blond hair is 1/4. If there are 6 children in the family, what is the probability that half of them will have blond hair? (b) If the probability of hitting a target on a single shot is 0.3, what is the probability that in 4 shots the target will be hit at least 3 times?

(a) Here $n = 6$, $X = 3$, $p = 1/4$, and $1-p = 3/4$. Substituting these values into the binomial formula, we get

$$P(3) = \frac{6!}{3!(6-3)!} (1/4)^3 (3/4)^3 = \frac{6!}{3!3!} (1/64)(27/64) = \frac{6 \cdot 5 \cdot 4 \cdot 3 \cdot 2 \cdot 1}{3 \cdot 2 \cdot 1 \cdot 3 \cdot 2 \cdot 1} (27/4096)$$

$$= 20 \frac{27}{4096} = \frac{540}{4096} \cong 0.13$$

(b) Here $n = 4$, $X \geq 3$, $p = 0.3$, and $1-p = 0.7$:

$$P(X \geq 3) = P(3) + P(4)$$

$$P(3) = \frac{4!}{3!(4-3)!} (0.3)^3 (0.7)^1 = \frac{4 \cdot 3 \cdot 2 \cdot 1}{3 \cdot 2 \cdot 1 \cdot 1} (0.027)(0.7) = (4)(0.0189) = 0.0756$$

$$P(4) = \frac{4!}{4!(4-4)!} (0.3)^4 (0.7)^0 = (0.3)^4 = 0.0081$$

Thus $$P(X \geq 3) = P(3) + P(4) = 0.0756 + 0.0081 = 0.0837$$

3.24 (a) A quality inspector picks a sample of 10 tubes at random from a very large shipment of tubes known to contain 20% defective tubes. What is the probability that no more than 2 of the tubes picked are defective? (b) An inspection engineer picks a sample of 15 items at random from a manufacturing process known to produce 85% acceptable items. What is the probability that 10 of the items picked are acceptable?

(a) Here $n = 10$, $X \leq 2$, $p = 0.2$, and $1-p = 0.8$:

$$P(X \leq 2) = P(0) + P(1) + P(2)$$

$$P(0) = \frac{10!}{0!(10-0)!}(0.2)^0(0.8)^{10}$$

$$= 0.1074 \text{ (looking up } n = 10, X = 0, \text{ and } p = 0.2 \text{ in App. 1)}$$

$$P(1) = 0.2684 \text{ (looking up } n = 10, X = 1, \text{ and } p = 0.2 \text{ in App. 1)}$$

$$P(2) = 0.3020 \text{ (looking up } n = 10, X = 2, \text{ and } p = 0.2 \text{ in App. 1)}$$

Thus $$P(X \leq 2) = P(0) + P(1) + P(2) = 0.1074 + 0.2684 + 0.3020 = 0.6778$$

(b) Here $n = 15$, $X = 10$, $p = 0.85$, and $1 - p = 0.15$. Since App. 1 only gives binomial probabilities for up to 0.5, we should transform the problem. The probability of $X = 10$ acceptable items with $p = 0.85$ equals the probability of $X = 5$ defective items with $p = 0.15$. Using $n = 15$, $X = 5$ defective, p (of objective) $= 0.15$, we get 0.0449 (from App. 1).

3.25 (a) If 4 balanced coins are tossed simultaneously (or 1 balanced coin is tossed 4 times), compute the entire probability distribution and plot it. (b) Compute and plot the probability distribution for a sample of 5 items taken at random from a production process known to produce 30% defective items.

(a) Using $n = 4$; $X = 0H$, $1H$, $2H$, $3H$, or $4H$; $P = 1/2$; and App. 1, we get $P(0H) = 0.0625$, $P(1H) = 0.2500$, $P(2H) = 0.3750$, $P(3H) = 0.2500$, $P(4H) = 0.0625$, and

thus $$P(0H) + P(1H) + P(2H) + P(3H) + P(4H)$$
$$= 0.0625 + 0.2500 + 0.3750 + 0.2500 + 0.0625 = 1$$

See Fig. 3-9. Note that $p = 0.5$ and the probability distribution in Fig. 3-9 is symmetrical.

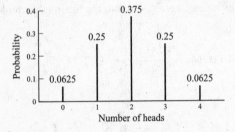 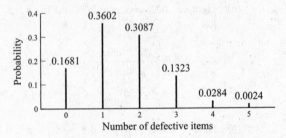

Fig. 3-9 Probability Distribution of Heads in Tossing Four Balanced Coins

Fig. 3-10 Probability Distribution of Defective Items

(b) Using $n = 5$; $X = 0, 1, 2, 3, 4,$ or 5 defective; and $p = 0.3$, we get $p(0) = 0.1681$, $P(1) = 0.3602$, $P(2) = 0.3087$, $P(3) = 0.1323$, $P(4) = 0.0284$, $P(5) = 0.0024$. Therefore

$$P(0) + P(1) + P(2) + P(3) + P(4) + P(5)$$
$$= 0.1681 + 0.3602 + 0.3087 + 0.1323 + 0.0284 + 0.0024 = 1$$

See Fig. 3-10. Note that $p < 0.5$ and the probability distribution in Fig. 3-10 is skewed to the right.

3.26 Calculate the expected value and standard deviation and determine the symmetry or asymmetry of the probability distribution of (a) Prob. 3.23(a), (b) Prob. 3.23(b), (c) Prob. 3.24(a), and (d) Prob. 3.24(b).

(a) $$E(X) = \mu = np = (6)(1/4) = 3/2 = 1.5 \text{ blond children}$$
$$\text{SD } X = \sqrt{np(1-p)} = \sqrt{6(1/4)(3/4)} = \sqrt{18/16} = \sqrt{1.125} \cong 1.06 \text{ blond children}$$

Because $p < 0.5$, the probability distribution of blond children is skewed to the right.

(b)
$$E(X) = \mu = np = (4)(0.3) = 1.2 \text{ hits}$$
$$\text{SD } X = \sqrt{np(1-p)} = \sqrt{(4)(0.3)(0.7)} = \sqrt{0.84} \cong 0.92 \text{ hits}$$

Because $p < 0.5$, the probability distribution is skewed to the right.

(c)
$$E(X) = \mu = np = (10)(0.2) = 2 \text{ defective tubes}$$
$$\text{SD } X = \sqrt{np(1-p)} = \sqrt{(10)(0.2)(0.8)} = \sqrt{1.6} \cong 1.26 \text{ defective tubes}$$

Because $p < 0.5$, the probability distribution is skewed to the right.

(d)
$$E(X) = \mu = np = (15)(0.85) = 12.75 \text{ acceptable items}$$
$$\text{SD } X = \sqrt{np(1-p)} = \sqrt{15(0.85)(0.15)} = \sqrt{1.9125} \cong 1.38 \text{ acceptable items}$$

Because $p > 0.5$, the probability distribution is skewed to the left.

3.27 When sampling is done from a finite population *without replacement*, the binomial distribution cannot be used because the events are not independent. Then the *hypergeometric* distribution is used. This is given by

$$P_{\text{H}} = \frac{\binom{N - X_t}{n - X}\binom{X_t}{X}}{\binom{N}{n}} \quad \text{hypergeometric distribution} \qquad (3.22)$$

It measures the number of successes X in a sample size n taken at random and without replacement from a population of size N, of which X_t items have the characteristic denoting success. (a) Using the formula, determine the probability of picking 2 men in a sample of 6 selected at random without replacement from a group of 10 people, 5 of which are men. (b) What would the result have been if we had (incorrectly) used the binomial distribution?

(a) Here $X = 2$ men, $n = 6$, $N = 10$, and $X_t = 5$:

$$P_{\text{H}} = \frac{\binom{10 - 5}{6 - 2}\binom{5}{2}}{\binom{10}{6}} = \frac{\binom{5}{4}\binom{5}{2}}{\binom{10}{6}} = \frac{\dfrac{5!}{4!1!}\dfrac{5!}{2!3!}}{\dfrac{10!}{6!4!}} = \frac{(5)(10)}{210} \cong 0.24$$

(b)
$$P(2) = \frac{n!}{X!(n-X)!}p^X(1-p)^{n-X} = \frac{6!}{2!4!}(1/2)^2(1/2)^4 = \frac{15}{64} = 0.23$$

It should be noted that when the sample is very small in relation to the population (say, less than 5% of the population), sampling without replacement has little effect on the probability of success in each trial and the binomial distribution (which is easier to use) is a good approximation for the hypergeometric distribution. This is the reason the binomial distribution was used in Prob. 3.24(a).

THE POISSON DISTRIBUTION

3.28 (a) What is the difference between the binomial and the Poisson distributions? (b) Give some examples of when we can apply the Poisson distribution. (c) Give the formula for the Poisson distribution and the meaning of the various symbols. (d) Under what conditions can the Poisson distribution be used as an approximation to the binomial distribution? Why can this be useful?

(a) Whereas the binomial distribution can be used to find the probability of a designated number of successes in n trials, the Poisson distribution is used to find the probability of a designated number of successes *per unit of time*. The other conditions required to apply the binomial distribution also are required to apply the Poisson distribution; that is (1) there must be only two mutually exclusive outcomes, (2) the events must be independent, and (3) the average number of successes per unit of time must remain constant.

(b) The Poisson distribution is often used in operations research in solving management problems. Some examples are the number of telephone calls to the police per hour, the number of customers arriving at a gasoline pump per hour, and the number of traffic accidents at an intersection per week.

(c) The probability of a designated number of successes per unit of time, $P(X)$, can be found by

$$P(X) = \frac{\lambda^X e^{-\lambda}}{X!}$$

where X = designated number of successes

λ = the average number of successes over a specific time period

e = the base of the natural logarithm system, or 2.71828

Given the value of λ, we can find $e^{-\lambda}$ from App. 2, substitute it into the formula, and find $P(X)$. Note that λ is the mean and variance of the Poisson distribution.

(d) We can use the Poisson distribution as an approximation to the binomial distribution when n, the number of trials, is large and p or $1 - p$ is small (rare events). A good rule of thumb is to use the Poisson distribution when $n \geq 30$ and np or $n(1 - p) < 5$. When n is large, it can be very time-consuming to use the binomial distribution and tables for binomial probabilities, for very small values of p may not be available. If $n(1 - p) < 5$, success and failure should be redefined so that $np < 5$ to make the approximation accurate.

3.29 Past experience indicates that an average number of 6 customers per hour stop for gasoline at a gasoline pump. (a) What is the probability of 3 customers stopping in any hour? (b) What is the probability of 3 customers or less in any hour? (c) What is the expected value, or mean, and standard deviation for this distribution?

(a) $$P(3) = \frac{6^3 e^{-6}}{3!} = \frac{(216)(0.00248)}{3 \cdot 2 \cdot 1} = \frac{0.53568}{6} = 0.08928$$

(b) $$P(X \leq 3) = P(0) + P(1) + P(2) + P(3)$$

$$P(0) = \frac{6^0 e^{-6}}{0!} = \frac{(1)(0.00248)}{1} = 0.00248$$

$$P(1) = \frac{6^1 e^{-6}}{1!} = \frac{(6)(0.00248)}{1} = 0.01488$$

$$P(2) = \frac{6^2 e^{-6}}{2!} = \frac{(36)(0.0248)}{2.1} = 0.04464$$

$$P(3) = 0.08928 \text{ (from part } a)$$

Thus $$P(X \leq 3) = 0.00248 + 0.01488 + 0.04464 + 0.08928 = 0.15128$$

(c) The expected value, or mean, of this Poisson distribution is $\lambda = 6$ customers, and the standard deviation is $\sqrt{\lambda} = \sqrt{6} \cong 2.45$ customers.

3.30 Past experience shows that 1% of the lightbulbs produced in a plant are defective. Find the probability that more than 1 bulb is defective in a random sample of 30 bulbs, using (a) the binomial distribution and (b) the Poisson distribution.

(a) Here $n = 30$, $p = 0.01$, and we are asked to find $P(X > 1)$. Using App. 1, we get

$$P(2) + P(3) + P(4) + \cdots = 0.0328 + 0.0031 + 0.0002 = 0.0361, \text{ or } 3.61\%$$

(b) Since $n = 30$ and $np = (30)(0.01) = 0.3$, we can use the Poisson approximation of the binomial distribution. Letting $\lambda = np = 0.3$, we have to find $P(X > 1) = 1 - P(X \leq 1)$, where X is the number of defective bulbs. Using Eq. (3.13), we get

$$P(1) = \frac{0.3^1 e^{-0.3}}{1!} = (0.3)(0.74082) = 0.222246$$

$$P(0) = \frac{0.3^0 e^{-0.3}}{0!} = e^{-0.3} = 0.74082$$

$$P(X \le 1) = P(1) + P(0) = 0.222246 + 0.74082 = 0.963066$$

Thus $P(X > 1) = 1 - P(X \le 1) = 1 - 0.963066 = 0.036934$, or 3.69%

As n becomes larger, the approximation becomes even closer.

CONTINUOUS PROBABILITY DISTRIBUTIONS: THE NORMAL DISTRIBUTION

3.31   (a) Define what is meant by a continuous variable and give some examples. (b) Define what is meant by a continuous probability distribution. (c) Derive the formula for the expected value and variance of a continuous probability distribution.

(a) A *continuous variable* is one that can assume any value within any given interval. A continuous variable can be measured with any degree of accuracy simply by using smaller and smaller units of measurement. For example, if we say that a production process takes 10 h, this means anywhere between 9.5 and 10.4 h (10 h rounded to the nearest hour). If we used minutes as the unit of measurement, we could have said that the production process takes 10 h and 20 min. This means anywhere between 10 h and 19.5 min and 10 h and 20.4 min, and so on. Time is thus a continuous variable, and so are weight, distance, and temperature.

(b) A *continuous probability distribution* refers to the range of all possible values that a continuous random value can assume, together with the associated probabilities. The probability distribution of a continuous random variable is often called a *probability density function*, or simply a *probability function*. It is given by a smooth curve such that the total area (probability) under the curve is 1. Since a continuous random variable can assume an infinite number of values within any given interval, the probability of a *specific* value is 0. However, we can measure the probability that a continuous random variable X assumes any value within a given interval (say, between X_1 and X_2) by the area under the curve within that interval:

$$P(X_1 < X < X_2) = \int_{X_1}^{X_2} f(X) \, dX \tag{3.23}$$

where $f(X)$ is the equation of the probability density function, and the integration sign, $\int$, is analogous to the summation sign $\sum$ for discrete variables. Probability tables for some of the most used continuous probability distributions are given in the appendixes, thus eliminating the need to perform the integration ourselves.

(c) The expected value, or mean, and variance for continuous probability distributions can be derived by substituting $\int$ for $\sum$ and $f(X)$ for $P(X)$ into the formula for the expected value and variance for discrete probability distributions [Eqs. (3.19) and (3.20)]:

$$E(X) = \mu = \int X f(X) \, dX \tag{3.24}$$

$$\text{Var } X = \sigma^2 = \int [X - E(X)]^2 f(X) \, dX \tag{3.25}$$

3.32   (a) What is a normal distribution? (b) What is its usefulness? (c) What is the standard normal distribution? What is its usefulness?

(a) The *normal distribution* is a continuous probability function that is bell-shaped, symmetrical about the mean, and mesokurtic (defined in Sec. 2.4). As we move further away from the mean in both directions, the normal curve approaches the horizontal axis (but never quite touches it). The equation of the normal probability function is given by

$$f(X) = \frac{1}{\sqrt{2\pi\sigma^2}} \exp\left[-\frac{1}{2}\left(\frac{X-\mu}{\sigma}\right)^2\right] \qquad (3.26)$$

where $f(X)$ = height of the normal curve

$\quad$ exp = constant 2.7183

$\quad \pi$ = constant 3.1416

$\quad \mu$ = mean of the distribution

$\quad \sigma$ = standard deviation of the distribution

$$\int_{-\infty}^{\infty} \frac{1}{\sqrt{2\pi\sigma^2}} \exp\left[-\frac{1}{2}\left(\frac{X-\mu}{\sigma}\right)^2\right] dX = 1 \qquad \text{(the total area under the normal curve from minus infinity to plus infinity)}$$

(b) The normal distribution is the most commonly used of all probability distributions in statistical analysis. Many distributions actually found in nature and industry are normal. Some examples are the IQs (intelligence quotients), weights, and heights of a large number of people and the variations in dimensions of a large number of parts produced by a machine. The normal distribution often can be used to approximate other distributions, such as the binomial and the Poisson distributions (see Probs. 3.37 and 3.38). Distributions of sample means and proportions are often normal, regardless of the distribution of the parent population (see Sec. 4.2).

(c) The *standard* normal distribution is a normal distribution with $\mu = 0$ and $\sigma^2 = 1$. Any normal distribution (defined by a particular value for μ and σ^2) can be transformed into a standard normal distribution by letting $\mu = 0$ and expressing deviations from μ in standard deviation units. We often can find areas (probabilities) by converting X values into corresponding z values [that is, $z = (X - \mu)/\sigma$] and looking up these z values in App. 3.

3.33 Find the area under the standard normal curve   (a) between $z \pm 1$, $z \pm 2$, and $z \pm 3$;   (b) from $z = 0$ to $z = 0.88$;   (c) from $z = -1.60$ to $z = 2.55$;   (d) to the left of $z = -1.60$;   (e) to the right of $z = 2.55$;   (f) to the left of $z = -1.60$ and to the right of $z = 2.55$.

(a) The area (probability) included under the standard normal curve between $z = 0$ and $z = 1$ is obtained by looking up the value of 1.0 in App. 3. This is accomplished by moving down the z column in the table to 1.0 and then across until we are below the column headed .00. The value that we get is 0.3413. This means that 34.13% of the total area (of 1, or 100%) under the curve lies between $z = 0$ and $z = 1.00$. Because of symmetry, the area between $z = 0$ and $z = -1$ is also 0.3413, or 34.13%. Therefore, the area between $z = -1$ and $z = 1$ is 68.26% (see Fig. 3-4). Similarly, the area between $z = 0$ and $z = 2$ is 0.4772, or 47.72% (by looking up $z = 2.00$ in the table), so that the area between $z = \pm 2$ is 95.44% (see Fig. 3-4). The area between $z \pm 3 = 99.74\%$ (see Fig. 3-4). Note that the table only gives detailed z values for up to 2.99 because the area under the curve outside $z \pm 3$ is negligible.

(b) The area between $z = 0$ and $z = 0.88$ is obtained by looking up 0.88 in the table. This is 0.3106.

(c) The area between $z = 0$ and $z = -1.60$ is obtained by looking up $z = 1.60$ in the table. This is 0.4452. The area between $z = 0$ and $z = 2.55$ is obtained by looking up $z = 2.55$ in the table. This is 0.4946. Thus the area under the standard normal curve from $z = -1.60$ and $z = 2.55$ equals 0.4452 plus 0.4946. This is 0.9398, or 93.98% (see Fig. 3-11). In all problems of this nature it is helpful to sketch a figure.

(d) We know that the total area under the normal curve is equal to 1. Because of symmetry, 0.5 of the area lies on either side of $\mu = 0$. Since 0.4452 extends from $z = 0$ to $z = -1.60$, $0.5 - 0.4452 = 0.0548$, or 5.48%, is the area in the left tail, to the left of -1.60 (see Fig. 3-11).

(e) $0.5 - 0.4946 = 0.0054$, or 0.54%, is the area in the right tail, to the right of $z = 2.55$ (see Fig. 3-11).

(f) The area to the left of $z = -1.60$ and to the right of $z = 2.55$ is equal to $1 - 0.9398$ (see part c). This is 0.0602, or 6.02% of the total.

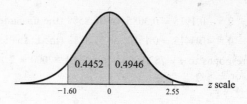

Fig. 3-11

3.34 The lifetime of lightbulbs is known to be normally distributed with $\mu = 100$ h and $\sigma = 8$ h. What is the probability that a bulb picked at random will have a lifetime between 110 and 120 burning hours?

We are asked here to find $P(110 < X < 120)$, where X refers to time measured in hours of burning time. Given $\mu = 100$ h and $\sigma = 8$ h, and letting $X_1 = 110$ h and $X_2 = 120$ h, we get

$$z_1 = \frac{X_1 - \mu}{\sigma} = \frac{110 - 100}{8} = 1.25 \quad \text{and} \quad z_2 = \frac{X_2 - \mu}{\sigma} = \frac{120 - 100}{8} = 2.50$$

Thus we want the area (probability) between $z_1 = 1.25$ and $z_2 = 2.50$ (the shaded area in Fig. 3-12). Looking up $z_2 = 2.50$ in App. 3, we get 0.4938. This is the area from $z = 0$ to $z_2 = 2.50$. Looking up $z_1 = 1.25$, we get 0.3944. This is the area from $z = 0$ to $z_1 = 1.25$. Subtracting 0.3944 from 0.4938, we get 0.0994, or 9.94%, for the shaded area that gives $P(110 < X < 120)$.

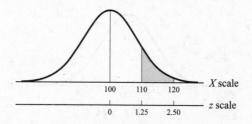

Fig. 3-12

3.35 Assume that family incomes are normally distributed with $\mu = \$16,000$, and $\sigma = \$2000$. What is the probability that a family picked at random will have an income: (*a*) Between \$15,000 and \$18,000? (*b*) Below \$15,000? (*c*) Above \$18,000? (*d*) Above \$20,000?

(*a*) We want $P(\$15,000 < X < \$18,000)$, where X is family income:

$$z_1 = \frac{X_1 - \mu}{\sigma} = \frac{\$15,000 - \$16,000}{\$2000} = -0.5 \quad \text{and} \quad z_2 = \frac{X_2 - \mu}{\sigma} = \frac{\$18,000 - \$16,000}{\$2000} = 1$$

Thus we want the area (probability) between $z_1 = -0.5$ and $z_2 = 1$ (the shaded area in Fig. 3-13). Looking up $z = 0.5$ in App. 3, we get 0.1915 for the area from $z = 0$ to $z = -0.5$. Looking up $z = 1$, we get 0.3413 for the area from $z = 0$ to $z = 1$. Thus, $P(\$15,000 \le X \le \$18,000) = 0.1915 + 0.3413 = 0.5328$, or 53.28%.

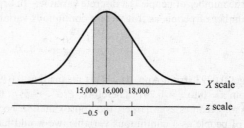

Fig. 3-13

(b) $P(X < \$15{,}000) = 0.5 - 0.1915 = 0.3085$, or 30.85% (the unshaded area in the left tail of Fig. 3-13).

(c) $P(X > \$18{,}000) = 0.5 - 0.3413 = 0.1587$, or 15.87% (the unshaded area in the right tail of Fig. 3-13).

(d) $X = \$20{,}000$ corresponds to $z = (\$20{,}000 - \$16{,}000)/\$2000 = 2$. Therefore, $P(X > \$20{,}000) = 0.5 - 0.4772 = 0.0228$, or 2.28%.

3.36 The grades on the midterm examination in a large statistics section are normally distributed with a mean of 78 and a standard deviation of 8. The professor wants to give the grade of A to 10% of the students. What is the lowest grade point that can be designated an A on the midterm?

In this problem we are asked to find the point grade such that 10% of the students will have higher grades. This involves finding the grade point X such that 10% of the area under the normal curve will be to the right of X (the shaded area in Fig. 3-14). Since the total area under the curve to the right of 78 is 0.5, the *unshaded* area in Fig. 3-14 to the *right* of 78 must be 0.4. We must look *into the body* of App. 3 for the value closest to 0.4. This is 0.3997, which corresponds to the z value of 1.28. The X value (the grade point) that corresponds to the z value of 1.28 is obtained by substituting the known values into $z = (X - \mu)/\sigma$ and solving for X:

$$1.28 = \frac{X - 78}{8}$$

This gives $10.24 = X - 78$. Therefore $X = 78 + 10.24 = 88.24$, or 88 to the nearest whole number.

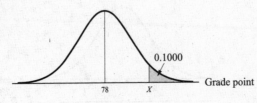

Fig. 3-14

3.37 Experience indicates that 30% of the people entering a store make a purchase. Using (a) the binomial distribution and   (b) the normal approximation to the binomial, find the probability that out of 30 people entering the store, 10 or more will make a purchase.

(a) Here $n = 30$, $p = 0.3$, and $1 - p = 0.7$ and we are asked to find $P(X \geq 10)$. Using App. 1 (the table of binomial probabilities),

$$(X \geq 10) = P(10) + P(11) + P(12) + \cdots + P(30) = 0.1416 + 0.1103 + 0.0749 + 0.0444 + 0.0231$$
$$+ 0.0106 + 0.0042 + 0.0015 + 0.005 + 0.001$$
$$= 0.4112$$

(b) $\mu = np = (30)(0.3) = 9$ persons, and $\sigma = \sqrt{np(1-p)} = \sqrt{(30)(0.3)(0.7)} = \sqrt{6.3} \cong 2.51$ persons. Since $n = 30$ and both np and $n(1 - p) > 5$, we can approximate the binomial probability with the normal. However, the number of people is a discrete variable. In order to use the normal distribution, we must treat the number of people as if it were a continuous variable and find $P(X \geq 9.5)$. Thus

$$z = \frac{X - \mu}{\sigma} = \frac{9.5 - 9}{2.51} = \frac{0.5}{2.51} \cong 0.20$$

From $z = 0.20$, we get 0.0793 (from App. 3). This means that 0.0793 of the area under the standard normal curve lies from $z = 0$ to $z = 0.20$. Therefore, $P(X \geq 9.5) = 0.5 - 0.0793 = 0.4207$ (the normal approximation). As n becomes even larger, the approximation becomes even closer. [If we had not treated the number of people as a continuous variable, we would have found that $P(X \geq 10) = 0.34$, and the approximation would not have been as close.]

3.38 A production process produces 10 defective items per hour. Find the probability that 4 or less items are defective out of the output of a randomly chosen hour using (a) the Poisson distribution and (b) the normal approximation of the Poisson.

(a) Here $\lambda = 10$ and we are asked to find $P(X \leq 4)$, where X is the number of defective items from the output of a randomly chosen hour. The value of e^{-10} from App. 2 is 0.00005. Therefore

$$P(0) = \frac{\lambda^0 e^{-10}}{0!} = e^{-10} = 0.00005$$

$$P(1) = \frac{\lambda^1 e^{-10}}{1!} = \frac{10(0.00005)}{1} = 0.0005$$

$$P(2) = \frac{\lambda^2 e^{-10}}{2!} = \frac{10^2(0.00005)}{2} = 0.0025$$

$$P(3) = \frac{\lambda^3 e^{-10}}{3!} = \frac{10^3(0.00005)}{6} = 0.0083335$$

$$P(4) = \frac{\lambda^4 e^{-10}}{4!} = \frac{10^4(0.00005)}{24} = 0.0208335$$

$$P(X \leq 4) = P(0) + P(1) + P(2) + P(3) + P(4)$$
$$= 0.00005 + 0.0005 + 0.0025 + 0.0083335 + 0.0208335$$
$$= 0.032217, \text{ or about } 3.22\%$$

(b) Treating the items as continuous [see Prob. 3.37(b)], we are asked to find $P(X \leq 4.5)$, where X is the number of defective items, $\mu = \lambda = 10$, and $\sigma = \sqrt{\lambda} = \sqrt{10} \cong 3.16$. Thus

$$z = \frac{X - \mu}{\sigma} = \frac{4.5 - 10}{3.16} = \frac{-5.5}{3.16} = -1.74$$

For $z = 1.74$ in App. 3, we get 0.4591. This means that $0.5 - 0.4591 = 0.0409$ of the area (probability) under the standard normal curve lies to the left of $z = -1.74$. Thus $P(X \leq 4.5) = 0.0409$, or 4.09%. As λ becomes larger, we get a better approximation. [If we had not treated the number of defective items as a continuous variable, we would have found that $P(X \leq 4) = 0.287$.]

3.39 If events or successes follow a Poisson distribution, we can determine the probability that the first event occurs within a designated period of time, $P(T \leq t)$, by the *exponential probability distribution*. Because we are dealing with time, the exponential is a continuous probability distribution. This is given by

$$P(T \leq t) = 1 - e^{-\lambda} \qquad (3.27)$$

where λ is the mean number of occurrences for the *interval of interest* and $e^{-\lambda}$ can be obtained from App. 2. The expected value and variance are

$$E(T) = \frac{1}{\lambda} \qquad (3.28)$$

$$\text{Var } T = \frac{1}{\lambda^2} \qquad (3.29)$$

(a) For the statement of Prob. 3.29, find the probability that starting at a random point in time, the first customer stops at the gasoline pump within a half hour. (b) What is the probability that no customer stops at the gasoline pump within a half hour? (c) What is the expected value and variance of the exponential distribution, where the continuous variable is time T?

(a) Since an average of 6 customers stop at the pump per hour, λ = average of 3 customers per half hour. The probability that the first customer will stop within the first half hour is

$$1 - e^{-\lambda} = 1 - e^{-3} = 1 - 0.04979 \text{ (from App. 2)} = 0.9502, \text{ or } 95.02\%$$

(b) The probability that no customer stops at the pump within a half hour is

$$e^{-\lambda} = e^{-3} = 0.04979$$

(c) $E(T) = 1/\lambda = 1/6 \cong 0.17\,\mathrm{h}$ per car, and var $T = 1/\lambda^2 = 1/36 \cong 0.03\,\mathrm{h}$ per car squared.   The exponential distribution also can be used to calculate the time between two successive events.

3.40 The mean level of schooling for a population is 8 years and the standard deviation is 1 year. What is the probability that a randomly selected individual from the population will have had between 6 and 10 years of schooling? Less than 6 years or more than 10 years?

Since we have not been told the form of the distribution, we can use Chebyshev's theorem, which applies to any distribution. With $\mu = 8$ years and $\sigma = 1$ year, 6 years of schooling is 2 standard deviations below μ and 10 years of schooling is 2 standard deviations above μ. Using Chevyshev's theorem or inequality we obtain

$$P(|\bar{X} - \mu| \le K\sigma) \ge 1 - \frac{1}{K^2} \qquad (3.30)$$

The probability of an individual picked at random from the population will be within 2 standard deviations from the mean is

$$1 - \frac{1}{K^2} = 1 - \frac{1}{2^2} = \frac{3}{4}, \text{ or } 75\%$$

Therefore, the probability that the individual will have had either less than 6 or more than 10 years of schooling is 25%.

Supplementary Problems

PROBABILITY OF A SINGLE EVENT

3.41 What approach to probability is involved in the following statements? (a) The probability of a head in the toss of a balanced coin is 1/2. (b) The relative frequency of a head in 100 tosses of a coin is 0.53. (c) The probability of rain tomorrow is 20%.
Ans. (a) The classical or a priori approach. (b) The relative frequency or empirical approach. (c) The subjective or personalistic approach.

3.42 What is the probability that in tossing a balanced coin we get (a) a tail, (b) a head, (c) not a tail, or (d) a tail or not a tail?
Ans. (a) $P(\mathrm{T}) = 1/2$ (b) $P(\mathrm{H}) = 1/2$ (c) $P(\mathrm{T}') = 1/2$ (d) $P(\mathrm{T}) + P(\mathrm{T}') = 1$

3.43 What is the probability that in one roll of a fair die we get (a) a 1, (b) a 6, (c) not a 1, or (d) a 1 or not a 1?
Ans. (a) $P(1) = 1/6$ (b) $P(6) = 1/6$ (c) $P(1') = 5/6$ (d) $P(1) + P(1') = 1$

3.44 What is the probability that in a single pick from a standard deck of cards we pick (a) a club, (b) an ace, (c) the ace of clubs, (d) not a club, or (e) a club or not a club?
Ans. (a) $P(\mathrm{C}) = 13/52 = 1/4$ (b) $P(\mathrm{A}) = 4/52 = 1/13$ (c) $P(\mathrm{A}_\mathrm{C}) = 1/52$ (d) $P(\mathrm{C}') = 3/4$ (e) $P(\mathrm{C}) + P(\mathrm{C}') = 1$

3.45 An urn contains 12 balls that are exactly alike except that 4 are blue, 3 are red, 3 are green, and 2 are white. What is the probability that by picking a single ball we pick (a) A blue ball? (b) A red ball? (c) A green ball? (d) A white ball? (e) A nonred ball? (f) A nonwhite ball? (g) A white or nonwhite ball? Also

(*h*) What are the odds of picking a green ball?   (*i*) What are the odds of picking a nongreen ball?
Ans. (*a*) $P(B) = 1/3$ or 0.33 (*b*) $P(R) = 1/4$ or 0.25 (*c*) $P(G) = 1/4$ or 0.25 (*d*) $P(W) = 1/6$ or 0.167
(*e*) $P(R') = 0.75$ (*f*) $P(W') = 0.833$ (*g*) $P(W) + P(W') = 1$ (*h*) $3 : 9$ (*i*) $9 : 3$

3.46 Suppose that a card is picked from a well-shuffled standard deck.   The card is then replaced, the deck reshuffled, and another card is picked.   As this process is repeated 520 times, we obtain 136 spades. (*a*) What is the relative frequency or empirical probability of getting a spade?   (*b*) What is the classical or a priori probability of getting a spade?   (*c*) What would you expect the relative frequency or empirical probability of getting a spade to be if the process is repeated many more times?
Ans. (*a*) $136/520 \cong 0.26$ (*b*) $P(S) = 1/4$ (*c*) To approach 1/4 or 0.25

3.47 An insurance company found that from a sample of 10,000 men between the ages of 30 and 40, 87 become seriously ill during a 1-year period.   (*a*) What is the relative frequency or empirical probability of men between 30 and 40 becoming seriously ill during a 1-year period?   (*b*) Why is the insurance company interested in these results?   (*c*) Suppose that the company subsequently sells health insurance to 1,387,684 men in the 30 to 40 age group.   How many claims can the company expect during a 1-year period? *Ans.* (*a*) The relative frequency or empirical probability is $87/10,000 = 0.0087$.   (*b*) The insurance company is interested in the relative frequency or empirical probability in order to determine its insurance premiums.   (*c*) 12,073, to the nearest person

PROBABILITY OF MULTIPLE EVENTS

3.48 What types of events are the following?   (*a*) Picking hearts or clubs on a single pick from a deck. (*b*) Picking diamonds or a queen on a single pick from a deck.   (*c*) Two successive flips of a balanced coin.   (*d*) Two successive tosses of a fair die.   (*e*) Picking two cards from a deck with replacement. (*f*) Picking two cards from a deck without replacement.   (*g*) Picking two balls from an urn without replacement.
Ans.   (*a*) Mutually exclusive   (*b*) Not mutuall exclusive   (*c*) Independent   (*d*) Independent   (*e*) Independent (*f*) Dependent (*g*) Dependent

3.49 What is the probability of getting   (*a*) Four or more on a single toss of a fair die?   (*b*) Ace or king on a single pick from a well-shuffled standard deck of cards?   (*c*) A green or white ball from the urn of Prob. 3.45?
Ans. (*a*) 1/2   (*b*) 8/52 or 2/13   (*c*) 5/12

3.50 What is the probability of getting   (*a*) A diamond or a queen on a single pick from a deck of cards?   (*b*) A diamond, a queen, or a king?   (*c*) An African-American or a woman president of the United States if the probability of an African-American president is 0.25, of a woman is 0.15, and of an African-American woman is 0.07?
Ans. (*a*) 16/52 or 4/13   (*b*) 19/52   (*c*) 0.33

3.51 What is the probability of   (*a*) Two ones in 2 rolls of a die?   (*b*) Three tails in 3 flips of a coin?   (*c*) A *total* of 6 in rolling 2 dice simultaneously?   (*d*) A total of less than 5 in rolling 2 dice simultaneously?   (*e*) A total of 10 or more in rolling 2 dice simultaneously?
Ans. (*a*) 1/36   (*b*) 1/8   (*c*) 5/36   (*d*) 1/6   (*e*) 1/6

3.52 What is the probability of obtaining the following from a deck of cards: (*a*) A diamond on the second pick when the first card picked and not replaced was a diamond?   (*b*) A diamond on the second pick when the first card picked and not replaced was not a diamond?   (*c*) A king on the third pick when a queen and a jack were already obtained on the first and second pick and not replaced?
Ans. (*a*) 12/51   (*b*) 13/51   (*c*) 4/50

3.53 What is the probability of picking   (*a*) the king of clubs and a diamond *in that order* in 2 picks from a deck without replacement?   (*b*) A white ball and a green ball *in that order* in 2 picks without replacement from the urn of Prob. 3.45?   (*c*) A green ball and a white ball *in that order* in 2 picks without replacement from the

urn of Prob. 3.45? (d) A green and a white ball *in that order* in 2 picks without replacement from the same urn? (e) Three green balls in 3 picks without replacement from the urn?
Ans. (a) 13/2652 or 1/204 (b) 6/132 or 1/22 (c) 1/22 (d) 1/11 (e) 6/1320 or 1/220

3.54 Suppose that the probability of rain on a given day is 0.1 and the probability of my having a car accident is 0.005 on any day and 0.012 on a rainy day. (a) What rule should I use to calculate the probability that on a given day it will rain and I will have a car accident? (b) State the rule asked for in part a, letting A signify accident and R signify rain. (b) Calculate the probability asked for in part a.
Ans. (a) The rule of multiplication for dependent events (b) $P(\text{R and A}) = P(\text{R}) \cdot P(\text{A}/\text{R})$ (c) 0.0012

3.55 (a) What rule or theorem should I use to calculate for the statement in Prob. 3.54 the probability that it was raining when I had a car accident? (b) State the rule or theorem applicable to part a. (c) Answer the question in part c.
Ans. (a) Bayes' theorem (b) $P(\text{R}/\text{A}) = P(\text{R}) \cdot P(\text{A}/\text{R})/P(\text{A})$ (c) 0.24

3.56 In how many different ways can 6 qualified individuals be assigned to (a) Three trainee positions available if the positions are identical? (b) Three trainee positions eventually if the positions differ? (c) Six trainee positions available if the positions differ?
Ans. (a) 20 (b) 120 (c) 720

DISCRETE PROBABILITY DISTRIBUTIONS: THE BINOMIAL DISTRIBUTION

3.57 The probability distribution of lunch customers at a restaurant is given in Table 3.5. Calculate (a) the expected number of lunch customers, (b) the variance, and (c) the standard deviation.

Table 3.5 Probability Distribution of Lunch Customers at a Restaurant

Number of Customers, X	Probability, $P(X)$
100	0.2
110	0.3
118	0.2
120	0.2
125	0.1
	1.0

Ans. (a) 113.1 customers (b) 65.69 customers squared (c) 8.10 customers

3.58 What is the probability of (a) Getting exactly 4 heads and 2 tails in 6 tosses of a balanced coin? (b) Getting 3 sixes in 4 rolls of a fair die?
Ans. (a) 0.23 (b) 0.0154321

3.59 (a) If 20% of the students entering college drop out before receiving their diplomas, find the probability that out of 20 students picked at random from the very large number of students entering college, less than 3 drop out. (b) If 90% of the bulbs produced in a plant are acceptable, what is the probability that out of 10 bulbs picked at random from the very large output of the plant, 8 are acceptable?
Ans. (a) 0.206 (b) 0.1937

3.60 Calculate the expected value and standard deviation and determine the symmetry or asymmetry of the probability distribution of (a) Prob. 3.58(a), (b) Prob. 3.59(a), and (c) Prob. 3.59(b).
Ans. (a) $E(X) = 3$ heads, SD $X = 1.22$ heads, and the distribution is symmetrical. (b) $E(X) = 4$ students, SD $X = 1.79$ students, and the distribution is skewed to the right. (c) $E(X) = 9$ bulbs, SD $X = 0.95$ bulbs, and the distribution is skewed to the left.

3.61   What is the probability of picking   (a) Two women in a sample of 5 drawn at random and without replacement from a group of 9 people, 4 of whom are women?   (b) Eight men in a sample of 10 drawn at random and without replacement from a population of 1000, half of which are men.
   Ans.   (a) About 0.71 (using the hypergeometric distribution)   (b) About 0.0439 (using the binomial approximation to the hypergeometric probability)

THE POISSON DISTRIBUTION

3.62   Past experience shows that there are two traffic accidents at an intersection per week.   What is the probability of: (a) Four accidents during a randomly selected week?   (b) No accidents?   (c) What is the expected value and standard deviation of the distribution?
   Ans.   (a) About 0.36   (b) About 0.14   (c) $E(X) = \lambda = 2$ accidents, and SD $X = \sqrt{\lambda} = 1.41$ accidents

3.63   Past experience shows that 0.003 of the national labor force get seriously ill during a year.   If 1000 persons are randomly selected from the national labor force: (a) What is the expected number of workers that will get sick during a year?   (b) What is the probability that 5 workers will get sick during the year?
   Ans.   (a) 3 workers   (b) About 0.1 (using the Poisson approximation to the binomial distribution)

CONTINUOUS PROBABILITY DISTRIBUTIONS: THE NORMAL DISTRIBUTION

3.64   Give the formulas: (a) the probability that continuous variable X falls between X_1 and X_2,   (b) the normal distribution,   (c) the expected value and variance of the normal distribution, and   (d) the standard normal distribution,   (e) what is the mean and variance of the standard normal distribution?
   Ans.   (a) $P(X_1 < X < X_2) = \int_{X_1}^{X_2} f(x)\, dX$ (b) $f(X) = (1/\sqrt{2\pi\sigma^2})\exp\{-(1/2)[(X-\mu)/\sigma]^2\}$ (c) $E(X) = \int_{-\infty}^{\infty} X f(X)\, dX$ and $\sigma^2 = \int_{-\infty}^{\infty} [X - E(X)]^2 f(X)\, dX$ (d) $f(X) = (1/\sqrt{2\pi})\exp[-(1/2)z^2]$ (e) $E(X) = \mu = 0$ and $\sigma^2 = 1$

3.65   Find the area under the standard normal curve   (a) within $z \pm 1.64$,   (b) within $z = \pm 1.96$,   (c) within $z = \pm 2.58$,   (d) between $z = 0.90$ and $z = 2.10$,   (e) to the left of $z = 0.90$,   (f) to the right of $z = 2.10$,   (g) to the left of $z = 0.90$ and to the right of $z = 2.10$.
   Ans.   (a) 0.899, or 89.90%   (b) 0.95   (c) 0.9902   (d) 0.1662   (e) 0.8159   (f) 0.0179   (g) 0.8338

3.66   A random variable is normally distributed with $\mu = 67$ and $\sigma = 3$.   What is the probability that this random variable will assume a value   (a) Between 67 and 70?   (b) Between 60 and 70?   (c) Between 60 and 65?   (d) Below 60?   (e) Above 65?
   Ans.   (a) 0.3413, or 34.13%   (b) 0.8334   (c) 0.2415   (d) 0.0099   (e) 0.7486

3.67   The mean weight of a large group of people is 180 lb and the standard deviation is 15 lb.   If the weights are normally distributed, find the probability that a person picked at random from the group will weigh   (a) between 160 and 180 lb,   (b) above 200 lb,   (c) below 150 lb.
   Ans.   (a) 0.4082, or 40.82%   (b) 0.0918   (c) 0.228

3.68   The IQs of army volunteers in a given year are normally distributed with $\mu = 110$ and $\sigma = 10$.   The army wants to give advanced training to the 25% of those recruits with the highest IQ scores.   What is the lowest IQ score acceptable for the advanced training?
   Ans.   117, to the nearest whole number

3.69   Past experience indicates that 60% of the students entering college get their degrees.   Using   (a) the binomial distribution and   (b) the normal approximation to the binomial, find the probability that out of 30 students picked at random from the entering class, more than 20 will receive their degrees.
   Ans.   (a) 0.1762   (b) 0.1762

3.70   An average of 10 cars per minute pass through a toll booth during rush hour.   Using   (a) the Poisson distribution and   (b) the normal approximation to the Poisson, find the probability that less than 6 cars pass through the toll booth during a randomly chosen minute.

Ans. (*a*) 0.0749, or 7.49%   (*b*) 0.0778, or 7.78%

3.71   A manufacturing process produces on the average two defective items per hour.   What is the probability that after a defective item: (*a*) One hour will pass before the next defective item?   (*b*) One-half hour will pass?   (*c*) Fifteen minutes will pass?   (*d*) What is the expected value and standard deviation of this distribution?

Ans. (*a*) 0.13534, or 13.53%   (*b*) 0.36788   (*c*) 0.60653   (*d*) $E(T) = \sigma = 1/\lambda = 0.5$ h per defective item

3.72   If a student has a grade point average 3 standard deviations above the mean in her school, what proportion of the other students in the school have: (*a*) A higher grade point average?   (*b*) A lower grade point average?

Ans. (*a*) < 0.11, or 11% (using Chevyshev's theorem)   (*b*) at least 0.89, or 89%

3.73   According to Chebyshev's theorem, at least what proportion of the observations fall within   (*a*) 1.5 standard deviations from the mean,   (*b*) 2.5 standard deviations from the mean?

Ans. (*a*) 0.56, or 56%   (*b*) 0.84, or 84%

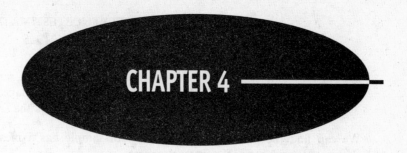

CHAPTER 4

Statistical Inference: Estimation

4.1 SAMPLING

Statistical inference is one of the most important and crucial aspects of the decision making process in economics, business, and science. *Statistical inference* refers to estimation and hypothesis testing (Chap. 5). *Estimation* is the process of inferring or estimating a population *parameter* (such as its mean or standard deviation) from the corresponding *statistic* of a sample drawn from the population.

To be valid, estimation (and hypothesis testing) must be based on a *representative* sample. This can be obtained by *random sampling*, whereby each member of the population has an equal chance of being included in the sample.

EXAMPLE 1. A random sample of 5 out of the 80 employees of a plant can be obtained by recording the name of each employee on a separate slip of paper, mixing the slips of paper thoroughly, and then picking 5 at random. A less cumbersome method is to use a table of random numbers (App. 4). To do this, we first assign each employee a number from 1 to 80. Then starting at random (say, from the third column and eleventh row) in App. 4, we can read 5 numbers (as pairs) either horizontally or vertically (eliminating all numbers exceeding 80). For example, reading vertically we get 13, 54, 19, 59, and 71.

4.2 SAMPLING DISTRIBUTION OF THE MEAN

If we take repeated random samples from a population and measure the mean of each sample, we find that most of these sample means, $\bar{X}$s, differ from each other. The probability distribution of these sample means is called the *sampling distribution of the mean*. However, the sampling distribution of the mean itself has a mean, given by the symbol $\mu_{\bar{X}}$, and a standard deviation or *standard error*, $\sigma_{\bar{X}}$.

Two important theorems relate the sampling distribution of the mean to the parent population.

1. If we take repeated random samples of size n from a population

$$\mu_{\bar{X}} = \mu \tag{4.1}$$

and

$$\sigma_{\bar{X}} = \frac{\sigma}{\sqrt{n}} \quad \text{or} \quad \sigma_{\bar{X}} = \frac{\sigma}{\sqrt{n}}\sqrt{\frac{N-n}{N-1}} \tag{4.2a, b}$$

where Eq. (*4.2b*) is used for finite populations of size N when $n \geq 0.05\,N$ [see Prob. 4.5(*b*)].

2. As the samples' size is increased (i.e., as $n \to \infty$), the sampling distribution of the mean approaches the normal distribution regardless of the shape of the parent population.   The approximation is sufficiently good for $n \geq 30$.   This is the *central-limit theorem*.

We can find the probability that a random sample has a mean $\bar{X}$ in a given interval by first calculating the z values for the interval, where

$$z = \frac{\bar{X} - \mu_{\bar{X}}}{\sigma_{\bar{X}}} \qquad\qquad (4.3)$$

and then looking up these values in App. 3, as explained in Sec. 3.5.

EXAMPLE 2.   In Fig. 4-1, the mean of the sampling distribution of the mean $\mu_{\bar{X}}$ is equal to the mean of the parent population μ regardless of the samples' size n.   However, the greater is n, the smaller is the spread or standard error of the mean, $\sigma_{\bar{X}}$.   If the parent population is normal, the sampling distributions of the mean are also normally distributed, even in small samples.   According to the central-limit theorem, even if the parent population is not normally distributed, the sampling distributions of the mean are approximately normal for $n \geq 30$.

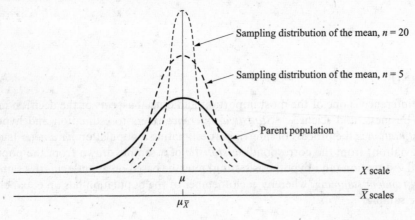

Fig. 4-1

EXAMPLE 3.   Assume that a population is composed of 900 elements with a mean of 20 units and a standard deviation of 12.   The mean and standard error of the sampling distribution of the mean for a sample size of 36 is

$$\mu_{\bar{X}} = \mu = 20 \text{ units}$$
$$\sigma_{\bar{X}} = \frac{\sigma}{\sqrt{n}} = \frac{12}{\sqrt{36}} = 2$$

If n had been 64 instead of 36 (so that $n > 0.05N$), then

$$\sigma_{\bar{X}} = \frac{\sigma}{\sqrt{n}} \sqrt{\frac{N-n}{N-1}} = \frac{12}{\sqrt{64}} \sqrt{\frac{900-64}{900-1}} = \frac{12}{8} \sqrt{\frac{836}{899}} = (1.5)(0.96) = 1.44$$

instead of $\sigma_{\bar{X}} = 1.5$, without the *finite correction factor*.

EXAMPLE 4.   The probability that the mean of a random sample $\bar{X}$ of 36 elements from the population in Example 3 falls between 18 and 24 units is computed as follows:

$$z_1 = \frac{\bar{X}_1 - \mu_{\bar{X}}}{\sigma_{\bar{X}}} = \frac{18 - 20}{2} = -1 \qquad \text{and} \qquad z_2 = \frac{\bar{X}_2 - \mu_{\bar{X}}}{\sigma_{\bar{X}}} = \frac{24 - 20}{2} = 2$$

Looking up z_1 and z_2 in App. 3, we get

$$P(18 < \bar{X} < 24) = 0.3413 + 0.4772 = 0.8185, \text{ or } 81.85\%$$

See Fig. 4-2.

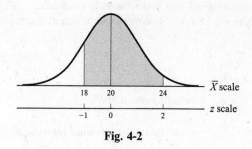

Fig. 4-2

4.3   ESTIMATION USING THE NORMAL DISTRIBUTION

We can get a point or an interval estimate of a population parameter.   A *point estimate* is a single number.   Such a point estimate is *unbiased* if in repeated random samplings from the population, the expected or mean value of the corresponding statistic is equal to the population parameter.   For example, $\bar{X}$ is an unbiased (point) estimate of μ because $\mu_{\bar{X}} = \mu$, where $\mu_{\bar{X}}$ is the expected value of $\bar{X}$.   The sample standard deviation s [as defined in Eqs. (*2.10b*) and (*2.11b*)] is an unbiased estimate of σ [see Prob. 4.13(*b*)], and the sample proportion $\bar{p}$ is an unbiased estimate of p (the proportion of the population with a given characteristic).

An *interval estimate* refers to a range of values together with the probability, or *confidence level*, that the interval includes the unknown population parameter.   Given the population standard deviation or its estimate, and given that the population is normal or that a random sample is equal to or larger than 30, we can find the 95% confidence interval for the unknown population mean as

$$P(\bar{X} - 1.96\sigma_{\bar{X}} < \mu < \bar{X} + 1.96\sigma_{\bar{X}}) = 0.95 \tag{4.4}$$

This states that in repeated random sampling, we expect that 95 out of 100 intervals such as Eq. (*4.4*) include the unknown population mean and that our confidence interval (based on a single random sample) is one of these.

A confidence interval can be constructed similarly for the *population proportion* (see Example 7) where

$$\mu_{\bar{p}} = \frac{\mu}{n} = p \qquad \text{(the proportion of successes in the population)} \tag{4.5}$$

$$\sigma_{\bar{p}} = \sqrt{\frac{p(1-p)}{n}} \qquad \text{(the standard error of the proportion)} \tag{4.6}$$

EXAMPLE 5.   A random sample of 144 with a mean of 100 and a standard deviation of 60 is taken from a population of 1000.   The 95% confidence interval for the unknown population mean is

$$\mu = \bar{X} \pm 1.96\sigma_{\bar{X}} \qquad \text{since } n > 30$$

$$= \bar{X} \pm 1.96\frac{\sigma}{\sqrt{n}}\sqrt{\frac{N-n}{N-1}} \qquad \text{since } n > 0.05N$$

$$= 100 \pm 1.96\frac{60}{\sqrt{144}}\sqrt{\frac{1000-144}{1000-1}} \qquad \text{using } s \text{ as an estimate of } \sigma$$

$$= 100 \pm 1.96(5)(0.93)$$

$$= 100 \pm 9.11$$

Thus μ is between 90.89 and 109.11 with a 95% degree of confidence.   Other frequently used confidence intervals are the 90 and 99% levels, corresponding to the z values of 1.64 and 2.58, respectively (see App. 3).

EXAMPLE 6. A manager wishes to estimate the mean number of minutes that workers take to complete a particular manufacturing process within ± 3 min and with 90% confidence. From past experience, the manager knows that the standard deviation σ is 15 min. The minimum required sample size ($n > 30$) is found as follows:

$$z = \frac{\bar{X} - \mu}{\sigma_{\bar{X}}}$$

$$z\sigma_{\bar{X}} = \bar{X} - \mu$$

$$1.64 \frac{\sigma}{\sqrt{n}} = \bar{X} - \mu \qquad \text{assuming } n < 0.05N$$

$$1.64 \frac{15}{\sqrt{n}} = 3 \qquad \text{since the total confidence interval, } \bar{X} - \mu, \text{ is 3 min}$$

$$1.64 \frac{15}{3} = \sqrt{n}$$

$$n = 67.24, \text{ or } 68 \text{ (rounded to the next higher integer)}$$

EXAMPLE 7. A state education department finds that in a random sample of 100 persons who attended college, 40 received a college degree. To find the 99% confidence interval for the proportion of college graduates out of all the persons who attended college, we proceed as follows. First, we note that this problem involves the binomial distribution (see Sec. 3.3). Since $n > 30$ and both $np > 5$ and $n(1 - p) > 5$, the binomial distribution approaches the normal distribution (which is simpler to use; see Sec. 3.5). Then

$$p = \bar{p} \pm z\sigma_{\bar{p}}$$

and

$$p = \bar{p} \pm z\sqrt{\frac{p(1-p)}{n}} \qquad \text{assuming } n < 0.05N$$

$$= 0.4 \pm 2.58\sqrt{\frac{(0.4)(0.6)}{100}} \qquad \text{using } \bar{p} \text{ as an estimate of } p$$

$$\cong 0.4 \pm 2.58(0.05)$$

$$\cong 0.4 \pm 0.13$$

Thus p is between 0.27 and 0.53 with a 99% level of confidence.

4.4 CONFIDENCE INTERVALS FOR THE MEAN USING THE *t* DISTRIBUTION

When the population is normally distributed but σ is not known and $n < 30$, we cannot use the normal distribution for determining confidence intervals for the unknown population mean, but we can use the *t* distribution. This is symmetrical about its zero mean but is flatter than the standard normal distribution, so that more of its area falls within the tails. While there is a single standard normal distribution, there is a different *t* distribution for each sample size, *n*. However, as *n* becomes larger, the *t* distribution approaches the standard normal distribution (see Fig. 4-3) until, when $n \geq 30$, they are approximately equal.

Appendix 5 gives the values of *t* *to the right of which* we find 10, 5, 2.5, 1, and 0.5% of the total area under the curve for various *degrees of freedom*. Degrees of freedom (df) are defined in this case as $n - 1$

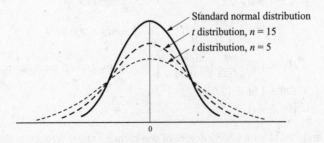

Fig. 4-3

(or the sample size minus 1 for the single parameter μ we wish to estimate). The 95% confidence interval for the unknown population mean when the t distribution is used is given by

$$P\left(\bar{X} - t\frac{s}{\sqrt{n}} < \mu < \bar{X} + t\frac{s}{\sqrt{n}}\right) = 0.95 \qquad (4.7)$$

where t refers to the t values such that 2.5% of the total area under the curve falls within each tail (for the degrees of freedom involved) and $s/\sqrt{n}$ is used instead of $\sigma_{\bar{X}} = \sigma/\sqrt{n}$.

EXAMPLE 8. A random sample of $n = 10$ flashlight batteries with a mean operating life $\bar{X} = 5\,\text{h}$ and a sample standard deviation $s = 1\,\text{h}$ is picked from a production line known to produce batteries with normally distributed operating lives. To find the 95% confidence interval for the unknown mean of the working life of the entire population of batteries, we first find the value of $\pm t_{0.025}$ so that 2.5% of the area is within each tail for $n - 1 = 9\,\text{df}$. This is obtained from App. 5 by moving down the column headed 0.025 to 9 df. The value we get is 2.262. Thus

$$\mu = \bar{X} \pm 2.262\frac{s}{\sqrt{n}} = 5 \pm 2.262\frac{1}{\sqrt{10}} \cong 5 \pm 2.262(0.316) \cong 5 \pm 0.71$$

and μ is between 4.29 and 5.71 h with 95% confidence (see Fig. 4-4). When $n < 30$ *and* the population is not normally distributed, we must use *Chebyshev's theorem* (see Prob. 4.27).

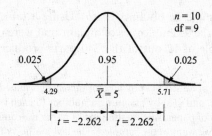

Fig. 4-4

Solved Problems

SAMPLING

4.1 (*a*) What is meant by statistical inference? What is its function and importance? (*b*) What is meant by and what is the relationship between a parameter and a statistic? (*c*) What is meant by estimation? Hypothesis testing?

(*a*) *Statistical inference* is the process of making inferences about populations from information provided by samples. A *population* is the collection of all the elements (people, parts produced by a machine, cars passing through a checkpoint, etc.) that we are describing. A *sample* is a portion chosen from the population. Analyzing an entire population may be impossible (if the population is infinite), it may destroy all the output (such as in testing all the flashbulbs produced), and it may be prohibitively expensive. These problems can be overcome by taking a (representative) sample from a population and making inferences about the population from the sample.

(*b*) A *parameter* is a descriptive characteristic (such as the mean and the standard deviation) of a population. A *statistic* is a descriptive characteristic of a sample. In statistical inference, we make inferences about parameters from their corresponding statistics.

(c) Statistical inference is of two kinds: estimation and hypothesis testing. *Estimation* is the process of inferring or estimating a parameter from the corresponding statistic. For example, we may estimate the mean and the standard deviation of a population from the mean and standard deviation of a sample drawn from the population. *Hypothesis testing* is the process of determining, on the basis of sample information, whether to accept or reject a hypothesis or assumption with regard to the value of a parameter. We deal with estimation in this chapter and with hypothesis testing in Chap. 5.

4.2   What is meant by random sampling? What is its importance?

Random sampling is a sampling procedure by which each member of a population has an equal chance of being included in the sample. Random sampling ensures a *representative* sample. There are several types of random sampling. In *simple* random sampling, not only each *item* in the population but each *sample* has an equal probability of being picked. In *systematic sampling*, items are selected from the population at uniform intervals of time, order, or space (as in picking every one-hundredth name from a telephone directory). Systematic sampling can be biased easily, such as, for example, when the amount of household garbage is measured on Mondays (which includes the weekend garbage). In *stratified* and *cluster* sampling, the population is divided into strata (such as age groups) and clusters (such as blocks of a city) and then a proportionate number of elements is picked at random from each stratum and cluster. Stratified sampling is used when the variations within each stratum are small in relation to the variations between strata. Cluster sampling is used when the opposite is the case. In what follows, we assume simple random sampling. Sampling can be from a finite population (as in picking cards from a deck without replacement) or from an infinite population (as in picking parts produced by a continuous process or cards from a deck *with replacement*).

4.3   (a) How can a random sample be obtained? (b) Using a table of random numbers, obtain a random sample of 10 from the 95 employees of a plant that were out sick during a particular day. (c) Obtain a random sample of 12 out of the 240 parts produced by a machine during its first hour of operation.

(a) A random sample can be obtained (1) by a computer programmed to assemble numbers, (2) from a table of random numbers, and (3) by assigning a number to each item in a population, recording each number on a separate slip of paper, mixing the slips of paper thoroughly, and then picking as many slips of paper and numbers as we want in the sample. The last method of obtaining a random sample is very cumbersome with large populations and may not give a representative sample because of the difficulty of thoroughly scrambling the pieces of paper.

(b) To obtain a random sample of 10 from the 95 employees, we assign each employee a number from 1 to 95 and then consult App. 4 (the table of random numbers of digits). Appendix 4 lists 1600 digits in sets of 5 digits generated by a completely random process and such that each digit and sequence of digits has the same probability of occurring as every other digit and sequence of digits. Starting at an arbitrary point in App. 4 (say, the fourteenth column and fifth row) and reading 10 numbers in pairs (say, vertically and omitting all numbers above 95), we get the following random sample: 60, 39, 4, 34, 76, 43, 52, 14, 8, and 95.

(c) Starting, say, from the third row and eighth line in App. 4 and reading 8 numbers horizontally (three digits at a time and eliminating numbers exceeding 240), we get the following random sample: 215, 182, 51, 9, 127, 177, 53, and 186 (the last four numbers were obtained from the ninth line after reaching the end of the eighth line).

SAMPLING DISTRIBUTION OF THE MEAN

4.4   (a) What does sampling distribution mean and how is a sampling distribution of the mean obtained? (b) What is meant by the mean and standard error of the sampling distribution of the mean?

(a) If we take repeated (or all possible) random samples, each of size n, from a population of values of the variable X and find the mean of each of these samples $\bar{X}$, we find that most of the sample means differ from each other. The probability distribution of these sample means is called the *theoretical sampling distribution of the mean*. Similarly, we could get the theoretical sampling distribution of a proportion, of the difference between two means, and of the difference between two proportions. For example, we could have found the proportion of defective items in each sample and obtained the theoretical sampling distribution of the proportion of defective items. For simplicity, this section deals only with the sampling distribution of the mean.

(b) Just as in other probability distributions (see Secs. 3.3 to 3.5), the theoretical sampling distribution of the mean can be described by its mean and standard deviation. The mean of the sampling distribution of the mean is given by the symbol $\mu_{\bar{X}}$ (read "mu sub X bar"). This is the mean of the $\bar{X}$s and is to be distinguished from μ (the mean of the parent population). The standard deviation of the sampling distribution of the mean is given by the symbol $\sigma_{\bar{X}}$ (read "sigma sub X bar"). This is the standard deviation of the $\bar{X}$s and is to be clearly distinguished from σ (which is the standard deviation of the parent population). The smaller is $\sigma_{\bar{X}}$, the more accurate is a sample mean $\bar{X}$ as an estimate of the (unknown) population mean μ. For this reason, $\sigma_{\bar{X}}$ is usually referred to as the *standard error* of the mean.

4.5 How can we find (a) The mean of the sampling distribution of the mean $\mu_{\bar{X}}$? (b) The standard deviation of the sampling distribution of the mean or standard error $\sigma_{\bar{X}}$?

(a) The mean of the theoretical sampling distribution of the mean $\mu_{\bar{X}}$ is equal to the mean of the parent population μ; that is, $\mu_{\bar{X}} = \mu$. Note that for this to be true, either we must take *all* the different samples of size n possible from the finite population or, if we are dealing with an infinite population (or a finite population with replacement), we must continue to take repeated random samples of size n *indefinitely*. Moreover, $\mu_{\bar{X}}$ is also equal to $E(\bar{X})$ (see Probs. 3.20 and 3.31).

(b) The standard error of the mean $\sigma_{\bar{X}}$ is given by the standard deviation of the parent population σ divided by the square root of the samples' size $\sqrt{n}$; that is, $\sigma_{\bar{X}} = \sigma/\sqrt{n}$. For finite populations of size N, a *finite correction factor* must be added, and $\sigma_{\bar{X}} = (\sigma/\sqrt{n})\sqrt{(N-n)/(N-1)}$. However, if the sample size is very small in relation to the population size, $\sqrt{(N-n)/(N-1)}$ is close to 1 and can be dropped from the formula. By convention, this is done whenever $n < 0.05N$. Independently of this finite correction factor, $\sigma_{\bar{X}}$ is directly related to σ and inversely related to $\sqrt{n}$ [see Eq. (4.2a,b)]. Thus increasing the samples size 4 times increases the accuracy of $\bar{X}$ as an estimate of μ by cutting $\sigma_{\bar{X}}$ in half. Note also that $\sigma_{\bar{X}}$ is always smaller than σ. The reason for this is that the sample means, as *averages* of the sample observations, exhibit less variability or spread than the population values. Furthermore, the larger are the sample sizes, the more the values of $\sigma_{\bar{X}}$ are averaged down with respect to the value of σ (see Fig. 4-1).

4.6 For a population composed of the following 5 numbers: 1, 3, 5, 7, and 9, find (a) μ and σ, (b) the theoretical sampling distribution of the mean for the sample size of 2, and (c) $\mu_{\bar{X}}$ and $\sigma_{\bar{X}}$.

(a)
$$\mu = \frac{\sum X}{N} = \frac{1+3+5+7+9}{5} = \frac{25}{5} = 5$$

$$\sigma = \sqrt{\frac{\sum(X-\mu)^2}{N}} = \sqrt{\frac{(1-5)^2 + (3-5)^2 + (5-5)^2 + (7-5)^2 + (9-5)^2}{5}}$$

$$= \sqrt{\frac{16+4+0+4+16}{5}} = \sqrt{\frac{40}{5}} = \sqrt{8} \cong 2.83$$

(b) The theoretical sampling distribution of the sample mean for the sample size of 2 from the given finite population n is given by the mean of *all the possible different samples* that can be obtained from this population. The number of *combinations* of 5 numbers taken 2 at a time *without concern for the order* is $5!/2!3! = 10$ (see Prob. 3.18). These 10 samples are 1, 3; 1, 5; 1, 7; 1, 9; 3, 5; 3, 7; 3, 9; 5, 7; 5, 9; and 7, 9. The mean, $\bar{X}$, of the preceding 10 samples is 2, 3, 4, 5, 4, 5, 6, 6, 7, 8. The theoretical sampling distribution of the mean is given in Table 4.1. Note that the variability or spread of the sample means (from 2 to 8) is less than the variability or spread of the values in the parent population (from 1 to 9), confirming the statement made at the end of Prob. 4.5(b).

(c) By applying theorem 1 (Sec. 4.2), $\mu_{\bar{X}} = \mu = 5$. Since the sample size of 2 is greater than 5% of the population size (that is, $n > 0.05N$),

$$\sigma_{\bar{X}} = \frac{\sigma}{\sqrt{n}}\sqrt{\frac{N-n}{N-1}} = \frac{\sqrt{8}}{\sqrt{2}}\sqrt{\frac{5-2}{5-1}} = \sqrt{4}\sqrt{\frac{3}{4}} = \sqrt{3} \cong 1.73$$

Table 4.1 Theoretical Sampling Distribution of the Mean

Values of the Mean	Possible Outcomes	Probability of Occurrence
2	2	0.1
3	3	0.1
4	4, 4	0.2
5	5, 5	0.2
6	6, 6	0.2
7	7	0.1
8	8	0.1
		Total 1.0

4.7 For the theoretical sampling distribution of the sample mean found in Prob. 4.6(*b*)   (*a*) find the mean and the standard error of the mean *using the formulas for the population mean and standard deviation given in Secs. 2.2 and 2.3.*   (*b*) What do the answers to part *a* show?

(*a*)
$$\mu_{\bar{X}} = \frac{\sum \bar{X}}{N} = \frac{2+3+4+5+4+5+6+6+7+8}{10} = \frac{50}{10} = 5$$

$$\sigma_{\bar{X}} = \sqrt{\frac{\sum (\bar{X} - \mu_{\bar{X}})^2}{N}}$$

$$= \sqrt{\frac{\begin{array}{c}(2-5)^2 + (3-5)^2 + (4-5)^2 + (5-5)^2 + (4-5)^2 \\ + (5-5)^2 + (6-5)^2 + (6-5)^2 + (7-8)^2 + (8-5)^2\end{array}}{10}}$$

$$= \sqrt{\frac{9+4+1+0+1+0+1+1+4+9}{10}} = \sqrt{\frac{30}{10}} = \sqrt{3} \cong 1.73$$

(*b*) The answers to part *a* confirm the results obtained in Prob. 4.5(*c*) by the application of *theorem 1* (Sec. 4.2), namely, that $\mu_{\bar{X}} = \mu$ and $\sigma_{\bar{X}} = (\sigma/\sqrt{n})\sqrt{(N-n)/(N-1)}$ for the finite population where $n > 0.05N$.   Note that we took *all the possible different samples of size 2 that we could take from our finite population of 5 numbers.*   Sampling from an infinite parent population (or from a finite parent population with replacement) would have required taking an *infinite* number of random samples of size n from the parent population (an obviously impossible task).   By taking only a limited number of random samples, theorem 1 would hold only approximately (i.e., $\mu_{\bar{X}} \approx \mu$ and $\sigma_{\bar{X}} \approx \sigma/\sqrt{n}$), with the approximation becoming better as the number of random samples taken is increased.   In this case, the sampling distribution of the sample mean generated is referred to as the *empirical sampling distribution of the mean.*

4.8 A population of 12,000 elements has a mean of 100 and a standard deviation of 60.   Find the mean and standard error of the sampling distribution of the mean for sample sizes of   (*a*) 100 and   (*b*) 900.

(*a*)
$$\mu_{\bar{X}} = \mu = 100$$

$$\sigma_{\bar{X}} = \frac{\sigma}{\sqrt{n}} = \frac{60}{\sqrt{100}} = 6$$

(*b*)
$$\mu_{\bar{X}} = \mu = 100$$

Since a sample of 900 is more than 5% of the population size, the finite correction factor must be used in the formula for the standard error:

$$\sigma_{\bar{X}} = \frac{\sigma}{\sqrt{n}}\sqrt{\frac{N-n}{N-1}} = \frac{60}{\sqrt{900}}\sqrt{\frac{12,000-900}{12,000-1}} = \frac{60}{30}\sqrt{\frac{11,100}{11,999}} \cong 2\sqrt{0.925} \cong 2(0.962) \cong 1.92$$

Without the correction factor, $\sigma_{\bar{X}}$ would have been equal to 2 instead of 1.92.

4.9   (a) What is the shape of the theoretical sampling distribution of the mean if the parent population is normal? If the parent population is not normal? (b) What is the importance of the answer to part a?

   (a) If the parent population is normally distributed, the theoretical sampling distributions of the mean are also normally distributed, *regardless of sample size*. According to the *central limit theorem*, even if the parent population is not normal, the theoretical sampling distributions of the sample mean approach normality as sample size increases (i.e., as $n \rightarrow \infty$). This approximation is sufficiently good for samples of at least 30.

   (b) The central-limit theorem is perhaps the most important theorem in all of statistical inference. It allows us to use sample statistics to make inferences about population parameters without knowing anything about the shape of the parent population. This will be done in this chapter and in Chap. 5.

4.10   (a) How can we calculate the probability that a random sample has a mean that falls within a given interval if the theoretical sampling distribution of the mean is normal or approximately normal? How is this different from the process of finding the probability that a normally distributed random variable assumes a value within a given interval? (b) Draw a normal curve in the $\bar{X}$ and z scales and show the percentage of the area under the curve within 1, 2, and 3 standard deviation units of its mean.

   (a) If the theoretical sampling distribution of the mean is normal or approximately normal, we can find the probability that a random sample has a mean $\bar{X}$ that falls within a given interval by calculating the corresponding z values in App. 3. This is analogous to what was done in Sec. 3.5, where the normal and the standard normal curves were introduced. The only difference is that now we are dealing with the distribution of the $\bar{X}$s rather than with the distribution of the Xs. In addition, before $z = (X - \mu)/\sigma$, while now $z = (\bar{X} - \mu_{\bar{X}})/\sigma_{\bar{X}} = (\bar{X} - \mu)/\sigma_{\bar{X}}$, since $\mu_{\bar{X}} = \mu$.

   (b) In Fig. 4-5, we have a normal curve in the $\bar{X}$ scale and a standard normal curve in the z scale. The area under the curve within 1, 2, and 3 standard deviation units from the mean is 68.26, 95.44, and 99.74%, respectively. Note the great similarity and important difference between Figs. 4-5 and 3-4.

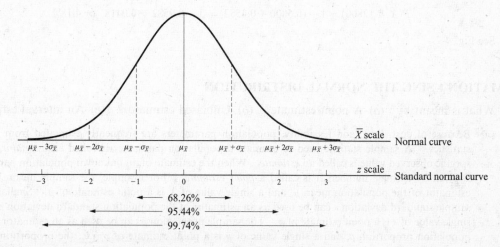

Fig. 4-5

4.11 Find the probability that the mean of a random sample of 25 elements from a normally distributed population with a mean 90 and a standard deviation of 60 is larger than 100.

Since the parent population is normally distributed, the theoretical sampling distribution of the mean is also normally distributed and $\sigma_{\bar{X}} = \sigma/\sqrt{n}$ because $n < 0.05N$. For $\bar{X} = 100$

$$z = \frac{\bar{X} - \mu_{\bar{X}}}{\sigma_{\bar{X}}} = \frac{\bar{X} - \mu}{\sigma/\sqrt{n}} = \frac{100 - 90}{60/\sqrt{25}} = \frac{10}{12} \cong 0.83$$

Looking up this value in App. 3, we get

$$P(\bar{X} > 100) = 1 - (0.5000 + 0.2967) = 1 - 0.7967 = 0.2033, \text{ or } 20.33\%$$

See Fig. 4-6.

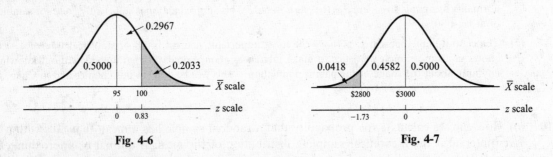

Fig. 4-6   Fig. 4-7

4.12 A small local bank has 1450 individual savings accounts with an average balance of $3000 and a standard deviation of $1200. If the bank takes a random sample of 100 accounts, what is the probability that the average savings for these 100 accounts will be below $2800?

Since $n = 100$, the theoretical sampling distribution of the mean is approximately normal, but since $n > 0.05N$, the finite correction factor must be used to find $\sigma_{\bar{X}}$. For $\bar{X} = \$2800$

$$z = \frac{\bar{X} - \mu_{\bar{X}}}{\sigma_{\bar{X}}} = \frac{\bar{X} - \mu}{\dfrac{\sigma}{\sqrt{n}}\sqrt{\dfrac{N-n}{N-1}}} = \frac{2800 - 3000}{\dfrac{1200}{\sqrt{100}}\sqrt{\dfrac{1450-100}{1450-1}}} = \frac{-200}{120\sqrt{\dfrac{1350}{1449}}} \cong \frac{-200}{120(0.965)} \cong -1.73$$

Looking up $z = 1.73$ in App. 3, we get

$$P(\bar{X} < \$2800) = 1 - (0.5000 + 0.4582) = 1 - 0.9582 = 0.0418, \text{ or } 4.18\%$$

See Fig. 4-7.

ESTIMATION USING THE NORMAL DISTRIBUTION

4.13 What is meant by (a) A point estimate? (b) Unbiased estimator? (c) An interval estimate?

(a) Because of cost, time, and feasibility, population parameters are frequently estimated from sample statistics. A sample statistic used to estimate a population parameter is called an *estimator*, and a specific observed value is called an *estimate*. When the estimate of an unknown population parameter is given by a single number, it is called a *point estimate*. For example, the sample mean $\bar{X}$ is an estimator of the population mean μ, and a single value of $\bar{X}$ is a point estimate of μ. Similarly, the sample standard deviation s can be used as an estimator of the population standard deviation σ and a single value of s is a point estimate of σ. The sample proportion $\bar{p}$ can be used as an estimator for the population proportion p, and a single value of $\bar{p}$ is a point estimate of p (i.e., the proportion of the population with a given characteristic).

(b) An estimator is *unbiased* if in repeated random sampling from the population the corresponding statistic from the theoretical sampling distribution is equal to the population parameter.   Another way of stating this is that an estimator is unbiased if its expected value (see Probs. 3.20 and 3.31) is equal to the population parameter being estimated.   For example, $\bar{X}$, s [defined in Eqs. (*2.10b*) and (*2.11b*)], and $\bar{p}$ are unbiased estimators of μ, σ, and p, respectively.   Other important criteria for a good estimator are discussed in Sec. 6.4.

(c) An *interval estimate* refers to the range of values used to estimate an unknown population parameter together with the probability, or *confidence level*, that the interval does include the unknown population parameter.   This is known as a *confidence interval* and is usually centered around the unbiased point estimate.   For example, the 95% confidence interval for μ is given by

$$P(\bar{X} - 1.96\sigma_{\bar{X}} < \mu < \bar{X} + 1.96\sigma_{\bar{X}}) = 0.95$$

The two numbers defining a confidence interval are called *confidence limits*.   Because an interval estimate also expresses the degree of accuracy or confidence we have in the estimate, it is superior to a point estimate.

4.14   A random sample of 64 with a mean of 50 and a standard deviation of 20 is taken from a population of 800.   (*a*) Find an interval estimate for the population mean such that we are 95% confident that the interval includes the population mean.   (*b*) What does the result of part *a* tell us?

(*a*) Since $n > 30$, we can use the z value of 1.96 from the standard normal distribution to construct the 95% confidence interval for the unknown population and we can use s as an estimate for the unknown σ:

$$\hat{\sigma} = s \tag{4.8}$$

where the "hat" (^) indicates an estimate, and

$$\hat{\sigma}_{\bar{X}} = \frac{\hat{\sigma}}{\sqrt{n}} = \frac{s}{\sqrt{n}} \quad \text{or} \quad \hat{\sigma}_{\bar{X}} = \frac{\hat{\sigma}}{\sqrt{n}}\sqrt{\frac{N-n}{N-1}} = \frac{s}{\sqrt{n}}\sqrt{\frac{N-n}{N-1}} \quad \text{when } n > 0.05N \tag{4.9a, b}$$

In this problem

$$\hat{\sigma}_{\bar{X}} = \frac{s}{\sqrt{n}}\sqrt{\frac{N-n}{N-1}} = \frac{20}{\sqrt{64}}\sqrt{\frac{800-64}{800-1}} \approx \frac{20}{8}0.96 \cong 2.4$$

Then $\mu \cong \bar{X} \pm z\sigma_{\bar{X}} \cong 50 \pm 1.96(2.4) \cong 50 \pm 4.70$.   Thus μ is between the lower confidence limit of 45.3 and the upper confidence limit of 54.7 with a 95% level of confidence.

(*b*) The result of part *a* tells us that if we take from the population repeated random samples, each of size $n = 64$, and construct the 95% confidence interval for each of the sample means, 95% of these confidence intervals will contain the true unknown population mean.   By assuming that our confidence interval (based on the single random sample that we have actually taken) is one of these 95% confidence intervals that includes μ, we take the calculated risk of being wrong 5% of the time.

4.15   A random sample of 25 with a mean 80 is taken from a population of 1000 that is normally distributed with a standard deviation of 30.   Find (*a*) the 90%, (*b*) the 95%, and (*c*) the 99% confidence intervals for the unknown population mean.   (*d*) What does the difference in the results to parts *a*, *b*, and *c* indicate?

(*a*) $\quad\quad\quad\quad \mu = \bar{X} \pm 1.64\sigma_{\bar{X}} \quad\quad$ since the population is normally distributed

$\quad\quad\quad\quad\quad \mu = \bar{X} \pm 1.64\frac{\sigma}{\sqrt{n}} \quad\quad$ since $n < 0.05N$ and σ is known

$\quad\quad\quad\quad\quad\quad = 80 \pm 1.64\frac{30}{\sqrt{25}}$

$\quad\quad\quad\quad\quad\quad = 80 \pm 1.64(6)$

$\quad\quad\quad\quad\quad\quad = 80 \pm 9.84$

Thus μ is between 70.16 and 89.94 with 90% confidence.

(b) $$\mu = 80 \pm 1.96(6) = 80 \pm 11.76$$

Thus μ is between 68.24 and 91.76 with 95% level of confidence.

(c) $$\mu = 80 \pm 2.58(6) = 80 \pm 15.48$$

Thus μ is between 64.52 and 95.48 with 99% level of confidence.

(d) The results of parts a, b, and c indicate that as we increase the degree of confidence required, the size of the confidence interval increases and the interval estimate becomes more vague (i.e., less precise). However, the degree of confidence associated with a very narrow confidence interval may be so low as to have little meaning. By convention, the most frequently used confidence interval is 95%, followed by 90 and 99%.

4.16 A random sample of 36 students is taken out of the 500 students from a high school taking the college entrance examintion. The mean test score for the sample is 380, and the standard deviation for the entire population of 500 students is 40. Find the 95% confidence interval for the unknown population mean score.

Since $n > 30$, the theoretical sampling distribution of the mean is approximately normal. Also, since $n > 0.05N$

$$\sigma_{\bar{X}} = \frac{\sigma}{\sqrt{n}} \sqrt{\frac{N-n}{N-1}} = \frac{40}{\sqrt{36}} \sqrt{\frac{500-36}{500-1}} \cong \frac{40}{6}(0.96) \cong 6.4$$

Then $$\mu = \bar{X} \pm z\sigma_{\bar{X}} = 380 \pm 1.96(6.4) = 380 \pm 12.54$$

Thus μ is between 367.46 and 392.54 with a 95% level of confidence.

4.17 A researcher wishes to estimate the mean weekly wage of the several thousands of workers employed in a plant within plus or minus $20 and with a 99% degree of confidence. From past experience, the researcher knows that the weekly wages of these workers are normally distributed with a standard deviation of $40. What is the minimum sample size required?

$$z = \frac{\bar{X} - \mu}{\sigma_{\bar{X}}}$$

$$z\sigma_{\bar{X}} = \bar{X} - \mu$$

$$z\frac{\sigma}{\sqrt{n}} = \bar{X} - \mu \quad \text{(presumably, } n < 0.05N\text{)}$$

$$2.58\frac{40}{\sqrt{n}} = 20$$

$$2.58\frac{40}{20} = \sqrt{n}$$

$$n = 5.16^2 = 26.63, \text{ or } 27 \text{ (rounded to the nearest higher integer)}$$

4.18 (a) Solve Prob. 4.17 by first getting an expression for n and then substituting the values from the problem into the expression obtained. (b) Why is the question of sample size important? (c) What is the size of the total confidence interval in Prob. 4.17? (d) What would have to be the sample size in Prob. 4.17 if we had not been told that the population was normally distributed? (e) What would have happened if we had not been told the population standard deviation?

(a) Starting with $z\sigma/\sqrt{n} = \bar{X} - \mu$ (see Prob. 4.17), we get $z\sigma/(\bar{X} - \mu) = \sqrt{n}$. Thus

$$n = \left(\frac{z\sigma}{\bar{X} - \mu}\right)^2 \tag{4.10}$$

Substituting the values from Prob. 4.17, we get

$$n = \left[\frac{(2.58)(40)}{20}\right]^2 = 26.63, \text{ or } 27 \text{ (the same as in Prob. 4.17)}$$

(b) The question of sample size is important because if the sample is too small, we fail to achieve the objectives of the analysis, and if the sample is too large, we waste resources because it is more expensive to collect and evaluate a larger sample.

(c) The size of the total confidence interval in Prob. 4.17 is \$40, or twice $\bar{X} - \mu$. Since we are using $\bar{X}$ as an estimate of μ, $\bar{X} - \mu$ is sometimes referred to as the *error of the estimate*. Because in Prob. 4.17 we want the error of the estimate to be "within plus or minus \$20," we get $\bar{X} - \mu = \pm\$20$, or a range of \$40 for the total confidence interval.

(d) If we had not been told that the population was normally distributed, we would have had to increase the sample to at least 30 in Prob. 4.17 in order to justify the use of the normal distribution.

(e) If we had not been told the value of σ, we could not have solved the problem. (Since we were deciding on what sample size to take in Prob. 4.17, we could not possibly have known the s to use as an estimate of σ.) The only way we could estimate σ (and thus approximate n) would be if we knew the range of wages from the highest to the lowest. Since $\pm 3\sigma$ includes 99.7% of all the area under the normal curve, we could have equated 6σ with the range of wages and thus estimate σ (and solve the problem).

4.19 With reference to a binomial distribution, indicate the relationship between (a) μ and $\mu_{\bar{p}}$, (b) p and $\bar{p}$, and   (c) σ, $\sigma_{\bar{p}}$, and $\hat{\sigma}_{\bar{p}}$.

(a) $\mu = np$ = mean *number* of successes in n trials, where p is the probability of success in any of the trials (see Sec. 3.3). $\mu_{\bar{p}} = \mu/n = p$ = the *proportion* of successes of the sampling distribution of the proportion.

(b) p = the proportion of successes *in the population*, and $\bar{p}$ = the proportion of successes *in the sample* (and an unbiased estimator of p).

(c) $\sigma = \sqrt{np(1-p)}$ = *standard deviation* of the number of successes in the population, and

$$\sigma_{\bar{p}} = \sqrt{\frac{p(1-p)}{n}} = \text{standard error of } p \tag{4.6a}$$

or $\qquad$ $$\sigma_{\bar{p}} = \sqrt{\frac{p(1-p)}{n}}\sqrt{\frac{N-n}{N-1}} \qquad\qquad \text{when } n > 0.05N \tag{4.6b}$$

$$\hat{\sigma}_{\bar{p}} = \sqrt{\frac{\bar{p}(1-\bar{p})}{n}} \quad \text{or} \quad \hat{\sigma}_{\bar{p}} = \sqrt{\frac{\bar{p}(1-\bar{p})}{n}}\sqrt{\frac{N-n}{N-1}} \qquad \text{when } n > 0.05N \tag{4.11a,b}$$

4.20 For a random sample of 100 workers in a plant employing 1200, 70 prefer providing for their own retirement benefits over belonging to a company-sponsored plan. Find the 95% confidence interval for the proportion of all the workers in the plant who prefer their own retirement plans.

$$\bar{p} = \frac{70}{100} = 0.7$$

$p = \bar{p} \pm z\sigma_{\bar{p}}$ since $n > 30$ and $np > 5$ and $n(1 - p) > 5$

$$= \bar{p} \pm z\sqrt{\frac{p(1-p)}{n}}\sqrt{\frac{N-n}{N-1}} \quad\quad \text{since } n > 0.05N$$

$$= 0.7 \pm 1.96\sqrt{\frac{(0.7)(0.3)}{100}}\sqrt{\frac{1200 - 100}{1200 - 1}} \quad\quad \text{using } \bar{p} \text{ as an estimate for } p$$

$$\cong 0.7 \pm 1.96(0.05)(0.96)$$

$$\cong 0.7 \pm 0.09$$

Thus p (the proportion of all the workers in the plant who prefer their own retirement plans) is between 0.61 and 0.79 with 95% degree of confidence.

4.21 A polling agency wants to estimate with 90% level of confidence the proportion of voters who would vote for a particular candidate within ±0.06 of the true (population) proportion of voters. What is the minimum sample size required if other polls indicate that the proportion voting for this candidate is 0.30?

$$z = \frac{\bar{p} - p}{\sigma_{\bar{p}}}$$

$$z\sigma_{\bar{p}} = \bar{p} - p$$

$$z\sqrt{\frac{p(1-p)}{n}} = \bar{p} - p \quad\quad \text{presumably } n < 0.05N$$

$$1.64\sqrt{\frac{(0.3)(0.7)}{n}} = 0.06$$

$$\frac{2.6896(0.3)(0.7)}{n} = 0.0036 \quad\quad \text{by squaring both sides}$$

$$n = \frac{(2.6896)(0.3)(0.7)}{0.0036} \cong 156.89, \text{ or } 157$$

4.22 (*a*) Solve Prob. 4.21 by first getting an expression for n and then substituting the values from the problem into the expression obtained. (*b*) How could we still have solved Prob. 4.21 if we had not been told that the proportion voting for the candidate was 0.30?

(*a*) Starting with $z\sqrt{p(1-p)/n} = \bar{p} - p$ (see Prob. 4.21), we get

$$\frac{z^2 p(1-p)}{n} = (\bar{p} - p)^2 \quad\quad \text{and} \quad\quad n = \frac{z^2 p(1-p)}{(\bar{p} - p)^2} \quad\quad\quad (4.12)$$

Substituting the values from Prob. 4.21, we get

$$n = \frac{(1.64)^2(0.3)(0.7)}{0.06^2} = \frac{(2.6896)(0.21)}{0.0036} \cong 156.89, \text{ or } 157$$

(the same as in Prob. 4.21).

(*b*) If we had not been told that the proportion voting for the candidate was 0.30, we could estimate the largest value of n to achieve the precision required *no matter what the actual value of p is*. This is done by letting $p = 0.5$ (so that $1 - p = 0.5$ also). Since $p(1-p)$ appears in the numerator of the formula for n (see part *a*) and this product is greatest when p and $1 - p$ both equal 0.5, the value of n is greatest. Thus

$$n = \frac{z^2 p(1-p)}{(\bar{p}-p)^2} = \frac{1.64^2(0.5)(0.5)}{0.06^2} = \frac{(2.6896)(0.25)}{0.0036} \cong 186.8, \text{ or } 187$$

(instead of $n = 157$ when we were told that $p = 0.30$). In this and similar cases, trying to get an actual estimate of p does not greatly reduce the size of the required sample. When p is taken to be 0.5, the formula for n can be simplified to

$$n = \left[\frac{z}{2(\bar{p}-p)}\right]^2 \tag{4.13}$$

Using this, we get

$$n = \left[\frac{1.64}{2(0.06)}\right]^2 = \left(\frac{1.64}{0.12}\right)^2 \cong 186.8, \text{ or } 187 \text{ (the same as above)}$$

CONFIDENCE INTERVALS FOR THE MEAN USING THE t DISTRIBUTION

4.23 (*a*) Under what conditions can we not use the normal distribution but can use the t distribution to find confidence intervals for the unknown population mean? (*b*) What is the relationship between the t distribution and the standard normal distribution? (*c*) What is the relationship between the z and t statistics for the theoretical sampling distribution of the mean? (*d*) What is meant by *degrees of freedom*?

(*a*) When the population is normally distributed but the population standard deviation σ is not known and the sample size n is smaller than 30, we cannot use the normal distribution for determining confidence intervals for the unknown population mean but we can use the Student t (or simply, the t) distribution.

(*b*) Like the standard normal distribution, the t distribution is bell-shaped and symmetrical about its zero mean, but it is platykurtic (see Sec. 2.4) or flatter than the standard normal distribution so that more of its area falls within the tails. While there is only one standard normal distribution, there is a different t distribution for each sample size n. However, as n becomes larger, the t distribution approaches the standard normal distribution until, when $n \geq 30$, they are approximately equal.

(*c*)
$$z = \frac{\bar{X} - \mu_{\bar{X}}}{\sigma_{\bar{X}}} = \frac{\bar{X} - \mu}{\sigma/\sqrt{n}}$$

and is found in App. 3.

$$t = \frac{\bar{X} - \mu}{s/\sqrt{n}} \tag{4.14}$$

and is found in App. 5 for the degrees of freedom involved.

(*d*) *Degrees of freedom* (df) refer to the number of values we can choose freely. For example, if we deal with a sample of 2 and we know that the sample mean for these two values is 10, we can freely assign the value to only one of these two numbers. If one number is 8, the other number must be 12 (to get the mean of 10). Then we say that we have $n - 1 = 2 - 1 = 1$ df. Similarly, if $n = 10$, this means that we can freely assign a value to only 9 of the 10 values if we want to estimate the population mean, and so we have $n - 1 = 10 - 1 = 9$ df.

4.24 (*a*) How can you find the t value for 10% of the area in each tail for 9 df? (*b*) In what way are t values interpreted differently from z values? (*c*) Find the t value for 5, 2.5, and 0.5% of the area within each tail for 9 df. (*d*) Find the t value for 5, 2.5, and 0.5% of the area within each tail for a sample size, n, that is very large or infinite. How do these t values compare with their corresponding z values?

(*a*) The t value for 10% of the area *within each tail* is obtained by moving down the column headed 0.10 in App. 5 to 9 df. This gives the t value of 1.383. By symmetry, 10% of the area under the t distribution with 9 df also lies within the left tail, to the left of $t = -1.383$.

(b) The t values given in App. 5 refer to the areas (probabilities) *within the tail(s)* of the t distribution indicated by the degrees of freedom. However, z values given in App. 3 refer to the areas (probabilities) under the standard normal curve *from the mean to the specified z values* (compare Example 4 with Example 8).

(c) Moving down the columns headed 0.05, 0.025, and 0.005 in App. 5 to 9 df, we get t values of 1.833, 2.262, and 3.250, respectively. Because of symmetry, 5, 2.5, and 0.5% of the area within the left tail of the t distribution for 9 df lie to the left of $t = -1.833$, $t = -2.262$, and $t = -3.250$, respectively.

(d) For sample sizes (and df) that are very large or infinite, $t_{0.05} = 1.645$, $t_{0.025} = 1.960$, and $t_{0.005} = 2.576$ (from the last row of App. 5). These coincide with the corresponding z values in App. 3. Specifically, $t_{0.025} = 1.960$ means that 2.5% of the area under the t distribution with ∞ df lies within the right tail, to the right of $t = 1.96$. Similarly, $z = 1.96$ gives (from App. 3) 0.4750 of the area under the standard normal curve from $\mu = 0$ to $z = 1.96$. Thus, for df $= n - 1 = \infty$, the t distribution is identical to the standard normal curve.

4.25 A random sample of 25 with a mean of 80 and a standard deviation of 30 is taken from a population of 1000 that is normally distributed. Find (a) the 90%, (b) the 95%, and (c) the 99% confidence intervals for the unknown population mean. (d) How do these results compare with those in Prob. 4.15?

(a) $t_{0.05} = 1.711$ for 24 df

$$\mu = \bar{X} \pm t \frac{s}{\sqrt{n}} = 80 \pm 1.711 \frac{30}{\sqrt{25}} = 80 \pm 10.266$$

Thus μ is between 69.734 and 90.266 with a 90% level of confidence.

(b) $t_{0.025} = 2.064$ for 24 df

$$\mu = \bar{X} \pm t \frac{s}{\sqrt{n}} = 80 \pm 2.064 \frac{30}{\sqrt{25}} = 80 \pm 12.384$$

Thus μ is between 67.616 and 92.384 with a 95% level of confidence.

(c) $t_{0.005} = 2.797$ for 24 df

$$\mu = \bar{X} \pm t \frac{s}{\sqrt{n}} = 80 \pm 2.797 \frac{30}{\sqrt{25}} = 80 \pm 16.782$$

Thus μ is between 63.218 and 96.782 with 99% degree of confidence.

(d) The 90, 95, and 99% confidence intervals, as anticipated, are larger in this problem, where the t distribution was used, than in Prob. 4.15, where the standard normal distribution was used. However, the differences are not great because when $n = 25$, the t distribution and the standard normal distribution are fairly similar. Note that in this problem we had to use the t distribution because s was given (and not σ, as in Prob. 4.15).

4.26 A random sample of $n = 9$ lightbulbs with a mean operating life of 300 h and a standard deviation s of 45 h is picked from a large shipment of lightbulbs known to have a normally distributed operating life. (a) Find the 90% confidence interval for the unknown mean operating life of the entire shipment. (b) Sketch a figure for the results of part a.

(a) $t_{0.05} = 1.860$ for 8 df

$$\mu = \bar{X} \pm t \frac{s}{\sqrt{n}} = 300 \pm 1.860 \frac{45}{\sqrt{9}} = 300 \pm 27.9$$

Thus μ is approximately between 272 and 328 h with a 90% level of confidence.

(b)   See Fig. 4-8.

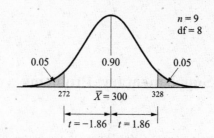

Fig. 4-8

4.27   A random sample of $n = 25$ with $\bar{X} = 80$ is taken from a population of 1000 with $\sigma = 30$. Suppose that we know that the population from which the sample is taken is not normally distributed.   (a) Find the 95% confidence interval for the unknown population mean. (b) How does this result compare with the results of Probs. 4.15(b) and 4.25(b)?

(a)   Since we know that the population from which the sample is taken is not normally distributed and $n < 30$, we can use neither the normal nor the t distributions.   We can apply *Chebyshev's theorem*, which states that regardless of the shape of the distribution, the proportion of observations (or area falling within K standard deviations of the mean) is at least $1 - (1/K^2)$, for $K \geq 1$ (see Prob. 3.40). Setting $1 - (1/K^2) = 0.95$ and solving for K, we get

$$\frac{1}{K^2} = 1 - 0.95$$

$$1 = 0.05K^2$$

$$K^2 = 20$$

$$K \cong 4.47$$

Then
$$\mu = \bar{X} + K\frac{\sigma}{\sqrt{n}} = 80 \pm 4.47\frac{30}{\sqrt{25}} \cong 80 \pm 26.82$$

Thus μ is approximately between 53 and 107 with a 95% level of confidence.

(b)   The 95% confidence interval using Chebyshev's theorem is much wider than that found when we could use the normal distribution [Prob. 4.15(b)] or the t distribution [Prob. 4.25(b)].   For this reason, Chebyshev's theorem is seldom used to find confidence intervals for the unknown population mean. However, it represents the only possibility short of increasing the sample size to at least 30 (so that the normal distribution can be used).

4.28   Under what conditions can we construct confidence intervals for the unknown population mean from a random sample drawn from a population using   (a) The normal distribution?   (b) The t distribution?   (c) Chebyshev's theorem?

(a)   We can use the normal distribution   (1) if the parent population is normal, $n \geq 30$, and σ or s are known; (2) if $n \geq 30$ (by invoking the central-limit theorem) and using s as an estimate for σ; or (3) if $n < 30$ but σ is given *and* the population from which the random sample is taken is known to be normally distributed.

(b)   We can use the t distribution (for the given degrees of freedom) when $n < 30$ but σ is not given *and* the population from which the sample is taken is known to be normally distributed.

(c)   If $n < 30$ but the population from which the random sample is taken is not known to be normally distributed, theoretically we should use neither the normal distribution nor the t distribution. In such cases, either we should use Chebyshev's theorem or we should increase the size of the random sample to

$n \geq 30$ (so as to be able to use the normal distribution). In reality, however, the t distribution is used even in these cases.

Supplementary Problems

SAMPLING

4.29   (a) What does *statistical inference* refer to?   (b) What are the names of the descriptive characteristics of populations and samples?   (c) How can representative samples be obtained?
Ans.   (a) Estimation and hypothesis testing   (b) Parameters and statistics   (c) By random sampling

4.30   (a) Starting from the third column and tenth row of App. 4 and reading horizontally, obtain a sample of 5 from 99 elements.   (b) Starting from the seventh column and first row of App. 4 and reading vertically, obtain a sample of 10 from 400 elements.
Ans.   (a) 31, 13, 33, 67, 68   (b) 24, 54, 290, 218, 385, 130, 24, 72, 313, 387

SAMPLING DISTRIBUTION OF THE MEAN

4.31   How can we obtain the theoretical sampling distribution of the mean from a population which is   (a) Finite? (b) Infinite?
Ans.   (a) By taking all possible different samples of size n from the population and then finding the mean of each sample   (b) By (hypothetically) taking an infinite number of samples of size n from the infinite population and then finding the mean of each sample

4.32   What is   (a) the mean and   (b) the standard error for a theoretical sampling distribution of the mean?
Ans.   (a) $\mu_{\bar{X}} = \mu$ where μ is the mean of the parent population   (b) $\sigma_{\bar{X}} = \sigma/\sqrt{n}$, where σ is the standard deviation of the parent population and n is the sample size; for finite populations of size N where $n > 0.05N$,
$\sigma_{\bar{X}} = (\sigma/\sqrt{n})\sqrt{(N-n)/(N-1)}$

4.33   For a population of 1000 items, $\mu = 50$ and $\sigma = 10$. What is the mean and standard error of the theoretical sampling distribution of the mean for sample sizes of   (a) 25 and   (b) 81?
Ans.   (a) $\mu_{\bar{X}} = 50$ units and $\sigma_{\bar{X}} = 2$ (b) $\mu_{\bar{X}} = 50$ units and $\sigma_{\bar{X}} = 1.07$

4.34   What is the shape of the theoretical sampling distribution of the mean for samples of   (a) 10 if the parent population is normal?   (b) 50 if the parent population is not normal?   (c) On what was the answer to part b based?
Ans.   (a) Nomal   (b) Approximately normal   (c) The central-limit theorem

4.35   What is the statistic for   (a) Random variable X?   (b) The theoretical sampling distribution of $\bar{X}$?
Ans.   (a) $z = (X - \mu)/\sigma$ (b) $z = (\bar{X} - \mu)/\sigma_{\bar{X}}$

4.36   What is the probability of $\bar{X}$ lying between 49 and 50 for a random sample of 36 from a population with $\mu = 48$ and $\sigma = 12$?
Ans.   0.1498, or 14.98%

4.37   What is the probability that the mean for a random sample of 144 accounts receivable drawn from a population of 2000 accounts with a mean of $10,000 and a standard deviation of $4000 will be between $9500 and $10,500?
Ans.   0.8812, or 88.12%

ESTIMATION USING THE NORMAL DISTRIBUTION

4.38   What are unbiased point estimators of μ, σ, and p, respectively?
Ans. $\bar{X}$, s [as defined in Eqs. (*2.10b*) and (*2.11b*)], and $\bar{p}$

4.39   Using the standardized normal distribution, state for μ (*a*) the 90%,   (*b*) the 95%, and   (*c*) the 99% confidence intervals.
Ans.   (*a*) $P(\bar{X} - 1.64\sigma_{\bar{X}} < \mu < \bar{X} + 1.64\sigma_{\bar{X}}) = 0.90$ (*b*) $P(\bar{X} - 1.96\sigma_{\bar{X}} < \mu < \bar{X} + 1.96\sigma_{\bar{X}}) = 0.95$ (*c*)
$P(\bar{X} - 2.58\sigma_{\bar{X}} < \mu < \bar{X} + 2.58\sigma_{\bar{X}}) = 0.99$

4.40   A random sample of 144 with a mean of 300 and a standard deviation of 100 is taken from a population of 5000.   Find an interval estimate for μ such that we are 90% confident that the interval includes μ.
Ans.   286.34 to 313.66

4.41   For Prob. 4.40, find   (*a*) the 95% and   (*b*) the 99% confidence intervals.   (*c*) What do the answers to parts *a* and *b* indicate?
Ans. (*a*) 283.67 to 316.33 (*b*) 278.51 to 321.49 (*c*) The greater is the degree of confidence, the larger is the confidence interval

4.42   A random sample of 400 is taken out of the more than 100,000 army recruits in a particular year.   The average weight for the sample of army recruits is 170 lb, and the standard deviation of the entire population of army recruits is 40 lb.   Find the 90% confidence interval for the mean weight of the population of army recruits.
Ans.   166.7 to 173.3 lb

4.43   A firm wishes to estimate the mean number of operating hours of a particular type of lightbulb within 10 operating hours (plus or minus) and with 95% confidence.   On the basis of previous knowledge with this type of lightbulb, the firm knows $\sigma = 30$ h.   How large a sample would the firm take?
Ans.   35

4.44   (*a*) Write down the expression for n to solve Prob. 4.43.   (*b*) What is the size of the total confidence interval in Prob. 4.43?   (*c*) What would have happened in Prob. 4.43 if $n < 30$?
Ans. (*a*) $n = [z\sigma/(\bar{X} - \mu)]^2$ (*b*) 20 operating hours (*c*) n would have had to be increased to 30 to justify the use of the normal distribution

4.45   For the binomial distribution, write the formula for   (*a*) μ and σ,   (*b*) $\sigma_{\bar{p}}$ and $\hat{\sigma}_{\bar{p}}$ when $n < 0.05N$, and (*c*) $\hat{\sigma}_{\bar{p}}$ when $n > 0.05N$.
Ans.   (*a*) $\mu = np$ and $\sigma = \sqrt{np(1 - p)}$ (*b*) $\sigma_{\bar{p}} = \sqrt{p(1 - p)/n}$ and $\hat{\sigma}_{\bar{p}} = \sqrt{\bar{p}(1 - \bar{p})/n}$
   (*c*) $\hat{\sigma}_{\bar{p}} = \sqrt{\bar{p}(1 - \bar{p})/n} \times \sqrt{(N - n)/(N - 1)}$

4.46   For a random sample of 36 graduate students in economics in a graduate economics program with 880 students, 8 students have an undergraduate degree in mathematics.   Find the proportion of all graduate students at this university with an undergraduate major in mathematics at the 90% confidence level.
Ans.   0.11 to 0.33

4.47   A manufacturer of lightbulbs wants to estimate the proportion of defective lightbulbs within ± 0.1 with a 95% degree of confidence.   What is the minimum sample size required if previous experience indicates that the proportion of defective lightbulbs produced is 0.2.
Ans.   62

4.48   (*a*) Write down the expression for n to solve Prob. 4.47.   (*b*) How could we still have solved Prob. 4.47 if the manufacturer did not know that $p = 0.2$?
Ans.   (*a*) $n = z^2 p(1 - p)/(\bar{p} - p)^2$ (*b*) By letting $p = 0.5$ and $n = 97$

CONFIDENCE INTERVALS FOR THE MEAN USING THE t DISTRIBUTION

4.49   Find the t value for 29 df for the following areas falling within the (right) tail of the t distribution: (a) 10%, (b) 5%, (c) 2.5%, and (d) 0.05%.
   Ans. (a) $t_{0.10} = 1.311$ (b) $t_{0.05} = 1.699$ (c) $t_{0.025} = 2.045$ (d) $t_{0.005} = 2.756$

4.50   Find the z value for the following areas falling from the mean to the z value under the standard normal curve: (a) $z = 40\%$,   (b) $z = 45\%$,   (c) $z = 47.5\%$, and   (d) $z = 49.5\%$ (e) How do these z values compare with the corresponding t values found in Prob. 4.49?
   Ans. (a) $z = 1.28$ (b) $z = 1.65$ (c) $z = 1.96$ (d) $z = 2.58$ (e) Corresponding z and t values are very similar (compare $z = 1.28$ to $t = 1.311$, $z = 1.65$ to $t = 1.699$, $z = 1.96$ to $t = 2.045$, and $z = 2.58$ to $t = 2.756$)

4.51   A random sample of $n = 16$ with $\bar{X} = 50$ and $s = 10$ is taken from a very large population that is normally distributed. (a) Find the 95% confidence interval for the unknown population mean. (b) How would the answer have differed if $\sigma = 10$?
   Ans. (a) 44.67 to 55.33 (using the t distribution with 15 df) (b) 45.1 to 54.9 (using the standard normal distribution)

4.52   On a particular test for a very large statistics class, a random sample of $n = 4$ students has a mean grade $\bar{X} = 75$ and $s = 8$. The grades for the entire class are known to be normally distributed. For the unknown population mean of the grades, find (a) the 95% confidence interval and (b) the 99% confidence interval.
   Ans. (a) Approximately from 62 to 88 (b) Approximately from 52 to 98

4.53   A random sample of $n = 16$ with $\bar{X} = 50$ and $s = 10$ is taken from a very large population that is not normally distrributed.   (a) Find the 95% confidence interval for the unknown population mean. (b) How is the answer in part a different from those of Prob. 4.51?
   Ans. (a) 39 to 61 (using Chebyshev's theorem and s as a rough estimate of σ) (b) The 95% confidence interval here is much wider than those found in Prob. 4.51

4.54   Indicate which distribution to use in order to find confidence intervals for the unknown population mean from a random sample taken from the population in the following cases: (a) $n = 36$ and $s = 10$, (b) $n = 20$ and $s = 10$ and the population is normally distributed, and (c) $n = 20$ and $s = 10$ and the population is not normally distributed.
   Ans. (a) Normal distribution (invoking the central limit theorem and using s as an estimate of σ) (b) The t distribution with 19 df (c) Chebyshev's theorem

CHAPTER 5

Statistical Inference: Testing Hypotheses

5.1 TESTING HYPOTHESES

Testing hypotheses about population characteristics (such as μ and σ) is another fundamental aspect of statistical inference and statistical analysis. In testing a hypothesis, we start by making an assumption with regard to an unknown population characteristic. We then take a random sample from the population, and on the basis of the corresponding sample characteristic, we either accept or reject the hypothesis with a particular degree of confidence.

We can make two types of errors in testing a hypothesis. First, on the basis of the sample information, we could reject a hypothesis that is in fact true. This is called a *type I error*. Second, we could accept a false hypothesis and make a *type II error*.

We can control or determine the probability of making a type I error, α. However, by reducing α, we will have to accept a greater probability of making a type II error, β, unless the sample size is increased. α is called the *level of significance*, and $1 - \alpha$ is the *level of confidence* of the test.

EXAMPLE 1. Suppose that a firm producing lightbulbs wants to know if it can claim that its lightbulbs last 1000 burning hours, μ. To do this, the firm can take a random sample of, say, 100 bulbs and find their average lifetime $\overline{X}$. The smaller the difference is between $\overline{X}$ and μ, the more likely is acceptance of the hypothesis that $\mu = 1000$ burning hours at a specified level of significance, α. By setting α at 5%, the firm accepts the calculated risk of rejecting a true hypothesis 5% of the time. By setting α at 1%, the firm would face a greater probability of accepting a false hypothesis, β.

5.2 TESTING HYPOTHESES ABOUT THE POPULATION MEAN AND PROPORTION

The formal steps in testing hypotheses about the population mean (or proportion) are as follows:

1. Assume that μ equals some hypothetical value μ_0. This is represented by H_0: $\mu = \mu_0$ and is called the *null hypothesis*. The *alternative hypotheses* are then H_1: $\mu \neq \mu_0$ (read "μ is not equal to μ_0"), H_1: $\mu > \mu_0$, or H_1: $\mu < \mu_0$, depending on the problem.

2. Decide on the level of significance of the test (usually 5%, but sometimes 1%) and define the *acceptance region* and *rejection region* for the test using the appropriate distribution.

3. Take a random sample from the population and compute $\overline{X}$. If $\overline{X}$ (in standard deviation units) falls in the acceptance region, accept H_0; otherwise, reject H_0 in favor of H_1.

EXAMPLE 2. Suppose that the firm in Example 1 wants to test whether it can claim that the lightbulbs it produces last 1000 burning hours. The firm takes a random sample of $n = 100$ of its lightbulbs and finds that the sample mean $\overline{X} = 980\,h$ and the sample standard deviation $s = 80\,h$. If the firm wants to conduct the test at the 5% level of significance, it should proceed as follows. Since μ could be equal to, larger than, or smaller than 1000, the firm should set the null and alternative hypotheses as

$$H_0: \quad \mu = 1000 \qquad H_1: \quad \mu \neq 1000$$

Since $n > 30$, the sampling distribution of the mean is approximately normal (and we can use s as an estimate of σ). The acceptance region of the test at the 5% level of significance is within ± 1.96 under the standard normal curve and the rejection region is outside (see Fig. 5-1). Since the rejection region is in both tails, we have a *two-tail test*. The third step is to find the z value corresponding to $\overline{X}$:

$$z = \frac{\overline{X} - \mu_0}{\sigma_{\overline{X}}} = \frac{\overline{X} - \mu_0}{\sigma/\sqrt{n}} = \frac{\overline{X} - \mu_0}{s/\sqrt{n}} = \frac{980 - 1000}{80/\sqrt{100}} = \frac{-20}{8} = -2.5$$

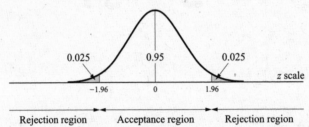

Fig. 5-1

Since the calculated z value falls in the rejection region, the firm should reject H_0, that $\mu = 1000$ and accept H_1, that $\mu \neq 1000$, at the 5% level of significance.

EXAMPLE 3. A firm wants to know with a 95% level of confidence if it can claim that the boxes of detergent it sells contain more than 500 g (about 1.1 lb) of detergent. From past experience the firm knows that the amount of detergent in the boxes is normally distributed. The firm takes a random sample of $n = 25$ and finds that $\overline{X} = 520\,g$ and $s = 75\,g$. Since the firm is interested in testing if $\mu > 500\,g$, we have

$$H_0: \quad \mu = 500 \qquad H_1: \quad \mu > 500$$

Since the population distribution is normal but $n < 30$ and σ is not known, we must use the t distribution (with $n - 1 = 24$ degrees of freedom) to define the critical, or rejection, region of the test at the 5% level of significance. This is found from App. 5 (see Sec. 4.4) and is given in Fig. 5-2. This is a *right-tail test*. Finally, since

$$t = \frac{\overline{X} - \mu}{s/\sqrt{n}} = \frac{520 - 500}{75/\sqrt{25}} = \frac{20}{15} = 1.33$$

and it falls within the acceptance region, we accept H_0, that $\mu = 500\,g$, at the 5% level of significance (or with a 95% level of confidence).

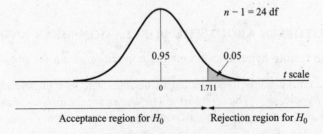

Fig. 5-2

EXAMPLE 4. In the past, 60% of the students entering a specialized college program received their degrees within 4 years. For the 1980 entering class of 36, only 15 received their degrees by 1984. To test if the 1980 class performed worse than previous classes, we first note that this problem involves the binomial distribution. However, since $n > 30$ and np and $n(1 - p) > 5$, we can use the normal distribution (see Sec. 3.5), with p (the proportion of successes) $= 0.60$. For the 1980 class, the proportion of successes $\bar{p} = 15/36 = 0.42$ and the standard error $\sigma_{\bar{p}} = p(1 - p)/n = (0.6)(0.4)/36 = 0.08$. Since we would like to test if the 1980 class performed worse, we have

$$H_0: \quad p = 0.60 \qquad H_1: \quad p < 0.60$$

Then
$$z = \frac{\bar{p} - p}{\sigma_{\bar{p}}} = \frac{0.42 - 0.60}{0.08} = -2.25$$

Since this is a *left-tail test* and 5% of the area under the standard normal curve lies to the left of -1.64 (see App. 3), we reject H_0 and conclude, at the 5% level of significance, that the 1980 class did perform worse than previous classes. However, if $\alpha = 1\%$, the critical region would be to the left of $z = -2.33$ and we would accept H_0. Problem 5.5 shows how to define the acceptance and rejection regions in the units of the problem instead of in standard deviation units. Problems 5.10 and 5.11 show how to find the *operating-characteristic curve* (OC curve), which gives the value of β for various values of $\mu > \mu_0$. Problem 5.12 then shows how to find the *power curve*, which gives the value of $(1 - \beta)$ for various values of $\mu > \mu_0$.

5.3   TESTING HYPOTHESES FOR DIFFERENCES BETWEEN TWO MEANS OR PROPORTIONS

In many decisionmaking situations, it is important to determine whether the means or proportions of two populations are the same or different. To do this, we take a random sample from each population and only if the *difference* in the sample means or proportions can be attributed to chance do we accept the hypothesis that the two populations have equal means or proportions.

If the two populations are normally distributed (or if both n_1 and $n_2 \geq 30$) and independent, then the sampling distribution of the difference between the sample means or proportions is also normal or approximately normal with standard error given by

$$\sigma_{\bar{X}_1 - \bar{X}_2} = \sqrt{\frac{\sigma_1^2}{n_1} + \frac{\sigma_2^2}{n_2}} \qquad \text{to test if } \mu_1 = \mu_2 \qquad (5.1)$$

and
$$\sigma_{\bar{p}_1 - \bar{p}_2} = \sqrt{\frac{\bar{p}(1 - \bar{p})}{n_1} + \frac{\bar{p}(1 - \bar{p})}{n_2}} \qquad \text{to test if } p_1 = p_2 \qquad (5.2)$$

where
$$\bar{p} = \frac{n_1 \bar{p}_1 + n_2 \bar{p}_2}{n_1 + n_2} \qquad \text{(a weighted average of } \bar{p}_1 \text{ and } \bar{p}_2) \qquad (5.3)$$

EXAMPLE 5. A manager wants to determine at the 5% level of significance if the hourly wages for semiskilled workers are the same in two cities. In order to do this, she takes a random sample of hourly wages in both cities and finds that $\bar{X}_1 = \$6.00$, $\bar{X}_2 = \$5.40$, $s_1 = \$2.00$, and $s_2 = \$1.80$ for $n_1 = 40$ and $n_2 = 54$. The hypotheses to be tested are

$$\begin{array}{llll} H_0: & \mu_1 = \mu_2 & \text{or} & H_0: \quad \mu_1 - \mu_2 = 0 \\ H_1: & \mu_1 = \mu_2 & \text{or} & H_1: \quad \mu_1 - \mu_2 \neq 0 \end{array}$$

This is a two-tail test and the *acceptance region* for H_0 lies within ± 1.96 under the standard normal curve (see Fig. 5-1).

$$\sigma_{\bar{X}_1 - \bar{X}_2} = \sqrt{\frac{\sigma_1^2}{n_1} + \frac{\sigma_2^2}{n_2}} \cong \sqrt{\frac{s_1^2}{n_1} + \frac{s_2^2}{n_2}} = \sqrt{\frac{2.00^2}{40} + \frac{1.80^2}{54}} = \sqrt{0.1 + 0.06} = \sqrt{0.16} = 0.4$$

$$z = \frac{(\bar{X}_1 - \bar{X}_2) - (\mu_1 - \mu_2)}{\sigma_{\bar{X}_1 - \bar{X}_2}} = \frac{(\bar{X}_1 - \bar{X}_2) - 0}{\sigma_{\bar{X}_1 - \bar{X}_2}} = \frac{0.6}{0.4} = 1.5$$

Since the calculated z value falls within the acceptance region, we accept H_0, that $\mu_1 = \mu_2$, at the 5% level of significance. However, if the two populations were known to be normally distributed but both n_1 and n_2 were less than 30 and it were assumed that $\sigma_1^2 = \sigma_2^2$ (but unknown), then the sampling distribution of the difference between the means would have a t distribution with $n_1 + n_2 - 2$ degrees of freedom (see Prob. 5.15).

EXAMPLE 6. A firm wants to determine at the 1% level of significance if the proportion of *acceptable* electronic components of a foreign supplier, p_1, is greater than for a domestic supplier, p_2. The firm takes a random sample from the shipment of each supplier and finds that $\bar{p}_1 = 0.9$ and $\bar{p}_2 = 0.7$ for $n_1 = 100$ and $n_2 = 80$. The firm sets up the following hypotheses:

$$H_0: \quad p_1 = p_2 \qquad H_1: \quad p_1 > p_2$$

This is a right-tail test and the *rejection region* for H_0 lies to the right of 2.33 under the standard normal curve.

$$\bar{p} = \frac{n_1\bar{p}_1 + n_2\bar{p}_2}{n_1 + n_2} = \frac{(100)(0.9) + (80)(0.7)}{180} = \frac{146}{180} = 0.8$$

$$\sigma_{\bar{p}_1 - \bar{p}_2} = \sqrt{\frac{\bar{p}(1-\bar{p})}{n_1} + \frac{\bar{p}(1-\bar{p})}{n_2}} = \sqrt{\frac{(0.8)(0.2)}{100} + \frac{(0.8)(0.2)}{80}} = \sqrt{0.0016 + 0.002} = \sqrt{0.0036} = 0.06$$

Since
$$z = \frac{(\bar{p}_1 - \bar{p}_2) - (p_1 - p_2)}{\sigma_{\bar{p}_1 - \bar{p}_2}} = \frac{0.2}{0.06} = 3.33$$

we *reject* H_0 and accept the hypothesis that $p_1 > p_2$ at the 1% level of significance.

5.4 CHI-SQUARE TEST OF GOODNESS OF FIT AND INDEPENDENCE

The χ^2 (chi-square) distribution is used to test whether (1) the observed frequencies differ "significantly" from expected frequencies when *more than two* outcomes are possible; (2) the sampled distribution is binomial, normal, or other; and (3) two variables are independent.

The χ^2 statistic calculated from the sample data is given by

$$\chi^2 = \sum \frac{(f_0 - f_e)^2}{f_e} \tag{5.4}$$

where f_0 denotes the frequencies and f_e, the expected frequencies.

If the calculated χ^2 is *greater* than the tabular value of χ^2 at the specified level of significance and degrees of freedom (from App. 6), the null hypothesis H_0 is *rejected* in favor of the alternative hypothesis H_1.

The degrees of freedom for *tests of goodness of fit* (1 and 2) are given by

$$df = c - m - 1 \tag{5.5}$$

where c represents the categories and m, the number of population parameters estimated from sample statistics.

The degrees of freedom for tests of independence, or *contingency-table tests* (3), are given by

$$df = (r - 1)(c - 1) \tag{5.6}$$

where r indicates the number of rows of the contingency table and c, the number of columns.

The expected frequency for each cell of a contingency table is

$$f_e = \frac{\sum_r f_0 \sum_c f_0}{n} \tag{5.7}$$

where $\sum_r$ and $\sum_c$ indicate sum over row and column, respectively, of the observed cell and n represents the overall sample size.

EXAMPLE 7. In the past, 30% of the TVs sold by a store were small-screen, 40% were medium, and 30% were large. In order to determine the inventory to maintain of each type of TV set, the manager takes a random sample of 100 recent purchases and finds that 20 were small-screen, 40 were medium, and 40 were large. To test at the 5%

level of significance the hypothesis that the past pattern of sales H_0 still prevails, the manager proceeds as follows (see Table 5.1):

$$\chi^2 = \sum \frac{(f_0 - f_e)^2}{f_e} = \frac{(20-30)^2}{30} + \frac{(40-40)^2}{40} + \frac{(40-30)^2}{30} = \frac{-10^2}{30} + \frac{0^2}{40} + \frac{10^2}{40} = \frac{100}{30} + \frac{100}{40} \cong 5.83$$

$$\mathrm{df} = c - m - 1 = 3 - 0 - 1 = 2$$

Because no population parameter was estimated, $m = 0$. df $= 2$ means that if we know the value of 2 of the 3 classes (and the total), the third class is not "free" to vary. Since the calculated value of $\chi^2 = 5.83$ is smaller than the tabular value of $\chi^2 = 5.99$ with $\alpha = 0.05$ and df $= 2$ (see App. 6), we cannot reject H_0, that the past sales pattern still prevails. When the expected frequency of a category is less than 5, the category should be combined with an adjacent one (see Prob. 5.18). For testing if the sampled distribution is binomial or normal, see Probs. 5.19 and 5.20.

Table 5.1   Observed and Expected Purchases of TV Sets by Screen Size

	Screen Size			Total
	Small	Medium	Large	
Observed pattern f_0	20	40	40	100
Past pattern f_e	30	40	30	100

EXAMPLE 8.   A car dealer has collected the data shown in Table 5.2 on the number of foreign and domestic cars purchased by customers under 30 years old and 30 and above. To test at the 1% level of significance if the type of car bought (foreign or domestic) is independent of the age of the buyer, the dealer constructs a table of expected frequencies (Table 5.3). For the first cell in row 1 and column 1, we obtain

$$f_e = \frac{\sum_r f_0 \sum_c f_0}{n} = \frac{(70)(50)}{170} \cong 21$$

The other three expected frequencies can be obtained by subtraction from row and column totals. Thus

Table 5.2   Contingency Table for Car Buyers

Age	Type of Car		Total
	Foreign	Domestic	
< 30	30	40	70
≥ 30	20	80	100
Total	50	120	170

Table 5.3   Table of Expected Frequencies for the Observed Frequencies in Table 5.2

Age	Type of Car		Total
	Foreign	Domestic	
< 30	21	49	70
≥ 30	29	71	100
Total	50	120	170

$$df = (r - 1)(c - 1) = (2 - 1)(2 - 1) = 1$$

$$\chi^2 = \sum \frac{(f_0 - f_e)^2}{f_e} = \frac{(30 - 21)^2}{21} + \frac{(40 - 49)^2}{49} + \frac{(20 - 29)^2}{29} + \frac{(80 - 71)^2}{71} = 9.44$$

Since the calculated value of χ^2 exceeds the tabular value of χ^2 with $\alpha = 0.01$ and df $= 1$ (see App. 6), we reject H_0, that age is not a factor in the type of car bought (and conclude that younger people seem more likely to buy foreign cars). When df $= 1$ but $n < 50$, a *correction for continuity* is made by using $(|f_0 - f_e| - 0.5)^2$ in the numerator of Eq. (5.4) (see Prob. 5.22).

5.5 ANALYSIS OF VARIANCE

The *analysis of variance* is used to test the null hypothesis that the means of *two or more* populations are equal versus the alternative that at least one of the means is different. The populations are assumed to be independently normally distributed, and of equal variance. The steps are as follows:

1. Estimate the population variance from the variance *between the sample means* (MSA in Table 5.4)
2. Estimate the population variance from the variance *within the samples* (MSE in Table 5.4)
3. Compute the F ratio (MSA/MSE in Table 5.4):

$$F = \frac{\text{variance between the sample means}}{\text{variance within the samples}}$$

Table 5.4 Analysis of Variance Table

Source of Variation	Sum of Squares	Degrees of Freedom	Mean Square	F Ratio
Between the means (explained by factor A)	$SSA = r \sum (\overline{X}_J - \overline{\overline{X}})^2$	$c - 1$	$MSA = \dfrac{SSA}{c - 1}$	$\dfrac{MSA}{MSE}$
Within the samples (error or unexplained)	$SSE = \sum \sum (\overline{X}_{iJ} - \overline{\overline{X}}_J)^2$	$(r - 1)c$	$MSE = \dfrac{SSE}{(r - 1)c}$	—
Total	$SST = \sum \sum (X_{iJ} - \overline{\overline{X}})^2 = SSA + SSE$	$rc - 1$	—	—

4. If the calculated F ratio is *greater* than the tabular value of F at the specified level of significance and degrees of freedom (from App. 7), the null hypothesis, H_0, of equal population means is *rejected* in favor of the alternative hypothesis, H_1. The preceding steps are formalized in Table 5.4.

where $\overline{X}_J$ = mean of sample J composed of r observations $= \left(\sum_i X_{iJ} \right) / r$ (5.8)

$\overline{\overline{X}}$ = grand mean of all c samples $= \left(\sum_i \sum_J X_{iJ} \right) / rc$ (5.9)

SSA = sum of squares explained by factor $A = r \sum (\overline{X}_J - \overline{\overline{X}})^2$ (5.10)

SSE = sum of squares of error unexplained by factor $A = \sum \sum (X_{iJ} - \overline{X}_J)^2$ (5.11)

SST = total sum of squares $= SSA + SSE = \sum \sum (X_{iJ} - \overline{\overline{X}})^2$ (5.12)

Appendix 7 gives F values for $\alpha = 0.05$ (the top number) and $\alpha = 0.01$ (the bottom or boldface number) for each *pair* of degrees of freedom:

$$\text{df of numerator} = c - 1 \qquad\qquad (5.13)$$

where c is the number of samples and

$$\text{df of denominator} = (r - 1)c \qquad\qquad (5.14)$$

where r is the number of observations in each sample.

EXAMPLE 9. A company sells identical soap in three different wrappings at the same price. The sales for 5 months are given in Table 5.5. Sales data are normally distributed with equal variance. To test at the 5% level of

Table 5.5   Five-Month Sales of Soap in Wrappings 1, 2, and 3

Wrapping 1	Wrapping 2	Wrapping 3
87	78	90
83	81	91
79	79	84
81	82	82
80	80	88
410	400	435

significance whether the mean soap sales for each wrapping is equal or not (i.e., H_0: $\mu_1 = \mu_2 = \mu_3$ versus H_1: $\mu_1, \mu_2,$ and μ_3 are not equal), the company proceeds as follows:

$$\overline{X}_1 = \frac{410}{5} = 82, \qquad \overline{X}_2 = \frac{400}{5} = 80, \qquad \overline{X}_3 = \frac{435}{5} = 87, \qquad \overline{X} = \frac{410 + 400 + 435}{(5)(3)} = 83$$

$$\text{SSA} = 5[(82 - 83)^2 + (80 - 83)^2 + (87 - 83)^2] = 130$$

$$\begin{aligned}
\text{SSE} = {} & (87 - 82)^2 + (83 - 82)^2 + (79 - 82)^2 + (81 - 82)^2 + (80 - 82)^2 + (78 - 80)^2 + (81 - 80)^2 + (79 - 80)^2 \\
& + (82 - 80)^2 + (80 - 80)^2 + (90 - 87)^2 + (91 - 87)^2 + (84 - 87)^2 + (82 - 87)^2 + (88 - 87)^2 \\
= {} & 110
\end{aligned}$$

$$\text{SST} = (87 - 83)^2 + (83 - 83)^2 + \cdots + (88 - 83)^2 = \text{SSA} + \text{SSE} = 240$$

The preceding data are used to construct Table 5.6 for the analysis of variance (ANOVA).

Table 5.6   ANOVA Table for Soap Wrappings

Variation	Sum of Squares	Degrees of Freedom	Mean Square	F Ratio
Explained by wrappings (between columns)	SSA = 130	$c - 1 = 2$	MSA = 130/2 = 65	MSA/MSE = 65/9.17 = 7.09
Error or unexplained (within columns)	SSE = 110	$(r - 1)c = 12$	MSE = 110/12 = 9.17	
Total	SST = 240	$rc - 1 = 14$	—	

Since the calculated value of $F = 7.09$ (from Table 5.6) exceeds the tabular value of $F = 3.88$ for $\alpha = 0.05$ and 2 and 12 degrees of freedom (see App. 7), we reject H_0, that the mean soap sales for each wrapping is the same, and accept H_1, that it is not the same. The preceding procedure is referred to as *one-way*, or *one-factor*, *analysis of variance*. For two-way analysis of variance, see Probs. 5.26 and 5.27.

5.6 NONPARAMETRIC TESTING

Nonparametric testing is used when one or more of the assumptions of the previous tests have not been met. Usually the assumption in question is the normality of the distribution (distribution of the data is unknown or the sample size is small). Nonparametric tests are often based on counting techniques that are easier to calculate and may be used for ordinal as well as quantitative data. These tests are inefficient if the distribution is known or the sample is large enough for a parametric test.

To test a hypothesis about the median of a population (analogous to test of population mean), the *Wilcoxon signed rank* test may be used:

1. For each observation, calculate the difference between the value and the hypothesized median.
2. Rank values according to the distance from the median, dropping zero differences.
3. The test statistic, W = the sum of the ranks of the positive differences. This is compared to the critical values in App. 9.

The signed rank test can be adjusted slightly to test equality of medians of more than two samples (analogous to ANOVA, but no assumption of normality) in the *Kruskal-Wallis* test:

1. Rank all data as if from a single sample.
2. Add ranks of each sample, $\sum R_j$.
3. The test statistic

$$H = \frac{12}{n(n+1)}\left(\frac{(\sum R_1^2)}{n_1} + \frac{(\sum R_2^2)}{n_2} + \cdots + \frac{(\sum R_c^2)}{n_c}\right) - 3(n+1)$$

If all sample sizes are at least 5, chi-square tables (App. 6) can be used with df $= c - 1$.

For a nonparametric test of goodness of fit, the *Kolmogorov-Smirnov* test compares cumulative probabilities of the data to a hypothesized distribution.

1. Arrange data from smallest value to largest value.
2. The proportion of data below each value is compared with cumulative probability below that value from the hypothesized distribution.
3. The test statistic is the maximum difference found in step 2, which can be compared to the critical value in App. 10.

EXAMPLE 10. A corporation has 8 subsidiaries with profits of 20, 35, 10, −5, −50, 5, 0, 13, respectively (in M\$), and wants to know with 95% confidence if the median firm is making profit of 5 M\$. Since we have a small sample (<30) and no assumption of normality, a t test cannot be used. We set the null and alternative hypotheses as

$$H_0: \text{ Med} = 5 \qquad H_1: \text{ Med} \neq 5$$

The steps for the signed rank test are listed in Table 5.7.

Since $4 < W < 32$, we accept H_0: Med $= 5$ at the 5% significance level.

EXAMPLE 11. A store owner wants to determine at the 5% significance level whether sales are normally distributed with mean of 10 units and standard deviation of 3 units. Sales for a week are observed of 2, 8, 4, 18, 9, 11, and 13 units.

The small sample precludes the use of the chi-square goodness-of-fit test but the nonparametric Kolmogorov-Smirnov test may be used to test H_0: normally distributed $\mu = 10$, $\sigma = 3$; H_1: not normally distributed $\mu = 10$, $\sigma = 3$.

Table 5.7   Signed Rank Test

$X - 5$	Ordered	Rank	Rank for Positive Differences
15	0	N/A	
30	5	1.5 (tie)	1.5
5	−5	1.5 (tie)	
−10	8	3	3
−55	−10	4	
0	15	5	5
−5	30	6	6
8	−55	7	

$$W = 15.5$$

Ordered data values	2	4	8	9	11	13	18
Proportion below, %	14.29	28.57	42.86	57.14	71.43	85.71	100
Normal cumulative probability, %	0.38	2.27	25.24	36.94	63.05	84.13	99.61
Difference, %	13.91	26.3	17.62	20.20	8.38	1.58	0.39

The maximum difference is 26.30% (0.2630), which is less than the critical value of 0.410; therefore we accept the null hypothesis that sales are normally distributed with a mean of 10 and standard deviation of 3.

Solved Problems

TESTING HYPOTHESES

5.1   (a) What is meant by *testing a hypothesis*?   What is the general procedure?   (b) What is meant by *type I* and *type II errors*?   (c) What is meant by the *level of significance*?   The *level of confidence*?

(a)   *Testing a hypothesis* refers to the acceptance or rejection of an assumption made about an unknown characteristic of a population, such as a parameter or the shape or form of the population distribution. The first step in testing a hypothesis is to make an assumption about an unknown population characteristic.   A random sample is then taken from the population, and on the basis of the corresponding sample characteristic, we accept or reject the hypothesis with a particular degree of confidence.

(b)   *Type I error* refers to the rejection of a true hypothesis.   *Type II error* refers to the acceptance of a false hypothesis.   In statistical analysis, we can control or determine the probability of type I or type II errors.   The probability of type I error is usually given by the Greek letter alpha (α), while the probability of type II error is represented by a beta (β).   By specifying a smaller type I error, we increase the probability of a type II error.   The only way to reduce both α and β is to increase the sample size.

(c)   The *level of significance* refers to the probability of rejecting a true hypothesis or committing type I error (α).   The *level of confidence* (given by $1 - \alpha$) refers to the probability of accepting a true hypothesis.   In statistical work, the level of significance, α, is usually set at 5%, so that the level of confidence, $1 - \alpha$, is 95%.   Sometimes $\alpha = 1\%$ (so that $1 - \alpha = 99\%$).

5.2   (*a*) How can we test the hypothesis that a particular coin is balanced?   (*b*) What is the meaning of type I and type II error in this case?

(*a*)   To test the hypothesis that a particular coin is balanced, we can toss the coin a number of times and record the number of heads and tails.   For example, we might toss the coin 20 times and get 9 heads instead of the expected 10.   This, however, does not necessarily mean that the coin is unbalanced. Indeed, since 9 is "so close" to 10, we are "likely" to be dealing with a balanced coin.   If, however, we get only 4 heads in 20 times, we are likely to be dealing with an unbalanced coin because the *probability* of getting 4 heads (and 16 tails) in 20 times with a balanced coin is very small indeed (see Sec. 3.3).

(*b*)   Even though 9 heads in 20 tosses indicates in all likelihood a balanced coin, there is always a small probability that the coin is unbalanced.   By accepting the hypothesis that the coin is balanced, we could thus be making a type I error.   However, 4 heads in 20 tosses is very likely to mean an unbalanced coin. But by *accepting* the hypothesis that the coin is *unbalanced*, we must face the small probability that the coin is instead balanced, which would mean that we made a type II error.   In testing a hypothesis, the investigator can set the probability of rejecting a true hypothesis, α, as small as desired.   However, by increasing the "region of acceptance" of the hypothesis, the investigator would necessarily increase the probability of accepting a false hypothesis or of making a type II error, β.

5.3   How can a producer of steel cables test that the breaking strength of the cables produced is (*a*) 5000 lb?   (*b*) Greater than 5000 lb?   (*c*) Less than 5000 lb?

(*a*)   The producer can test if the breaking strength of the steel cables produced is 5000 lb by taking a random sample of the cables and finding their mean breaking strength $\overline{X}$.   The closer $\overline{X}$ is to the hypothesized $\mu = 5000$ lb, the more likely the producer is to accept the hypothesis for the specified level of significance α.

(*b*)   The producer may instead by interested in testing if the breaking strength of the cable exceeds 5000 lb (i.e., $\mu > 5000$ lb).   To do this, once again, the producer takes a random sample of the cable produced and tests the mean breaking strength $\overline{X}$.   The more $\overline{X}$ *exceeds* the hypothesized $\mu = 5000$ lb, the more likely the producer is to accept the hypothesis at the specified level of significance, α.

(*c*)   To test that the breaking strength of the cable does not exceed 5000 lb, the producer finds the mean breaking strength of a random sample of the steel cables.   The more $\overline{X}$ *falls short* of 5000 lb, the more likely the producer is to accept the hypothesis that the breaking strength of the steel cables is less than the 5000 lb (i.e., $\mu < 5000$ lb), with a particular degree of confidence $1 - \alpha$.

TESTING HYPOTHESES ABOUT THE POPULATION MEAN AND PROPORTION

5.4   A producer of steel cables wants to test if the steel cables it produces have a breaking strength of 5000 lb.   A breaking strength of less than 5000 lb would not be adequate, and to produce steel cables with breaking strengths of more than 5000 lb would unnecessarily increase production costs.   The producer takes a random sample of 64 pieces and finds that the average breaking strength is 5100 lb and the sample standard deviation is 480 lb.   Should the producer accept the hypothesis that its steel cable has a breaking strength of 5000 lb at the 5% level of significance?

Since μ could be equal to, greater than, or smaller than 5000 lb, we set up the null and alternative hypotheses as follows:

$$H_0: \quad \mu = 5000\,\text{lb} \qquad H_1: \quad \mu \neq 5000\,\text{lb}$$

Since $n > 30$, the sampling distribution of the mean is approximately normal (and we can use s as an estimate of σ).   The acceptance region of the test at the 5% level of significance is within ± 1.96 under the standard normal curve and the rejection or critical region is outside (see Fig. 5-3).   Since the rejection region is in both tails, we have a *two-tail test*.   *The third step is to find the z value corresponding to $\overline{X}$:*

$$z = \frac{\overline{X} - \mu_0}{\sigma_{\overline{X}}} = \frac{\overline{X} - \mu_0}{\sigma/\sqrt{n}} = \frac{\overline{X} - \mu_0}{s/\sqrt{n}} = \frac{5100 - 5000}{480/\sqrt{64}} = \frac{100}{60} = 1.67$$

Since the calculated value of z falls within the acceptance region, the producer should accept the null hypothesis H_0 and reject H_1 at the 5% level of significance (or with a 95% level of confidence).   Note

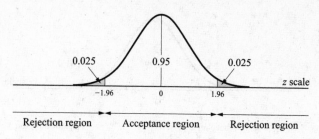

Fig. 5-3

that this does not "prove" that μ is indeed equal to 5000 lb. It only "proves" that *there is no statistical evidence that μ is not equal to 5000 lb* at the 5% level of significance.

5.5   Define the rejection and acceptance regions for Prob. 5.4 in terms of pounds, the units of the problem.

To find the acceptance region (at the 5% level of significance) in terms of pounds, we proceed as in Sec. 4.4 by finding the 95% confidence interval about μ_0:

$$\mu_0 \pm z\sigma_{\overline{X}} = \mu_0 \pm z\frac{\sigma}{\sqrt{n}} = \mu_0 \pm z\frac{s}{\sqrt{n}} = 5000 \pm 1.96\frac{480}{\sqrt{64}} = 5000 \pm 117.6$$

Thus, to accept H_0 at the 5% level of significance, $\overline{X}$ must have a value greater than 4882.4 lb and smaller than 5117.6 lb. The relationship between this and the result obtained in Prob. 5.4 is shown in Fig. 5-4.

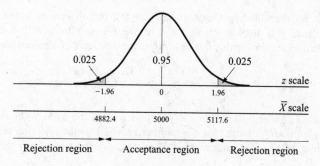

Fig. 5-4

5.6   An army recruiting center knows from past experience that the weight of army recruits is normally distributed with a mean μ of 80 kg (about 176 lb) and a standard deviation σ of 10 kg. The recruiting center wants to test, at the 1% level of significance, if the average weight of this year's recruits is above 80 kg. To do this, it takes a random sample of 25 recruits and finds that the average weight for this sample is 85 kg. How can this test be performed?

Since the center is interested in testing that $\mu > 80$ kg, it sets up the following hypotheses:

$$H_0: \quad \mu = 80\,\text{kg} \qquad H_1: \quad \mu > 80\,\text{kg}$$

(Some books state the null hypothesis as H_0: $\mu \leq 80$ kg, but the result is the same.) Since the parent population is normally distributed and σ is known, the standard normal distribution can be used to define the critical region, or rejection region, of the test. With H_1: $\mu > 80$ kg, we have a *right-tail test* with the critical region to the right of $z = 2.33$ at the 1% level of significance (see App. 3 and Fig. 5-5). Then

$$z = \frac{\overline{X} - \mu_0}{\sigma_{\overline{X}}} = \frac{\overline{X} - \mu_0}{\sigma/\sqrt{n}} = \frac{85 - 80}{10/\sqrt{25}} = 2.5$$

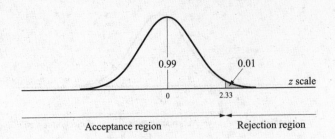

Fig. 5-5

Since the calculated value of z falls within the rejection region, we reject H_0 and accept H_1 (that $\mu > 80\,\text{kg}$). This means that if $\mu = 80\,\text{kg}$, the probability of getting a random sample from this population that gives $\overline{X} = 85\,\text{kg}$ is less than 1%. That would be an unusual sample indeed. Thus we reject H_0 at the 1% level of significance (i.e., we are 99% confident of making the right decision).

5.7   A government agency receives many consumer complaints that the boxes of detergent sold by a company contain less than the 20 oz of detergent advertised. To check the consumers' complaints, the agency purchases 9 boxes of the detergent and finds that $\overline{X} = 18\,\text{oz}$ and $s = 3\,\text{oz}$. How can the agency conduct the test at the 5% level of significance if it knows that the amount of detergent in the boxes is normally distributed?

The agency can set up H_0 and H_1 as follows:

$$H_0:\quad \mu = 20\,\text{oz} \qquad H_1:\quad \mu < 20\,\text{oz}$$

(Some books set up the null hypothesis as H_0: $\mu \geq 20$, but the result is the same.)   Since the parent population is normal, σ is not known, and $n < 30$, the t distribution (with 8 df and $\sigma = s$) must be used to define the rejection region for this *left-tail test* at the 5% level of significance (see Fig. 5-6).   Then

$$t = \frac{\overline{X} - \mu_0}{\sigma_{\overline{X}}} = \frac{\overline{X} - \mu_0}{\sigma/\sqrt{n}} = \frac{\overline{X} - \mu_0}{s/\sqrt{n}} = \frac{18 - 20}{3/\sqrt{9}} = -2.0$$

Since the calculated t value falls within the rejection region, the agency should reject H_0 and accept the consumers' complaints, H_1. Note that if α had been set at 1%, the rejection region would lie to the left of $t = -2.896$, leading to the acceptance of H_0. Thus it is important to specify the level of significance *before* the test.

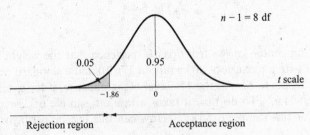

Fig. 5-6

5.8   A hospital wants to test that 90% of the dosages of a drug it purchases contain 100 mg ($1/1000\,\text{g}$) of the drug. To do this, the hospital takes a sample of $n = 100$ dosages and finds that only 85 of them contain the appropriate amount.   How can the hospital test this at   (*a*) $\alpha = 1\%$?   (*b*) $\alpha = 5\%$?   (*c*) $\alpha = 10\%$?

(*a*)   This problem involves the binomial distribution. However, since $n > 30$ and np and $n(1-p) > 5$, we can use the normal distribution with $p = 0.90$. For the sample

$$\bar{p} = \frac{85}{100} = 0.85 \qquad \text{and} \qquad \sigma_{\bar{p}} = \sqrt{\frac{p(1-p)}{100}} = \sqrt{\frac{(0.9)(0.1)}{100}} = 0.03$$

Since we are interested in finding if $p \gtrless 0.90$, we have $H_0: p = 0.90$ and $H_1: p \neq 0.90$. The acceptance region for H_0 at the 1% level of significance lies within ± 2.58 standard deviation units (see App. 3). Since

$$z = \frac{\bar{p} - p}{\sigma_{\bar{p}}} = \frac{0.85 - 0.90}{0.03} = 1.67$$

the hospital should accept H_0, that $p = 0.90$, at the 1% level of significance.

(b) At the 5% level of significance, the acceptance region for H_0 lies within ± 1.96 standard deviation units, and thus the hospital should accept H_0 and reject H_1 at the 95% level of confidence as well.

(c) At the 10% level of significance, the acceptance region for H_0 lies within ± 1.64 standard deviation units (see App. 3), and thus the hospital should *reject* H_0 and accept H_1, that $p \neq 0.90$. Note that larger values of α increase the rejection region for H_0 (i.e., increase the probability of acceptance of H_1). Furthermore, the greater is the value of α (i.e., the greater is the probability of rejecting H_0 when true), the smaller is β (the probability of accepting a false hypothesis).

5.9 The government antipollution spokesperson asserts that more than 80% of the plants in the region meet the antipollution standards. An antipollution advocate does not believe the government claim. She takes a random sample of published data on pollution emission for 64 plants in the area and finds that 56 plants meet the pollution standards. (a) Do the sample data support the government claim at the 5% level of significance? (b) Would the conclusion change if the sample had been 124, but with the sample proportion of the firms meeting the pollution standards the same as before?

(a) Here $H_0: p = 0.80$ and $H_1: p > 0.80$. The rejection region for H_0 lies to the right 1.64 standard normal deviation units for $\alpha = 5\%$. For the sample

$$\bar{p} = \frac{56}{64} = 0.88 \qquad \text{and} \qquad \sigma_{\bar{p}} = \sqrt{\frac{p(1-p)}{n}} = \sqrt{\frac{(0.8)(0.2)}{64}} = 0.05$$

Since

$$z = \frac{\bar{p} - p}{\sigma_{\bar{p}}} = \frac{0.88 - 0.80}{0.05} = 1.6$$

it falls within the acceptance region for H_0. This means that there is no statistical support for the government claim that $p > 0.8$ at the 5% level of significance.

(b) If the sample size had been 124 instead of 64, but $\bar{p} = 0.88$ as before,

$$\sigma_{\bar{p}} = \frac{(0.8)(0.2)}{124} = 0.04 \qquad \text{and} \qquad z = \frac{0.88 - 0.80}{0.04} = 2$$

and would fall in the rejection region for H_0 (so that there would be no evidence against the government claim that $p > 0.8$). Note that increasing n (and holding everything else the same) increases the probability of accepting the government claim.

5.10 Find the probability of accepting H_0 for Prob. 5.6 if   (a) $\mu = \mu_0 = 80\,\text{kg}$,   (b) $\mu = 82\,\text{kg}$,   (c) $\mu = 84\,\text{kg}$,   (d) $\mu = 85\,\text{kg}$,   (e) $\mu = 87\,\text{kg}$, and   (f) $\mu = 90\,\text{kg}$.

(a) If $\mu = \mu_0 = 80\,\text{kg}$, $\bar{X} = 85$, $\sigma = 10\,\text{kg}$, and $n = 25$, then

$$z = \frac{\bar{X} - \mu_0}{\sigma_{\bar{X}}} = \frac{\bar{X} - \mu}{\sigma/\sqrt{n}} = \frac{85 - 80}{10/\sqrt{25}} = \frac{5}{2} = 2.5$$

The probability of *accepting* H_0 when $\mu = \mu_0 = 80$ kg is 0.9938 (by looking up the value of $z = 2.5$ in App. 3 and adding 0.5 to it). Therefore, the probability of *rejecting* H_0 when H_0 is in fact true equals $1 - 0.9938$, or 0.0062.

(b) If $\mu = 82$ kg instead, then

$$z = \frac{\overline{X} - \mu}{\sigma/\sqrt{n}} = \frac{85 - 82}{10/\sqrt{25}} = \frac{3}{2} = 1.5$$

Therefore, the probability of accepting H_0 when H_0 is false equals 0.9332 (by looking up the value of $z = 1.5$ in App. 3 and adding 0.5 to it).

(c) If $\mu = 84$ kg, $z = (85 - 84)/2 = 1/2$ and $\beta = 0.6915$.

(d) If $\mu = 85$ kg, $z = 0$ and $\beta = 0.5$.

(e) If $\mu = 86$ kg, $z = (85 - 86)/2 = -1/2$ and $\beta = 0.5 - 0.1915 = 0.3085$.

(f) If $\mu = 87$ kg, $z = -1$ and $\beta = 0.5 - 0.3413 = 0.1587$.

5.11 (a) Draw a figure for the answers to Prob. 5.10 showing on the vertical axis the probability of accepting H_0 when $\mu = 80$ kg, 84 kg, 85 kg, 86 kg, and 88 kg. (b) What does this show? (c) What is the importance of knowing the value of β?

(a) See Fig. 5-7.

(b) The *operating-characteristic* (OC) *curve* in Fig. 5-7 shows the values of β for various values of $\mu > \mu_0$. Note that the more the actual value of μ exceeds μ_0, the smaller is β (or the probability of accepting H_0 when false).

(c) Knowing the value of β is important if accepting a false hypothesis (type II error) leads to very damaging results, such as, for example, when a drug is accepted as effective when it is not. In such cases, we want to keep β low, even if we have to accept a higher α (type I error). The only way to avoid this and reduce both α and β is to increase the sample size, n.

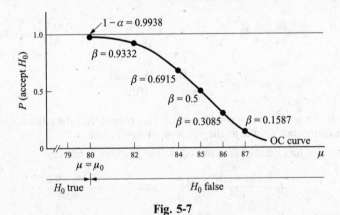

Fig. 5-7

5.12 (a) Draw a figure for the answers to Prob. 5.10 showing on the vertical axis the probability of *rejecting* H_0 for various values of $\mu > \mu_0$. What does this show? (b) How would the OC curve found in Prob. 5.11(a) and in part b of this problem look if the alternative hypothesis had been $H_1 : \mu < \mu_0$?

(a) For each value of $\mu > \mu_0$, the probability of rejecting H_0 when H_0 is false is given by $1 - \beta$, where β was found in Prob. 5.10(b) to part f. Joining these $1 - \beta$ points (starting with the value of α), we get the *power curve* (see Fig. 5-8). The power curve shows the probability of rejecting H_0 for various values of $\mu > \mu_0$. Note that the more μ exceeds μ_0, the greater is the power of the test (i.e., the greater is the probability of rejecting a false hypothesis).

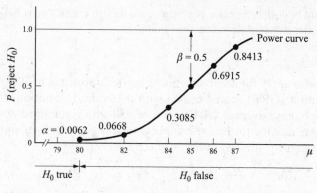

Fig. 5-8

(b) For H_1: $\mu < \mu_0$, the OC curve (for an actual value of $\overline{X}$ and for various alternative values of $\mu < \mu_0$) would look like the power curve in Fig. 5-8. However, the power curve would resemble the OC curve in Fig. 5-7.

TESTING HYPOTHESES FOR DIFFERENCES BETWEEN TWO MEANS OR PROPORTIONS

5.13 A large buyer of lightbulbs wants to decide, at the 5% level of significance, which of two equally priced brands to purchase. To do this, he takes a random sample of 100 bulbs of each brand and finds that brand 1 lasts 980 h on the average $\overline{X}_1$ with a sample standard deviation s_1 of 80 h. For brand 2, $\overline{X}_2 = 1010$ h and $s_2 = 120$ h.   Which brand should the buyer purchase to reach a decision at the significance level of   (a) 5%?   (b) 1%?

(a)
$$H_0:\ \mu_1 = \mu_2 \quad \text{or} \quad H_0:\ \mu_1 - \mu_2 = 0$$
$$H_1:\ \mu_1 \neq \mu_2 \quad \text{or} \quad H_1:\ \mu_1 - \mu_2 \neq 0$$

$$\overline{X}_1 = 980\,\text{h} \qquad s_1 = 80\,\text{h} \qquad n_1 = 100$$
$$\overline{X}_2 = 1010\,\text{h} \qquad s_2 = 120\,\text{h} \qquad n_2 = 100$$

This is a two-tail test with an acceptance region within ± 1.96 under the standard normal curve (see Fig. 5-1).   Therefore

$$\sigma_{\overline{X}_1 - \overline{X}_2} = \sqrt{\frac{\sigma_1^2}{n_1} + \frac{\sigma_2^2}{n_2}} \cong \sqrt{\frac{s_1^2}{n_1} + \frac{s_2^2}{n_2}} = \sqrt{\frac{80^2}{100} + \frac{120^2}{100}} = \sqrt{64 + 144} \cong 14.42$$

$$z = \frac{(\overline{X}_1 - \overline{X}_2) - (\mu_1 - \mu_2)}{\sigma_{\overline{X}_1 - \overline{X}_2}} = \frac{(\overline{X}_1 - \overline{X}_2) - 0}{\sigma_{\overline{X}_1 - \overline{X}_2}} = \frac{980 - 1010}{14.42} = \frac{-30}{14.42} = -2.08$$

Since the calculated value of z falls within the rejection region for H_0, the buyer should accept H_1, that $\mu_1 \neq \mu_2$, at the 5% level of significance (and presumably decide to purchase brand 2).

(b) At the 1% level of significance, the calculated z would fall within the acceptance region for H_0 (see Fig. 5-9). This would indicate that there is no significant difference between μ_1 and μ_2 at the 1% level, so

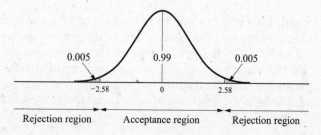

Fig. 5-9

the buyer could buy either brand. Note that even though brand 2 lasts longer than brand 1, brand 2 also has a greater standard deviation than brand 1.

5.14 The 65 students who apply for admission into a master's program in 1981 have average Graduate Record Examination (GRE) scores of 640 with a standard deviation of 20. In 1982, the 81 students who apply have average GRE scores of 650 with a standard deviation of 40. (*a*) Are the 1981 applicants inferior to the 1982 applicants at the 1% level of significance? (*b*) What is the acceptance region for the test *in terms of GRE scores*?

(*a*)
$$H_0: \ \mu_1 = \mu_2 \quad \text{and} \quad H_1: \ \mu_1 < \mu_2$$

$$\overline{X}_1 = 640 \qquad s_1 = 20 \qquad n_1 = 64$$
$$\overline{X}_2 = 650 \qquad s_2 = 40 \qquad n_2 = 81$$

This is a left-tail test with *acceptance* region for H_0 to the right of -2.33 under the standard normal curve. Therefore

$$\sigma_{\overline{X}_1 - \overline{X}_2} \cong \sqrt{\frac{s_1^2}{n_1} + \frac{s_2^2}{n_2}} = \sqrt{\frac{20^2}{64} + \frac{40^2}{81}} = \sqrt{6.25 + 19.75} = \sqrt{26} = 5.10$$

$$z = \frac{\overline{X}_1 - \overline{X}_2}{\sigma_{\overline{X}_1 - \overline{X}_2}} = \frac{640 - 650}{5.10} = \frac{-10}{5.10} = -1.96$$

Since the calculated value of z falls within the acceptance region, H_0 is accepted. This means that there is no statistical evidence at the 1% level of significance indicating that the applicants in the two years are of different quality.

(*b*) Since the hypothesized difference between the two *population* means in H_0 is 0, we can find the acceptance region for the test in terms of GRE scores as follows:

$$(\mu_1 - \mu_2)_0 - z\sigma_{\overline{X}_1 - \overline{X}_2} = 0 - (2.33)(5.10) = -11.88$$

Since $\overline{X}_1 - \overline{X}_2 = -10$, it falls within the acceptance region for H_0 (see Fig. 5-10).

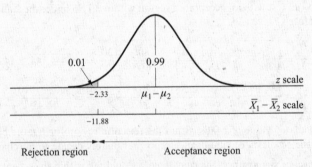

Fig. 5-10

5.15 The American Dental Association wants to test which of two toothpaste brands is better for fighting tooth decay. A random sample is taken of 21 persons using each toothpaste. The average number of cavities for the first group over a 10-year period is 25 with a standard deviation of 5. In the second group, the average number of cavities is 23 with a standard deviation of 4. Assuming that the distribution of cavities is normal for all the users of toothpastes 1 and 2 and that $\sigma_1^2 = \sigma_2^2$, determine if $\mu_1 = \mu_2$ at the 5% level of significance.

$$H_0: \ \mu_1 = \mu_2 \quad \text{and} \quad H_1: \ \mu_1 \neq \mu_2$$

$$\overline{X}_1 = 25 \qquad s_1 = 5 \qquad n_1 = 21$$
$$\overline{X}_2 = 23 \qquad s_2 = 4 \qquad n_2 = 21$$

Since the two populations are normally distributed but both n_1 and $n_2 < 30$ and it is assumed that $\sigma_1^2 = \sigma_2^2$ (but unknown), the sampling distribution of the difference between the means has a t distribution with $n_1 + n_2 - 2$ degrees of freedom. Since it is assumed that $\sigma_1^2 = \sigma_2^2$ (and we can use s_1^2 as an estimate of σ_1^2 and s_2^2 as an estimate of σ_2^2), we get

$$\sigma_{\overline{X}_1 - \overline{X}_2} \cong \sqrt{\frac{s^2}{n_1} + \frac{s^2}{n_2}} \qquad (5.1a)$$

where

$$s^2 = \frac{(n_1 - 1)s_1^2 + (n_2 - 1)s_2^2}{n_1 + n_2 - 2} \qquad (5.3a)$$

where s^2 is a weighted average of s_1^2 and s_2^2. The weights are $n_1 - 1$ and $n_2 - 1$, as in Eq. (2.8b) for s_1^2 and s_2^2, in order to get "unbiased" estimates for σ_1^2 and σ_2^2 (see Prob. 2.16). This is a two-tail test with the acceptance region for H_0 within ± 2.021 under the t distribution with $\alpha = 5\%$ and $n_1 + n_2 - 2 = 21 + 21 - 2 = 40$ df:

$$s^2 = \frac{20(5)^2 + 20(4)^2}{40} = \frac{500 + 320}{40} = 20.5$$

$$\sigma_{\overline{X}_1 - \overline{X}_2} \cong \sqrt{\frac{20.5}{21} + \frac{20.5}{21}} = \sqrt{\frac{42}{21}} = \sqrt{2} \cong 1.41$$

$$z = \frac{\overline{X}_1 - \overline{X}_2}{\sigma_{\overline{X}_1 - \overline{X}_2}} = \frac{25 - 23}{1.41} \cong 1.42$$

Since the calculated value of z falls within the acceptance region, we cannot reject H_0, that $\mu_1 = \mu_2$ (see Fig. 5-11).

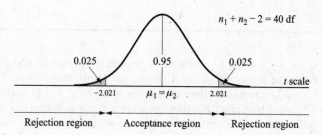

Fig. 5-11

5.16 Suppose that 50% of the 60 plants in region 1 abide by the antipollution standards but only 40% of the 40 plants in region 2 do so. Is the percentage of plants abiding by the antipollution standards significantly greater in region 1 as opposed to region 2 at: (a) the 5% level of significance? (b) the 10% level of significance?

(a)

$$H_0: \quad p_1 = p_2 \qquad \text{and} \qquad H_1: \quad p_1 > p_2$$
$$\overline{p}_1 = 0.50 \qquad \text{and} \qquad n_1 = 60$$
$$\overline{p}_2 = 0.40 \qquad \text{and} \qquad n_2 = 40$$

This is a right-tail test, and the acceptance region for H_0 with $\alpha = 0.05$ lies to the left of 1.64 under the standard normal curve:

$$\overline{p} = \frac{n_1 \overline{p}_1 + n_2 \overline{p}_2}{n_1 + n_2} = \frac{60(0.5) + 40(0.4)}{60 + 40} = \frac{30 + 16}{100} = 0.46$$

$$\sigma_{\bar{p}_1 - \bar{p}_2} = \sqrt{\frac{\bar{p}(1-\bar{p})}{n_1} + \frac{\bar{p}(1-\bar{p})}{n_2}} = \sqrt{\frac{(0.46)(0.54)}{60} + \frac{(0.46)(0.54)}{40}}$$

$$= 0.00414(0.00621) = 0.01035 = 0.10$$

Since $z = (\bar{p}_1 - \bar{p}_2)/\sigma_{\bar{p}_1 - \bar{p}_2} = (0.5 - 0.4)/0.1 = 0.10/0.10 = 1$, we accept H_0, that $p_1 = p_2$, with $\alpha = 0.05$.

(b) With $\alpha = 0.10$, the acceptance region for H_0 lies to the left of 1.28 under the standard normal curve. Since the calculated z falls within the acceptance region, we accept H_0 at $\alpha = 0.10$ as well.

CHI-SQUARE TEST OF GOODNESS OF FIT AND INDEPENDENCE

5.17 A plant manager takes a random sample of 100 sick days and finds that 30% of the plant labor force in the 20 to 29 age group took 26 of the 100 sick days, that 40% of the labor force in the 30 to 39 age group took 37 sick days, that 20% in the 40 to 49 age group took 24 sick days, and that 10% of the 50-and-over age group took 13 sick days. How can the manager test at the 5% level of significance the hypothesis that age is *not* a factor in taking sick days?

If age is not a factor in taking sick days, then the *expected* number of sick days taken by each age group should be the same as the proportion of the age group in the plant's labor force (see Table 5.8):

$$\chi^2 = \sum \frac{(f_0 - f_e)^2}{f_e} = \frac{(26-30)^2}{30} + \frac{(37-40)^2}{40} + \frac{(24-20)^2}{20} + \frac{(13-10)^2}{10}$$

$$= \frac{16}{30} + \frac{9}{40} + \frac{16}{20} + \frac{9}{10} \cong 2.46$$

Table 5.8   Observed and Expected Sick Days

Age Group	20–29	30–39	40–49	≥ 50	Total
f_0	26	37	24	13	100
f_e	30	40	20	10	100

where df $= c - m - 1 = 4 - 0 - 1 = 3$. Because no population parameter was estimated, $m = 0$. df $= 3$ means that if we know the value of 3 of the 4 classes, the fourth class is not "free" to vary. Since the calculated value of $\chi^2 = 2.46$ is smaller than the tabular value of $\chi^2 = 7.81$ with $\alpha = 0.05$ and df $= 3$ (see App. 6 and Fig. 5-12), we cannot reject H_0, that age is not a factor in taking sick days. Note that as in the case of the t distribution, there is a χ^2 distribution for each degree of freedom. However, the χ^2 test is used here as a right-tail test only.

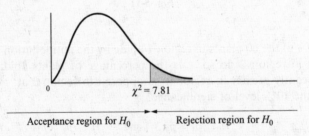

$\chi^2 = 7.81$

Acceptance region for H_0 Rejection region for H_0

Fig. 5-12

5.18 Table 5.9 indicates the observed and expected frequency of 4 rare diseases (A, B, C, and D) in a city. Is the difference between the observed and the expected frequency of the diseases significant at the 10% level?

Since for $f_e < 5$, diseases C and D, we combine these two classes (see Table 5.10):

Table 5.9   Observed and Expected Frequencies of Rare Diseases A, B, C, D

	Type of Disease				Total
	A	B	C	D	
f_0	3	5	6	3	17
f_e	6	6	3	2	17

Table 5.10   Observed and Expected Frequencies of Rare Diseases A, B, C, and D

	Type of Disease			Total
	A	B	C and D	
f_0	3	5	9	17
f_e	6	6	5	17

$$\chi^2 = \sum \frac{(f_0 - f_e)^2}{f_e} = \frac{(3-6)^2}{6} + \frac{(5-6)^2}{6} + \frac{(9-5)^2}{5} = \frac{9}{6} + \frac{1}{6} + \frac{16}{5} = 4.87$$

Since the calculated value of χ^2 exceeds the tabular value of $\chi^2 = 4.61$ for $\alpha = 0.10$ and df $= 2$, we reject H_0 and accept the alternative hypothesis H_1, that there is a significant difference between the observed and expected frequencies of occurrence of these rare diseases in this city. Note that if $f_0 = f_e$, $\chi^2 = 0$. The greater is the difference between f_0 and f_e, the larger is the calculated value of χ^2 and the more likely it is that H_0 would be rejected. Note also that because of the squaring, χ^2 can never be negative.

5.19   Table 5.11 gives the distribution of the number of acceptances of 100 students into 3 colleges. Test at the 5% level of significance that the distribution of acceptances is approximately binomial if the probability of a student's being accepted into college is 0.40.

Table 5.11   Distribution of Acceptances of 100 Students into 3 Colleges

Number of Acceptances	Number of Students
0	25
1	34
2	31
3	10
	100

The binomial probabilities given in Table 5.12 for 0, 1, 2, or 3 acceptances by any one student with $p = 0.4$ are obtained from App. 1. Therefore

$$\chi^2 = \sum \frac{(f_0 - f_e)^2}{f_e} = \frac{(25-22)^2}{22} + \frac{(34-43)^2}{43} + \frac{(31-29)^2}{29} + \frac{(10-6)^2}{6} = \frac{9}{22} + \frac{81}{43} + \frac{4}{29} + \frac{16}{6} = 5.10$$

Since the calculated value of $\chi^2 = 5.10$ is smaller than the tabular value of $\chi^2 = 7.81$ with $\alpha = 0.05$ and df $= 3$, we cannot reject H_0, that the distribution of acceptances follows a binomial distribution, with $p = 0.40$. Note that the χ^2 distribution is a continuous distribution (as are the normal and t distributions).

Table 5.12 Observed Frequencies, Binomial Probabilities, and Expected Frequencies of Acceptances

Number of Acceptances	Observed Frequency	Binomial Probabilities		Number of Applicants	Expected Frequency of Acceptance
0	25	0.216	×	100	22
1	34	0.432	×	100	43
2	31	0.288	×	100	29
3	10	0.064	×	100	6
		1.000	×		100

5.20 Table 5.13 gives the distribution of Scholastic Aptitude Test (SAT) scores for a random sample of 100 college students. Test at the 5% level of significance that the SAT scores are normally distributed.

Table 5.13 Frequency Distribution of SAT Scores

SAT Score	Number of Students
251–350	3
351–450	25
451–550	50
551–650	20
651–750	2
	100

To conduct this test, we must first calculate $\overline{X}$ and s for this distribution, as shown in Table 5.14:

$$\overline{X} = \frac{\sum fX}{n} = \frac{49,300}{100} = 493$$

and

$$s = \sqrt{\frac{\sum fX^2 - n\overline{X}^2}{n-1}} = \sqrt{\frac{24,950,000 - (100)(493)^2}{99}} \cong 80.72$$

If the SAT scores are normally distributed, then f_e is estimated as shown in Table 5.15:

Table 5.14 Calculation of $\overline{X}$ and s for SAT Scores

Class Interval	Frequency f_0	Midpoint X	fX	X^2	fX^2
251–350	3	300	900	90,000	270,000
351–450	25	400	10,000	160,000	4,000,000
451–550	50	500	25,000	250,000	12,500,000
551–650	20	600	12,000	360,000	7,200,000
651–750	2	700	1,400	490,000	980,000
	100		49,300		24,950,000

Table 5.15   Expected Frequencies for SAT Scores Using $\overline{X} = 493$ and $s = 80.72$

SAT Score, $x =$ Upper Class Limit	$z = \dfrac{X - 493}{80.72}$	Area to Left of X	Area of Class Interval	Expected Frequency f_e
≤ 350	-1.77	0.0384	$0.0384 \times 100 = 3.84$ ⎫	29.81
450	-0.53	0.2981	$0.2597 \times 100 = 25.97$ ⎭	
550	0.71	0.7612	$0.4631 \times 100 = 46.31$	46.31
650	1.94	0.9738	$0.2126 \times 100 = 21.26$ ⎫	23.88
> 750	3.18	1.0000	$\underline{0.0262 \times 100 = 2.62}$ ⎭	
			$1.0000 \qquad 100.00$	

$$\chi^2 = \sum \frac{(f_0 - f_e)^2}{f_e} = \frac{(28 - 29.81)^2}{29.81} + \frac{(50 - 46.31)^2}{46.31} + \frac{(22 - 23.88)^2}{23.88} \cong 0.54$$

Note that the first two and the last two classes of observed and expected frequencies were combined because $f_e < 5$. df $= c - m - 1 = 5 - 2 - 1 = 2$. Because two population parameters were estimated (μ and σ with $\overline{X}$ and s, respectively), $m = 2$. The tabular value of χ^2 with $\alpha = 0.05$ and df $= 2$ is 5.99. Since the calculated value of χ^2 is smaller than the tabular value, we cannot reject H_0. That is, we cannot reject the hypothesis that the random sample of SAT scores comes from a normal distribution with $\mu = 493$ and $\sigma = 80.72$.

5.21   The number of heart attacks suffered by males and females of various age groups in a city is given by contingency Table 5.16. Test at the 1% level of significance the hypothesis that age and sex are independent in the occurrence of heart attacks.

Table 5.16   Number of Heart Attacks of Males and Females in Various Age Groups in a City

Age Group	Male	Female	Total
< 30	10	10	20
30–60	50	30	80
> 60	$\underline{30}$	$\underline{20}$	$\underline{50}$
	90	60	150

To test this hypothesis, expected frequencies f_e must be estimated (see Table 5.17):

Table 5.17   Expected Frequencies of Heart Attacks

Age Group	Male	Female	Total
< 30	12	8	20
30–60	48	32	80
> 60	$\underline{30}$	$\underline{20}$	$\underline{50}$
	90	60	150

$$f_e = \frac{\sum_r f_0 \sum_c f_0}{n} = \frac{(20)(90)}{150} = 12 \qquad \text{for the cell in row 1, column 1}$$

$$= \frac{\sum_r f_0 \sum_c f_0}{n} = \frac{(80)(90)}{150} = 48 \qquad \text{for the cell in row 2, column 1}$$

All other expected frequencies can be obtained by subtraction from the appropriate row or column totals. Therefore

$$\chi^2 = \sum \frac{(f_0 - f_e)^2}{f_e} = \frac{(10 - 12)^2}{12} + \frac{(10 - 8)^2}{8} + \frac{(50 - 48)^2}{48}$$
$$+ \frac{(30 - 32)^2}{32} + \frac{(30 - 30)^2}{30} + \frac{(20 - 20)^2}{20} = 1.04$$

where df $= (r - 1)(c - 1) = (3 - 2)(2 - 1) = 2$ (corresponding to the two expected frequencies we had to calculate by formula). From App. 6, $\chi^2 = 9.21$ with $\alpha = 0.01$ and df $= 2$. Since the calculated χ^2 is smaller than the tabular χ^2, we accept the null hypothesis, H_0, that age is independent of sex in the occurrence of heart attacks. To be sure, males seem more likely to suffer heart attacks, but this tendency does not differ significantly with age at the 1% level of significance.

5.22   A random sample of 37 workers above the age of 65 in a town gives the results indicated by contingency Table 5.18. Test at the 10% level of significance the hypothesis that the number of male and female workers in the 66 to 70 and 71-plus age groups in the town is independent of sex.

Table 5.18   Male and Female Workers over 65 in a Town

Age Group	Male	Female	Total
66–70	17	9	26
≥ 71	3	8	11
	20	17	37

Table 5.19 gives the expected frequencies. For the first cell

$$f_e = \frac{\sum_r f_0 \sum_c f_0}{n} = \frac{(26)(20)}{37} = 14$$

Table 5.19   Expected Male and Female Workers over 65

Age Group	Male	Female	Total
66–70	14	12	26
≥ 71	6	5	11
	20	17	37

For the other cells, f_e is found by subtraction from the row and column totals.   df $= (r - 1)(c - 1) = (2 - 1)(2 - 1) = 1$. Since df $= 1$ and $n < 50$, a correction for continuity must be made to calculate χ^2, as indicated in Eq. (5.4a):

$$\chi^2 = \sum \frac{(|f_0 - f_e| - 0.5)^2}{f_e} \tag{5.4a}$$

Thus
$$\chi^2 = \frac{(|17 - 14| - 0.5)^2}{14} + \frac{(|9 - 12| - 0.5)^2}{12} + \frac{(|3 - 6| - 0.5)^2}{6} + \frac{(|8 - 5| - 0.5)^2}{5}$$
$$= \frac{2.5^2}{14} + \frac{2.5^2}{12} + \frac{2.5^2}{6} + \frac{2.5^2}{5} = 3.25$$

Since the calculated χ^2 is larger than the tabular value of χ^2 with $\alpha = 0.10$ and df $= 1$, we reject H_0, that males and females over 65 continue to work in this town independently of whether they are above or below 70 years of age. The proportion of workers is significantly higher for males in the 66 to 70 age group and for females in the 71-plus age group.   Note that the same adjustment indicated by Eq. (*5.4a*) is also made for tests of the goodness of fit when df $= 1$ and $n < 50$.

ANALYSIS OF VARIANCE

5.23   Table 5.20 gives the output for 8 years of an experimental farm that used each of 4 fertilizers. Assume that the outputs with each fertilizer are normally distributed with equal variance. (*a*) Find the *mean* output for each fertilizer and the *grand mean* for all the years and for all four fertilizers.   (*b*) Estimate the population variance from the variance *between* the means or columns.   (*c*) Estimate the population variance from the variance *within* the samples or columns.   (*d*) Test the hypothesis that the population means are the same at the 5% level of significance.

Table 5.20   Eight-Year Outputs with 4 Different Fertilizers

Fertilizer 1	Fertilizer 2	Fertilizer 3	Fertilizer 4
51	47	57	50
47	50	48	61
56	58	52	57
52	61	60	65
57	51	61	58
59	48	57	53
58	59	51	61
60	50	46	59
440	424	432	464

(*a*)
$$\overline{X}_1 = \frac{\sum_i X_{i1}}{r} = \frac{440}{8} = 55 \qquad \overline{X}_2 = \frac{\sum_i X_{i2}}{r} = \frac{424}{8} = 53$$

$$\overline{X}_3 = \frac{\sum_i X_{i3}}{r} = \frac{432}{8} = 54 \qquad \overline{X}_4 = \frac{\sum_i X_{i4}}{r} = \frac{464}{8} = 58$$

$$\overline{\overline{X}} = \frac{\sum_J \sum_i X_{iJ}}{rc} = \frac{440 + 424 + 432 + 464}{(8)(4)} = 55$$

(*b*)
$$\sigma^2 = \frac{\sigma_{\overline{X}}^2}{n} \cong \frac{\sum (X - \overline{X})^2/(n-1)}{n} \qquad \text{[from Eqs. (4.2a), (4.9a), and (2.8b)]}$$

Here
$$\sigma^2 = \frac{\sigma_{\overline{X}}^2}{n} \cong \frac{r \sum (\overline{X}_J - \overline{\overline{X}})^2}{c-1}$$

where $\overline{X}_J$ is a sample or column mean, $\overline{\overline{X}}$ is the grand mean, r is the number of observations in each sample, and c is the number of samples.   Then

$$\sum (\overline{X}_J - \overline{\overline{X}})^2 = (55 - 55)^2 + (53 - 55)^2 + (54 - 55)^2 + (58 - 55)^2 = 14$$

$$\sigma^2 = \frac{r \sum (\overline{X}_J - \overline{\overline{X}})^2}{c-1} = \frac{8(14)}{3} = \frac{112}{3} = 37.33$$

which is an estimate of population variance from the variance *between* the means or columns.

(*c*)   An estimate of the population variance from the variance *within* the samples or columns is obtained by averaging the four sample variances:

$$S_1^2 = \frac{\sum(X_{i1} - \overline{X}_1)^2}{r-1} = \frac{(51-55)^2 + (47-55)^2 + \cdots + (60-55)^2}{8-1} = \frac{144}{7} \cong 20.57$$

$$S_2^2 = \frac{\sum(X_{i2} - \overline{X}_2)^2}{r-1} = \frac{(47-53)^2 + (50-53)^2 + \cdots + (50-53)^2}{8-1} = \frac{208}{7} \cong 29.71$$

$$S_3^2 = \frac{\sum(X_{i3} - \overline{X}_3)^2}{r-1} = \frac{(57-54)^2 + (48-54)^2 + \cdots + (46-54)^2}{8-1} = \frac{216}{7} \cong 30.86$$

$$S_4^2 = \frac{\sum(X_{i4} - \overline{X}_4)^2}{r-1} = \frac{(50-58)^2 + (61-58)^2 + \cdots + (59-58)^2}{8-1} = \frac{158}{7} \cong 22.57$$

$$\sigma^2 \cong \frac{S_1^2 + S_2^2 + S_3^2 + S_4^2}{4} = \frac{20.57 + 29.71 + 30.86 + 22.57}{4} \cong 25.93$$

A more concise way of expressing the above is

$$\sigma^2 = \frac{\sum S_J^2}{c} = \frac{S_1^2 + S_2^2 + \cdots + S_c^2}{c}$$

$$= \frac{\dfrac{\sum(X_{i1} - \overline{X}_1)^2}{r-1} + \dfrac{\sum(X_{i2} - \overline{X}_2)^2}{r-1} + \dfrac{\sum(X_{i3} - \overline{X}_3)^2}{r-1} + \dfrac{\sum(X_{i4} - \overline{X}_4)^2}{r-1}}{c}$$

$$= \frac{\sum\sum(X_{iJ} - \overline{X}_J)^2}{(r-1)c} = \frac{144 + 208 + 216 + 158}{(7)(4)} = \frac{726}{28} = 25.93$$

(d)
$$F = \frac{\text{variance between sample means}}{\text{variance within samples}} = \frac{37.33}{25.93} = 1.44$$

The value of F from App. 7 for $\alpha = 0.05$ and $c - 1 = 3$ df in the numerator and $(r-1)c = 28$ df in the denominator is 2.95. Since the calculated value of F is smaller than the tabular value, we accept H_0, that the population means are the same.

5.24 (a) From the results obtained in Prob. 5.23, find the value of SSA, SSE, and SST; the degrees of freedom for SSA, SSE, and SST; and MSA, MSE, and the F ratio. (b) From the results in part a, construct an ANOVA table similar to Table 5.4. (c) Conduct the analysis of variance and draw a figure showing the acceptance and rejection regions for H_0.

(a)
$$\text{SSA} = r\sum(\overline{X}_J - \overline{\overline{X}})^2 = 112 \qquad \text{[from Prob. 5.23(b)]}$$

$$\text{SSE} = \sum\sum(X_{iJ} - \overline{X}_J)^2 = 726 \qquad \text{[from Prob. 5.23(c)]}$$

$$\text{SST} = \sum\sum(X_{iJ} - \overline{\overline{X}})^2 = (51-55)^2 + (47-55)^2 + \cdots + (59-55)^2 = 838$$

$$= \text{SSA} + \text{SSE} = 112 + 726 = 838$$

The df of SSA $= c - 1 = 4 - 1 = 3$; df of SSE $= (r-1)c = (8-1)(4) = 28$; and df of SST $= rc - 1 = 32 - 1 = 31$, which is the same as the df of SSA plus the df of SSE.

$$\text{MSA} = \frac{\text{SSA}}{c-1} = \frac{112}{3} = 37.33$$

$$\text{MSE} = \frac{\text{SSE}}{(r-1)c} = \frac{726}{28} = 25.93$$

$$F = \frac{\text{MSA}}{\text{MSE}} = \frac{37.33}{25.93} = 1.44$$

(b) See Table 5.21.

(c) The hypotheses to be tested are

$$H_0: \quad \mu_1 = \mu_2 = \mu_3 = \mu_4 \qquad \text{versus} \qquad H_1: \quad \mu_1, \mu_2, \mu_3, \mu_4 \text{ are not equal}$$

Since the calculated value of $F = 1.44$ is smaller than the tabular value of $F = 2.95$ with $\alpha = 0.05$ and df $= 3$ and 28, we accept H_0 (see Fig. 5-13); that is, we accept the null hypothesis, H_0, that $\mu_1 = \mu_2 = \mu_3 = \mu_4$. Since we were told (in Prob. 5.23) that the populations were normal with equal variance, we could view the four samples as coming from the *same* population. Note that the

Table 5.21   One-Way ANOVA Table for Fertilizer Experiment

Variation	Sum of Squares	Degrees of Freedom	Mean Square	F Ratio
Explained by fertilizer (between columns)	SSA = 112	$c - 1 = 3$	MSA = 37.33	
Error or unexplained (within columns)	SSE = 726	$(r - 1)c = 28$	MSE = 25.93	MSA/MSE = 1.44
Total	SST = 838	$rc - 1 = 31$	—	

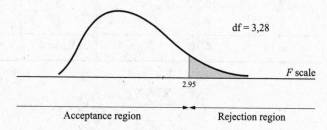

Fig. 5-13

MSE is a good estimate of σ^2 whether H_0 is true.   However, MSA is about equal to MSE only if H_0 is true (so that $F = 1$).   Note that the F distribution is continuous and is used here for a right-tail test only.

5.25   Table 5.22 gives the outputs of an experimental farm that used each of four fertilizers and three pesticides such that each plot of land had an equal probability of receiving each fertilizer-pesticide combination (*completely randomized design*).   (a) Find the average output for each fertilizer $\overline{X}_{\cdot J}$ for each pesticide $\overline{X}_i$. and for the sample as a whole $\overline{\overline{X}}$.   (b) Find the total sum of squares, SST, the sum of squares for fertilizer or factor A, SSA, for pesticides or factor B, SSB, and for the error or unexplained residual, SSE.   (c) Find the degrees of freedom for SSA, SSB, SSE, and SST. (d) Find MSA, MSB, MSE, MSA/MSE, and MSB/MSE.

Table 5.22   Output with 4 Fertilizers and 3 Pesticides

	Fertilizer 1	Fertilizer 2	Fertilizer 3	Fertilizer 4
Pesticide 1	21	12	9	6
Pesticide 2	13	10	8	5
Pesticide 3	8	8	7	1

(a)   The column mean for each fertilizer is given by

$$\overline{X}_{\cdot J} = \frac{\sum_i X_{iJ}}{r}$$

(5.8a)

The row mean for each pesticide is given by

$$\overline{X}_{i\cdot} = \frac{\sum_J X_{iJ}}{c} \qquad (5.8b)$$

The grand mean is given by

$$\overline{\overline{X}} = \frac{\sum \overline{X}_{i\cdot}}{r} = \frac{\sum \overline{X}_{\cdot J}}{c} \qquad (5.9a)$$

The subscripted dots signify that more than one factor is being considered.   The results are shown in Table 5.23.

Table 5.23   Output with 4 Fertilizers and 3 Pesticides (with Row, Column, and Grand Means)

	Fertilizer 1	Fertilizer 2	Fertilizer 3	Fertilizer 4	Sample Mean
Pesticide 1	21	12	9	6	$\overline{X}_{1\cdot} = 12$
Pesticide 2	13	10	8	5	$\overline{X}_{2\cdot} = 9$
Pesticide 3	8	8	7	1	$\overline{X}_{3\cdot} = 6$
Sample mean	$\overline{X}_{\cdot 1} = 14$	$\overline{X}_{\cdot 2} = 10$	$\overline{X}_{\cdot 3} = 8$	$\overline{X}_{\cdot 4} = 4$	$\overline{\overline{X}} = 9$

(b)
$$\text{SST} = \sum \sum (X_{iJ} - \overline{\overline{X}})^2$$

$$
\begin{array}{llll}
(21-9)^2 = 144 & (12-9)^2 = 9 & (9-9)^2 = 0 & (6-9)^2 = 9 \\
(13-9)^2 = 16 & (10-9)^2 = 1 & (8-9)^2 = 1 & (5-9)^2 = 16 \\
(8-9)^2 = \underline{1} & (8-9)^2 = \underline{1} & (7-9)^2 = \underline{4} & (1-9)^2 = \underline{64} \\
\qquad\quad 161 & \qquad\quad 11 & \qquad\quad 5 & \qquad\quad 89
\end{array}
$$

$$\text{SST} = 161 + 11 + 5 + 89 = 266$$

$$\text{SSA} = r \sum (\overline{X}_{\cdot J} - \overline{\overline{X}})^2 \quad \text{(between-column variations)}$$

$$= 3[(14-9)^2 + (10-9)^2 + (8-9)^2 + (4-9)^2]$$

$$= 3(25 + 1 + 1 + 25) = 156$$

$$\text{SSB} = c \sum (\overline{X}_{i\cdot} - \overline{\overline{X}})^2 \quad \text{(between-row variations)}$$

$$= 4[(12-9)^2 + (9-9)^2 + (6-9)^2]$$

$$= 4(9 + 0 + 9) = 72$$

$$\text{SSE} = \text{SST} - \text{SSA} - \text{SSB} = 266 - 156 - 72 = 38$$

(c)
$$\text{df of SSA} = c - 1 = 3 \qquad (5.13a)$$
$$\text{df of SSB} = r - 1 = 2 \qquad (5.13b)$$
$$\text{df of SSE} = (r-1)(c-1) = 6 \qquad (5.14a)$$
$$\text{df of SST} = rc - 1 = 11 \qquad (5.15)$$

(d)
$$\text{MSA} = \frac{\text{SSA}}{c-1} = \frac{156}{3} = 52 \qquad (5.16)$$

$$\text{MSB} = \frac{\text{SSB}}{r-1} = \frac{72}{2} = 36 \qquad (5.17)$$

$$\text{MSE} = \frac{\text{SSE}}{(r-1)(c-1)} = \frac{38}{6} = 6.33 \qquad (5.18)$$

$$\frac{\text{MSA}}{\text{MSE}} = \frac{52}{6.33} = 8.21 \qquad F \text{ ratio for factor A (fertilizer)} \qquad (5.19)$$

$$\frac{\text{MSB}}{\text{MSE}} = \frac{36}{6.33} = 5.69 \qquad F \text{ ratio for factor B (pesticide)} \qquad (5.20)$$

5.26 (a) From the results of Prob. 5.25, construct an ANOVA table similar to Table 5.4.   (b) Test at the 1% level of significance the hypothesis that the means for factor A populations (fertilizers) are identical.   (c) Test at the 1% level of significance the hypothesis that the means for factor B populations (pesticides) are identical.

(a)   See Table 5.24.

Table 5.24   Two-Factor ANOVA Table for Effect of Fertilizers and Pesticides on Output

Variation	Sum of Squares	Degrees of Freedom	Mean Square	F
Explained by fertilizer (between columns)	SSA = 156	$c - 1 = 3$	MSA = 52	$\frac{\text{MSA}}{\text{MSE}} = 8.21$
Explained by pesticide (between rows)	SSB = 72	$r - 1 = 2$	MSB = 36	$\frac{\text{MSB}}{\text{MSE}} = 5.69$
Error or unexplained	SSE = 38	$(r - 1)(c - 1) = 6$	MSE = 6.33	—
Total	SST = 266	$rc - 1 = 11$	—	—

(b)   The hypotheses to be tested are

$$H_0: \quad \mu_1 = \mu_2 = \mu_3 = \mu_4 \qquad \text{versus} \qquad H_1: \quad \mu_1, \mu_2, \mu_3, \mu_4 \text{ are not all equal}$$

where μ refers to the various means for factor A (fertilizer) populations.   For factor A, $F = 9.78$ (from App. 7) for degrees of freedom 3 (numerator) and 6 (denominator) and $\alpha = 0.01$.   Since the calculated value of $F = 8.21$ (from Table 5.24) is less than the tabular value of F, we accept H_0, that the means for factor A (fertilizer) populations are equal.

(c)   The second set of hypotheses to be tested consists of

$$H_0: \quad \mu_1 = \mu_2 = \mu_3 \qquad \text{versus} \qquad H_1: \quad \mu_1, \mu_2, \mu_3 \text{ are not all equal}$$

but now μ refers to the various means for factor B (pesticide) populations.   For factor B, $F = 10.92$ (from App. 7) for degrees of freedom 2 and 6 and $\alpha = 0.01$.   Since the calculated value of $F = 5.69$ (from Table 5.24) is less than the tabular value of F, we accept H_0, that the means for factor B (pesticide) populations are also equal.   Note that in two-factor analysis of variance (with an ANOVA table similar to Table 5.24) we can test two null hypotheses, one for factor A and one for factor B.

5.27 Table 5.25 gives the first-year earnings (in thousands of dollars) of students with master's degrees from 5 schools and for 3 class rankings at graduation.   Test at the 5% level of significance that the means are identical   (a) for school populations and   (b) for class-ranking populations.

(a)   The hypotheses to be tested are

$$H_0: \quad \mu_1 = \mu_2 = \mu_3 = \mu_4 = \mu_5 \qquad \text{versus} \qquad H_1: \quad \mu_1, \mu_2, \mu_3, \mu_4, \mu_5 \text{ are not equal}$$

where μ refers to the various means for factor A (school) populations.

$$\text{SST} = \sum \sum (X_{iJ} - \overline{\overline{X}})^2$$

Table 5.25   First-Year Earnings of MA Graduates of 5 Schools and 3 Class Ranks
(in Thousands of Dollars)

Class Ranks	School 1	School 2	School 3	School 4	School 5	Sample Mean
Top 1/3	20	18	16	14	12	$\overline{X}_{1.} = 16$
Middle 1/3	19	16	13	12	10	$\overline{X}_{2.} = 14$
Bottom 1/3	18	14	10	10	8	$\overline{X}_{3.} = 12$
Sample mean	$\overline{X}_{.1} = 19$	$\overline{X}_{.2} = 16$	$\overline{X}_{.3} = 13$	$\overline{X}_{.4} = 12$	$\overline{X}_{.5} = 10$	$\overline{\overline{X}} = 14$

$$
\begin{array}{lllll}
(20-14)^2 = 36 & (18-14)^2 = 16 & (16-14)^2 = 4 & (14-14)^2 = 0 & (12-14)^2 = 4 \\
(19-14)^2 = 25 & (16-14)^2 = 4 & (13-14)^2 = 1 & (12-14)^2 = 4 & (10-14)^2 = 16 \\
(18-14)^2 = \underline{16} & (14-14)^2 = \underline{0} & (10-14)^2 = \underline{16} & (10-14)^2 = \underline{16} & (8-14)^2 = \underline{36} \\
\qquad\quad 77 & \qquad\quad 20 & \qquad\quad 21 & \qquad\quad 20 & \qquad\quad 56
\end{array}
$$

$\text{SST} = 77 + 20 + 21 + 20 + 56 = 194$

$\text{SSA} = r \sum (\overline{X}_{.J} - \overline{\overline{X}})^2 \qquad \text{(between-column variations)}$

$\qquad = 3[(19-14)^2 + (16-14)^2 + (13-14)^2 + (12-14)^2 + (10-14)^2] = 3(25 + 4 + 1 + 4 + 16)$

$\qquad = 150$

$\text{SSB} = c \sum (\overline{X}_{i.} - \overline{\overline{X}})^2 = 5[(16-14)^2 + (14-14)^2 + (12-14)^2] = 5(4 + 0 + 4) = 40$

$\text{SSE} = \text{SST} - \text{SSA} - \text{SSB} = 194 - 150 - 40 = 4$

These results are summarized in Table 5.26. From App. 7, $F = 3.84$ for degrees of freedom 4 and 8 and $\alpha = 0.05$. Since the calculated $F = 70$, we reject H_0 and accept H_1, that the population means of first-year earnings for the 5 schools are different.

Table 5.26   Two-Factor ANOVA Table for First-Year Earnings

Variation	Sum of Squares	Degrees of Freedom	Mean Square	F
Explained by schools (A) (between columns)	$\text{SSA} = 150$	$c - 1 = 4$	$\text{MSA} = \dfrac{150}{4} = 37.5$	$\dfrac{\text{MSA}}{\text{MSE}} = \dfrac{37.5}{0.5} = 70$
Explained by ranking (B) (between rows)	$\text{SSB} = 40$	$r - 1 = 2$	$\text{MSB} = \dfrac{40}{2} = 20$	$\dfrac{\text{MSB}}{\text{MSE}} = \dfrac{20}{0.5} = 40$
Error or unexplained	$\text{SSE} = 4$	$(r-1)(c-1) = 8$	$\text{MSE} = \dfrac{4}{8} = 0.5$	
Total	$\text{SST} = 194$	$rc - 1 = 14$	—	

(b)   The hypotheses to be tested are

$$H_0: \quad \mu_1 = \mu_2 = \mu_3 \qquad \text{versus} \qquad H_1: \quad \mu_1, \mu_2, \mu_3 \text{ are not equal}$$

where μ refers to the various means for factor B (class-ranking) populations. From Table 5.26, we get that the calculated value of $F = \text{MSB}/\text{MSE} = 40$. Since this is larger than the tabular value of $F = 4.46$ for df 2 and 8 and $\alpha = 0.05$, we reject H_0 and accept H_1, that the population means of first-year earnings for the 3 class rankings are different. Thus the type of school and class ranking are both statistically significant at the 5% level in explaining differences in first-year earnings. The preceding analysis implicitly assumes that the effects of the two factors are *additive* (i.e., there is no *interaction* between them).

NONPARAMETRIC TESTING

5.28 (*a*) What are nonparametric tests?   (*b*) When would one want to use a nonparametric test? (*c*) What are the advantages and disadvantages of nonparametric tests?

(*a*) Nonparametric tests require fewer assumptions to establish the validity of their results. Parametric tests involve assumptions about the specific distribution that the data follows, as well as the structure of data-generating process. Nonparametric tests allow the researcher to relax the assumptions regarding the distribution of the data and/or the functional form of the underlying processes.

(*b*) Nonparametric tests should be used only when one is uncertain about the assumptions behind the parametric test. The usual situation for using a nonparametric test in statistics is a small sample size. If the values are not normally distributed, a small sample would invalidate the assumption that the sample mean is normally distributed with a mean of μ and a variance of σ^2/n.

(*c*) A nonparametric test is advantageous because of its ease of calculation and its flexibility. There are nonparametric tests appropriate for most scales of measurement, and for nonstandard functional forms and distributions. Also, the nonparametric goodness-of-fit test does not have the researcher choose class intervals to compare observed and expected values. The chi-square goodness-of-fit test is often not robust to changes in class specifications. The disadvantages of a nonparametric test focus around the loss of information. Nonparametric tests are based on counting rules, such as ranking, and therefore summarize magnitudes into a rank statistic. This only uses the relative position of values. If the standard assumptions hold, a parametric test will be more efficient, and therefore more powerful, for a given data set.

5.29 A marketing firm is deciding whether food additive B is better tasting than food additive A.   A focus group of 10 individuals rate the taste on a scale of 1 to 10.   Results of the focus group are listed in Table 5.27. Test at the 5% significance level the null hypothesis that food additive B is no better tasting than food additive A.

Table 5.27   Food Additive Taste Comparison

Individual ID No.	Additive A Rating	Additive B Rating
1	5.5	6
2	7	8
3	9	9
4	3	6
5	6	8
6	6	6
7	8	4
8	6.5	8
9	7	8
10	6	9

This is a small sample with ratings rather than quantitative variables; therefore the usual assumptions do not hold. We proceed with the nonparametric test. Since we have two samples with data that are paired (two ratings per person), we first take the difference of the two ratings for each person to test the hypotheses

$$H_0:\quad \text{Med}_A - \text{Med}_B \geq 0 \qquad H_1:\quad \text{Med}_A - \text{Med}_B < 0$$

The steps are shown in Table 5.28.   Since $W < 11$, we reject H_0: $\text{Med}_A - \text{Med}_B \geq 0$ and accept H_1: $\text{Med}_A - \text{Med}_B < 0$ at the 5% significance level.

Table 5.28   Ratings Signed Rank Test

$X_A - X_B$	Ordered	Rank	Rank for Positive Differences
−0.5	0	N/A	—
−1	0	N/A	—
0	−0.5	1	—
−3	−1	2.5	—
−2	−1	2.5	—
0	−1.5	4	—
4	−2	5	—
−1.5	−3	6.5	—
−1	−3	6.5	—
−3	4	8	8

$$W = 8$$

5.29   Data from the World Bank's World Development Indicators reports that for 9 Latin American countries, male illiteracy is as follows (in percent):

Argentina 3, Bolivia 8, Brazil, 15, Chile 4, Colombia 9, Ecuador 7, Peru 6, Uruguay 3, Venezuela 7

(a) Test the null hypothesis that the median illiteracy rate is 8% at the 10% significance level.
(b) Test the null hypothesis that the median illiteracy rate is greater than or equal to 8% at the 10% significance level.

(a)   Calculations for the signed rank test are given in Table 5.29 to test H_0: Med $= 8$ versus H_1: Med $\neq 8$. The critical values for the signed rank test (App. 9) for a two-tail test at the 10% significance level and $n = 9$ are 9 and 36.   Since $9 < W < 36$, we accept the null hypothesis that the median illiteracy rate for South American countries is equal to 8.

(b)   To test H_0: Med ≥ 8 versus H_1: Med < 8, we would expect a higher value of W for a higher population median.   Therefore this is a one-tail test with the rejection region in the left tail.   The critical value from App. 9 is 11.   Since $W < 11$, we reject the null hypothesis at the 10% significance level and conclude that the median illiteracy rate for males is less than 8. Note accepting the null in part a just means that we could not rule out a median of 8, but not that there was proof that the median was 8.   As with any

Table 5.29   Illiteracy Signed Rank Test

Illiteracy Rates (X)	$X - \mathrm{Med}_0$	Ordered	Rank	Rank for Positive Differences
3	−5	0	N/A	
8	0	1	2	2
15	7	−1	2	
4	−4	−1	2	
9	1	−2	4	
7	−1	−4	5	
6	−2	−5	6.5	
3	−5	−5	6.5	
7	−1	7	8	8

$$W = 10$$

test, the one-tail test with the signed rank statistic has a larger rejection region in the tail tested since the significance percentage is not split between two tails.

5.30 Continuing with the analysis from Prob. 5.29, the following is male illiteracy data from two other regions:

Asia: China 9, Hong Kong 4, Indonesia 9, Korea, Republic 1, Malaysia 9, Philippines 5, Singapore 4

Africa: Chad 50, Ivory Coast 46, Egypt, Arab Republic 34, Ethiopia 57, Morocco 39, Niger 77, Nigeria 29, Rwanda 27

(*a*) Perform a nonparametric test of equality of the median male illiteracy rate for South America, Asia, and Africa at the 1% significance level.   (*b*) Test the equality of the median illiteracy rate for South America and Asia at the 5% significance level.

(*a*)   Since we are testing more than one group, the Kruskal-Wallis rank test should be used.   Calculations are listed in Table 5.30.   SA indicates South America, As indicates Asia, and Af indicates Africa

Table 5.30   Kruskal-Wallis Rank Test

Illiteracy Rates (X)	Ordered	Rank	Totals
3 (SA)	1 (As)	1	For South America: $\Sigma R_{SA} = 76.5$
8 (SA)	3 (SA)	2.5	
15 (SA)	3 (SA)	2.5	
4 (SA)	4 (SA)	4	
9 (SA)	4 (As)	4	
7 (SA)	4 (As)	4	
6 (SA)	5 (As)	7	
3 (SA)	6 (SA)	8	
7 (SA)	7 (SA)	9.5	
9 (As)	7 (SA)	9.5	For Asia: $\Sigma R_{As} = 56.5$
4 (As)	8 (SA)	11	
9 (As)	9 (SA)	13.5	
1 (As)	9 (As)	13.5	
9 (As)	9 (As)	13.5	
5 (As)	9 (As)	13.5	
4 (As)	15 (SA)	16	
50 (Af)	27 (Af)	17	For Africa: $\Sigma R_{Af} = 164.0$
46 (Af)	29 (Af)	18	
34 (Af)	34 (Af)	19	
57 (Af)	39 (Af)	20	
39 (Af)	46 (Af)	21	
77 (Af)	50 (Af)	22	
29 (Af)	57 (Af)	23	
27 (Af)	77 (Af)	24	

$$H = \frac{12}{n(n+1)} \left(\frac{(\Sigma R_{SA})^2}{n_{SA}} + \frac{(\Sigma R_{As})^2}{n_{As}} + \frac{(\Sigma R_{Af})^2}{n_{Af}} \right) - 3(n+1)$$

$$= \frac{12}{24(24+1)} \left(\frac{(76.5)^2}{9} + \frac{(56.5)^2}{7} + \frac{(164)^2}{8} \right) - 3(24+1) = 14.36$$

The critical value for the chi-square distribution with 3 degrees of freedom at the 1% significance level is 11.34. Since $H = 14.36 > 11.34$, we reject the null hypothesis that the median male illiteracy rates of all three groups are equal.

(b)   For testing two samples, one can use the same ranking method, but can compare the sum of ranks of the smallest of the two groups with the critical values in the two-sample section of the Wilcoxon statistics in App. 9.   Since all African rankings fell above the rankings of South American and Asian countries, the rankings from Table 5.12 may be used.

$$W = \sum R_{As} = 56.5$$

From the table in App. 9, the critical values at the 5% significance level are 41 and 78.   Since $41 < W < 78$, we accept the null hypothesis that the median male illiteracy rates in South America are equal.

5.31   Using the African male illiteracy rates from Prob. 5.30, test the null hypothesis at the 10% significance level that the illiteracy rates in Africa follow the continuous uniform distribution   (a) between 25 and 80   (b) between 25 and 100.

(a)   The *continuous uniform distribution* has equal value of the density function at each point between 25 and 80.   To calculate the probability of being between values a and b, one can take the area under the density function: $P(a < X < b) = (b - a)/(80 - 25)$, where the denominator is the difference between the upper and lower bound.   Since we have a small sample size, we will use the Kolmogorov-Smirnov goodness-of-fit test.

Ordered data values	27	29	34	39	46	50	57	77
Proportion below, %	12.5	25.0	37.5	50.0	42.5	75.0	87.5	100.0
Uniform cumulative probability, %	3.6	7.3	16.4	25.5	38.2	45.5	58.2	94.5
Difference, %	8.9	17.7	21.1	24.5	4.3	29.5	29.3	5.5

The maximum difference is 29.5% (0.295), which is less than the critical value of 0.411 (App. 10); therefore we accept the null hypothesis that illiteracy rates in Africa follow the continuous uniform distribution between 25 and 80.

(b)   This continuous uniform distribution has equal value of the density function at each point between 25 and 100 under the null: $P(a < X < b) = (b - a)/(100 - 25)$.   The calculations are as follows

Ordered data values	27	29	34	39	46	50	57	77
Proportion below, %	12.5	25.0	37.5	50.0	42.5	75.0	87.5	100.0
Uniform cumulative probability, %	2.7	5.3	12.0	18.7	28.0	33.3	42.7	69.3
Difference, %	9.8	19.7	25.5	31.3	14.5	41.7	44.8	30.7

The maximum difference is 44.8% (0.448), which is greater than the critical value of 0.411; therefore we reject the null hypothesis that illiteracy rates in Africa follow the continuous uniform distribution between 25 and 100.

5.32   Repeat the test from Prob. 5.19 with the Kolmogorov-Smirnov goodness-of-fit test to test the H0: data are from the binomial distribution with probability of acceptance equal to 0.4.

Calculations are given in Table 5.31. The largest difference is 0.058 in absolute value. The critical value from the table for $n = 100$ is 0.136 at the 5% level of significance. Since $0.058 < 0.136$, we accept the null hypothesis that the distribution of college acceptances follows the binomial distribution with a probability of acceptance of 40%.

Table 5.31   Kolmogorov-Smirnov Goodness-of-Fit Test

Number of Acceptances	Frequency	Relative Frequency	Binomial Probabilities	Cumulative Relative Frequency (Observed)	Cumulative Probability (Expected)	Difference
0	25	0.25	0.216	0.25	0.216	0.034
1	34	0.34	0.432	0.59	0.648	−0.058
2	31	0.31	0.288	0.90	0.936	−0.036
3	10	0.10	0.064	1.00	1.00	0.00

Supplementary Problems

TESTING HYPOTHESIS

5.33 (a) What do we call the error of accepting a false hypothesis? Of rejecting a true hypothesis? (b) What symbol is usually used for the probability of type I error? What is another name for this? (c) What is the symbol conventionally used for the probability of type II error? (d) What is the level of confidence? (e) If α is reduced from 5 to 1%, what happens to β?
Ans. (a) Type II error; type I error (b) α; level of significance (c) β (d) $1 - \alpha$ (e) β increases

5.34 Having set $\alpha = 5\%$, when is a graduate school more likely to accept the hypothesis that the average Graduate Record Examination (GRE) scores of its entering class (a) Equal 600? (b) Are larger than 600? (c) Are smaller than 600?
Ans. (a) The closer the mean sample, $\overline{X}$, is to 600 (b) The more $\overline{X} > 600$ (c) The more $\overline{X} < 600$

TESTING HYPOTHESES ABOUT THE POPULATION MEAN AND PROPORTION

5.35 An aircraft manufacturer needs to buy aluminum sheets of 0.05 in thickness. Thinner sheets would not be appropriate, and thicker sheets would be too heavy. The aircraft manufacturer takes a random sample of 100 sheets from a supplier of aluminum sheets and finds that their average thickness is 0.048 in and their standard deviation is 0.01 in. Should the aircraft manufacturer buy the aluminum sheets from this supplier in order to make the decision at the 5% level of significance?
Ans. No

5.36 Define the acceptance region for Prob. 5.35 in inches.
Ans. 0.04804 to 0.05196 in

5.37 A navy recruiting center knows from past experience that the height of recruits is normally distributed with a mean μ of 180 cm (1 cm = 1/100 m) and a standard deviation σ of 10 cm. The recruiting center wants to test at the 1% level of significance the hypothesis that the average height of this year's recruits is above 180 cm. To do this, the recruiting officer takes a random sample of 64 recruits and finds that the average height for this sample is 182 cm. (a) Should the recruiting officer accept the hypothesis? (b) What is the rejection region for the test in centimeters?
Ans. (a) No (b) Greater than 182.9125

5.38 A purchaser of electronic components wants to test the hypothesis that they last less than 100 h. To do this she takes a random sample of 16 such components and finds that, on average, they last 96 h, with a standard deviation of 8 h. If the purchaser knows that the lifetime of the components is normally distributed, should she accept the hypothesis that they last less than 100 h at   (a) A 95% level of confidence? (b) A 99% level of confidence?
Ans. (a) Yes (b) No

5.39 In the past, 20% of applicants for admission into a master's program had GRE scores above 650. Of the 88 students applying to be admitted into the program in 1981, 22 had GRE scores above 650.   Do the 1981 applicants have greater GRE scores than previous applicants at the 5% level of significance?
Ans. No

5.40 Find the probability of accepting H_0 (that $\mu = 650$) for Prob. 5.39 if $\sigma_{\bar{p}} = 0.043$ and   (a) $p = 0.20$, (b) $p = 0.22$, (c) $p = 0.24$, (d) $p = 0.25$, (e) $p = 0.26$, and (f) $p = 0.28$.
Ans. (a) 0.877 (b) 0.758 (c) 0.591 (d) 0.5 (e) 0.409 (f) 0.242

5.41 (a) What is the value of α when $p = 0.20$ in Prob. 5.39 (b) How can the OC curve be derived for Prob. 5.39?
Ans. (a) 0.123 (b) By joining the value of $1 - \alpha$ for $p = 0.20$ with the values of β found in Prob. 5.40(b) to (f) for various values of $p > 0.20$

5.42 Find the probability of rejecting H_0 (that $\mu = 650$) for Prob. 5.39 if $\sigma_{\bar{p}} = 0.043$ and   (a) $p = 0.20$, (b) $p = 0.22$, (c) $p = 0.24$, (d) $p = 0.25$, (e) $p = 0.26$, and (f) $p = 0.28$.
Ans. (a) 0.123 (b) 0.242 (c) 0.409 (d) 0.5 (e) 0.591 (f) 0.758

5.43 How can we get the power curve for Prob. 5.39?
Ans. By joining the values found in Prob. 5.42(a) to (f) for various alternative values of $p > 0.2$.

TESTING HYPOTHESES FOR DIFFERENCES BETWEEN TWO MEANS OR PROPORTIONS

5.44 A consulting firm wants to decide at the 5% level of significance if the salaries of construction workers differ between New York and Chicago.   A random sample of 100 construction workers in New York has an average weekly salary of $400 with a standard deviation of $100.   In Chicago, a random sample of 75 workers has an average weekly salary of $375 with a standard deviation of $80.   Is there a significant difference between the salaries of construction workers in New York and Chicago at   (a) The 5% level? (b) The 10% level?
Ans. (a) No (b) Yes

5.45 A random sample of 21 AFC football players has a mean weight of 265 lb with a standard deviation of 30 lb, while a random sample of 11 NFC players has a mean weight of 240 lb with a standard deviation of 20 lb. Is the mean weight of all AFC football players greater than that for the NFC players at the 1% level of significance?
Ans. Yes

5.46 A random sample of 100 soldiers indicates that 20% are married in year 1, while 30% are married in year 2. Determine whether to accept the hypothesis that the proportion of married soldiers in year 1 is less than that in year 2 (a) at the 5% level of significance and   (b) at the 1% level of significance.
Ans. (a) Accept the hypothesis (b) Reject the hypothesis

CHI-SQUARE TEST OF GOODNESS OF FIT AND INDEPENDENCE

5.47 A die is rolled 60 times with the following results: a 1 came up 12 times, a 2 came up 8 times, a 3 came up 13 times, a 4 came up 12 times, a 5 came up 7 times, and a 6 came up 8 times. Is the die balanced at the 5% level of significance?
Ans. Yes

5.48 An urn contains balls of 4 colors: green, white, red, and blue. A ball is picked from the urn and its color is recorded.   The ball is then replaced in the urn, the balls are thoroughly mixed, and another ball is picked.

The process is repeated 18 times, and the result is that a green ball is picked 8 times, a white ball is picked 7 times, a red ball is picked once, and a blue ball is picked twice. Does the urn contain an equal number of green, white, red, or blue balls? Test the hypothesis at the 5% level of significance.

Ans. The hypothesis should be accepted at the 5% level of significance that the urn contains an equal number of balls of all four colors.

5.49 A random sample of 64 cities in the United States indicates the number of rainy days during the month of June given in Table 5.32. Do rainy days in U.S. cities follow a normal distribution with $\mu = 3$ and $\sigma = 2$ at the 10% level of significance?

Ans. No

Table 5.32 Number of Rainy Days during June for 64 U.S. Cities

Number of Rainy Days	Number of Cities
0	10
1	12
2	22
3	13
4	6
5	1
	64

5.50 Contingency Table 5.33 gives the number of acceptable and nonacceptable electronic components produced at various hours of the morning in a random sample from the output of a plant. Should the hypothesis be accepted or rejected at the 5% level of significance that the production of acceptable items is independent of the hour of the morning in which they are produced?

Ans. Accept H_0

Table 5.33 Acceptable and Nonacceptable Components Produced Each Hour of the Morning

	8–9 A.M.	9–10 A.M.	10–11 A.M.	11–12 A.M.	Total
Acceptable	60	75	80	65	280
Nonacceptable	30	25	30	35	120
	90	100	110	100	400

5.51 The number of people voting Democrat or Republican below the age of 40 and 40 plus in a random sample of 30 voters in a city is given in contingency Table 5.34. Is voting Democrat or Republican independent of the voter being below the age of 40 or 40 plus in this city at the 5% level of significance?

Ans. No

Table 5.34 Democrats and Republicans below and above Age 40

Age Group	Democrats	Republicans	Total
< 40	6	5	11
≥ 40	10	9	19
	16	14	30

ANALYSIS OF VARIANCE

5.52   Table 5.35 gives the miles per gallon for 4 different octanes of gasoline for 5 days.   Assume that the miles per gallon for each octane is normally distributed with equal variance.   Should the hypothesis of equal population means be accepted or rejected at the 5% level of significance?
Ans.   Rejected

Table 5.35   Miles per Gallon with 4 Types of Gasoline for 5 Days

Type 1	Type 2	Type 3	Type 4
12	12	16	17
11	14	14	15
12	13	15	17
13	15	13	16
11	14	14	18

5.53   Table 5.36 gives the miles per gallon for each of 4 different octanes of gasoline and 3 types of car (heavy, medium, and light) in a completely randomized design.   Should the hypothesis be accepted at the 1% level of significance that the population means are the same for each   (*a*) Octane of gasoline?   (*b*) Type of car?
Ans.   (*a*) Yes   (*b*) No

Table 5.36   Miles per Gallon for Each of 4 Octanes and 3 Types of Car

Type of Car	Octane 1	Octane 2	Octane 3	Octane 4
Heavy	8	9	9	10
Medium	16	15	18	17
Light	24	26	28	30

5.54   Table 5.37 gives sales data for soap with each of 3 different wrappings and 4 different formulas in a completely randomized design.   Should the hypothesis be accepted at the 5% level of significance that the population means are the same for each   (*a*) Wrappings?   (*b*) Formula?
Ans.   (*a*) No   (*b*) Yes

Table 5.37   Soap Sales for Each of 3 Wrappings and 4 Formulas

	Wrapping 1	Wrapping 2	Wrapping 3
Formula 1	87	78	90
Formula 2	79	79	84
Formula 3	83	81	91
Formula 4	85	83	89

NONPARAMETRIC TESTING

5.55   Using the data from Table 5.35, would the Wilcoxon signed rank test reject at the 10% significance level the null hypothesis that the median miles per gallon for type 1 gasoline is   (*a*) 12   (*b*) 15?
Ans.   (*a*) No ($W = 2$)   (*b*) Yes ($W = 0$)

5.56   Repeat the test from Prob. 5.52 using the Kruskal-Wallis rank test.   Is the null hypothesis of equality of medians accepted at the 5% level of significance?
Ans.   No, it is rejected ($H = 14.25$)

5.57   Repeat the test from Prob. 5.49 using the Kolmogorov-Smirnov goodness-of-fit test.   Are the data normally distributed with $\mu = 3$ and $\sigma = 2$ at the 10% level of significance?
Ans.   No (maximum difference = 0.391)

Statistics Examination

1. Table 1 gives the frequency distribution of the rate of unemployment in a sample of 20 large U.S. cities in 1980. (a) Find the mean, median, and mode of the unemployment rate. (b) Find the variance, standard deviation, and coefficient of variation. (c) Find the Pearson's coefficient of skewness and sketch the relative frequency histogram

Table 1 Frequency Distribution of Unemployment Rate

Unemployment Rate, %	Frequency
7.0–7.4	2
7.5–7.9	4
8.0–8.4	5
8.5–8.9	4
9.0–9.4	3
9.5–9.9	2
	$n = 20$

2. The lifetime of an electronic component is known to be normally distributed with a mean of 1000 h and a standard deviation of 80 h. What is the probability that a component picked at random from the production line will have a lifetime (a) Between 1120 and 1180 h? (b) Between 955 and 975 h? (c) Below 955 h? (d) Above 975 h? (e) Sketch the normal and the standard normal distribution for this problem and shade the area corresponding to part d.

3. The average IQ of a random sample of 25 students at a college is 110. If the distribution of the IQ at the college is known to be normal with a standard deviation of 10 (a) Find the 95% confidence interval for the unknown mean IQ for the entire student body at the college. (b) Answer the same question if the population standard deviation had not been known, but the sample standard deviation was calculated to be 8. (c) Specify all possible cases when the normal distribution, the t distribution, or Chebyshev's inequality can be used.

4. A firm sells detergent packed in two plants. From past experience, the firm knows that the amount of detergent in the boxes packed in the two plants is normally distributed. The firm takes a random sample of 25 boxes from the output of each plant and finds that the mean weight and standard deviation of the detergent in the boxes from plant 1 is 1064 g (2.34 lb) and 100 g, respectively. For the sample in plant 2, the mean is 1024 g and the standard deviation is 60 g. (a) Can the firm claim with a 95% level of confidence that the boxes of detergent from plant 1 contain more than 1000 g? (b) Test at the 95% level of confidence that the amount of detergent in the boxes of both plants is the same.

Answers
1. (a) See Table 2.

$$\overline{X} = \frac{\sum fX}{n} = \frac{168.0}{20} = 8.4\%$$

$$\text{Med} = L + \frac{n/2 - F}{f_m} c = 8.0 + \frac{20/2 - 6}{5} 0.4 = 8.32\%$$

$$\text{Mode} = L + \frac{d_1}{d_1 + d_2} c = 8.0 + \frac{1}{1 + 1} 0.4 = 8.2\%$$

(b) See Table 3.

Table 2   Calculations to Find Sample Mean, Median, and Mode

Unemployment Rate, %	Class Midpoint X	Frequency f	fX
7.0–7.4	7.2	2	14.4
7.5–7.9	7.7	4	30.8
8.0–8.4	8.2	5	41.0
8.5–8.9	8.7	4	34.8
9.0–9.4	9.2	3	27.6
9.5–9.9	9.7	2	19.4
		$\sum f = n = 20$	$\sum fX = 168.0$

Table 3   Calculations to Find the Variance, Standard Deviation, and Coefficient of Variation

Unemployment Rate, %	Class Midpoint X	Frequency f	Mean $\overline{X}$	$(X - \overline{X})$	$(X - \overline{X})^2$	$f(X - \overline{X})^2$
7.0–7.4	7.2	2	8.4	−1.2	1.44	2.88
7.5–7.9	7.7	4	8.4	−0.7	0.49	1.96
8.0–8.4	8.2	5	8.4	−0.2	0.04	0.20
8.5–8.9	8.7	4	8.4	0.3	0.09	0.36
9.0–9.4	9.2	3	8.4	0.8	0.64	1.92
9.5–9.9	9.7	2	8.4	1.3	1.69	3.38
		$\sum f = n = 20$				$\sum f(X - \overline{X})^2 = 10.70$

$$s^2 = \frac{\sum f(X - \overline{X})^2}{n - 1} = \frac{10.70}{19} \cong 0.56\% \text{ squared}$$

$$s = \sqrt{s^2} \cong 0.75\%$$

$$V = \frac{s}{\overline{X}} = \frac{0.75\%}{8.4\%} \cong 0.09$$

(c) $$\text{Sk} = 3\frac{\overline{X} - \text{med}}{s} = 3\frac{8.40 - 8.32}{0.75} \cong 0.32 \quad \text{(see Fig. 1)}$$

Fig. 1

2. (a) The problem asks to find $P(1120 < X < 1180)$, where X refers to time measured in hours of lifetime for electronic component. Given $\mu = 1000$ h and $\sigma = 80$ h and letting $X_1 = 1120$ h and $X_2 = 1180$ h, we get

$$z_1 = \frac{X_1 - \mu}{\sigma} = \frac{1120 - 1000}{80} = 1.5 \quad \text{and} \quad z_2 = \frac{1180 - 1000}{80} = 2.25$$

Subtracting the value of $z_2 = 0.4878$ from the value of $z_1 = 0.4332$ (obtained from the table of the standard normal distribution), we get

$$P(1120 < X < 1180) = 0.0546, \text{ or } 5.46\%$$

(b) $$z_1 = \frac{955 - 1000}{80} = -0.5625 \quad \text{and} \quad z_2 = \frac{975 - 1000}{80} = -0.3125$$

Looking up $z_1 = 0.56$ in the table, we get 0.2123. For $z_2 = 0.31$, we get 0.1217. Thus $P(955 < X < 975) = 0.2123 - 0.1217 = 0.0906$, or 9.06%.

(c) $P(X < 955) = 0.5 - 0.2123 = 0.2877$, or 28.77%.

(d) $P(X > 975) = 0.1217 + 0.5 = 0.6217$ or 62.17%.

(e)   See Fig. 2

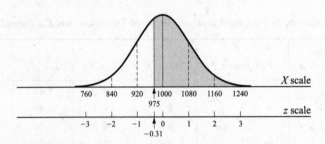

Fig. 2

3.   (a)   Since the population is normally distributed and σ is known, the normal distribution can be used:

$$\mu = \overline{X} \pm z\sigma_{\overline{X}} = \overline{X} \pm z\frac{\sigma}{\sqrt{n}} = 110 \pm 1.96\frac{10}{\sqrt{25}} = 110 \pm 3.92$$

Thus μ is between 106.08 and 113.92 with 95% confidence.

(b)   Since the distribution is normal, $n < 30$, and σ is not known, the t rather than the normal distribution must be used, with s as an estimate of σ:

$$\mu = \overline{X} \pm t_{0.025}\frac{s}{\sqrt{n}} \qquad t_{0.025} \text{ with } 25\,\text{df} = 2.064$$

$$= 110 \pm 2.064\frac{8}{\sqrt{25}}$$

$$= 110 \pm 3.30$$

Thus μ is between 106.70 and 113.30 with 95% confidence.

(c)   The *normal distribution* can be used   (1) if the parent population is normal, $n \geq 30$, and σ or s are known; (2) if $n \geq 30$ (by invoking the central-limit theorem) and using s as an estimate for σ; or   (3) if $n < 30$ but σ is given *and* the population from which the random sample is taken is known to be normally distributed.   The *t distribution* can be used (for the given degrees of freedom) when $n < 30$ but σ is not given *and* the population from which the sample is taken is known to be normally distributed.   If $n < 30$ but either σ is not given or the population from which the random sample is taken is not known to be normally distributed, we should use Chebyshev's inequality or increase the size of the random sample to $n \geq 30$ (to enable us to use the normal distribution).   In reality, however, the t distribution is used even when $n < 30$ and σ is not known, as long as the population is normally distributed.

4.   (a)   Since the firm is interested in testing if $\mu > 1000$ g in plant 1, we have a right-tail test:

$$H_0: \quad \mu_1 = 1000 \qquad \text{and} \qquad H_1: \quad \mu_1 > 1000$$

Since the population distribution is normal, but $n < 30$ and σ is not known, we must use the t distribution with $n - 1 = 24$ degrees of freedom:

$$t = \frac{\overline{X}_1 - \mu_1}{s_1/\sqrt{n_1}} = \frac{1064 - 1000}{100/\sqrt{25}} = 3.2$$

The calculated value of t exceeds the tabular value of $t_{0.05} = 1.71$ with 24 degrees of freedom. Thus H_0 is rejected and H_1 is accepted, so that the firm can claim at the 95% level of confidence that the boxes of detergent from plant 1 contain more than 1000 g of detergent.

(b)

$$H_0: \ \mu_1 = \mu_2 \quad \text{or} \quad H_0: \ \mu_1 - \mu_0 = 0$$
$$H_1: \ \mu_1 \neq \mu_2 \quad \text{or} \quad H_1: \ \mu_1 - \mu_0 \neq 0$$

$$\sigma_{\overline{X}_1 - \overline{X}_2} \cong \sqrt{\frac{s_1^2}{n_1} + \frac{s_2^2}{n_2}} = \sqrt{\frac{100^2}{25} + \frac{60^2}{25}} = \sqrt{544} \cong 23.32$$

$$t = \frac{(\overline{X}_1 - \overline{X}_2) - (\mu_1 - \mu_2)}{\sigma_{\overline{X}_1 - \overline{X}_2}} = \frac{\overline{X}_1 - \overline{X}_2 - 0}{\sigma_{\overline{X}_1 - \overline{X}_2}} = \frac{1064 - 1024}{23.32} \cong 1.72$$

This is a two-tail test with $n_1 + n_2 - 1 = 49$ degrees of freedom. Since the tabular value of $t_{0.025} > 2.00$ with 49 df, the firm can accept at the 95% level of confidence the hypothesis that there is no difference in the amount of detergent in the boxes from both plants.

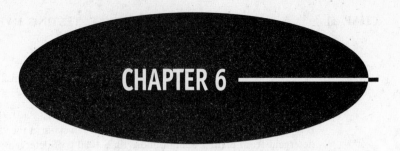

CHAPTER 6

Simple Regression Analysis

6.1 THE TWO-VARIABLE LINEAR MODEL

The two-variable linear model, or *simple regression analysis*, is used for testing hypotheses about the relationship between a dependent variable Y and an independent or explanatory variable X and for prediction. Simple *linear* regression analysis usually begins by plotting the set of XY values on a *scatter diagram* and determining by inspection if there exists an approximate linear relationship:

$$Y_i = b_0 + b_1 X_i \qquad (6.1)$$

Since the points are unlikely to fall precisely on the line, the exact linear relationship in Eq. (*6.1*) must be modified to include a *random disturbance*, *error*, or *stochastic term*, u_i (see Sec. 1.2 and Prob. 1.8):

$$Y_i = b_0 + b_1 X_i + u_i \qquad (6.2)$$

The error term is assumed to be (1) normally distributed, with (2) zero expected value or mean, and (3) constant variance, and it is further assumed (4) that the error terms are uncorrelated or unrelated to each other, and (5) that the explanatory variable assumes fixed values in repeated sampling (so that X_i and u_i are also uncorrelated).

EXAMPLE 1. Table 6.1 gives the bushels of corn per acre, Y, resulting from the use of various amounts of fertilizer in pounds per acre, X, produced on a farm in each of 10 years from 1971 to 1980. These are plotted in the scatter diagram of Fig. 6-1. The relationship between X and Y in Fig. 6-1 is approximately linear (i.e., the points would fall on or near a straight line).

6.2 THE ORDINARY LEAST-SQUARES METHOD

The *ordinary least-squares method* (OLS) is a technique for fitting the "best" straight line to the sample of XY observations. It involves minimizing the sum of the squared (vertical) deviations of points from the line:

$$\text{Min} \sum (Y_i - \hat{Y}_i)^2 \qquad (6.3)$$

where Y_i refers to the actual observations, and $\hat{Y}_i$ refers to the corresponding *fitted* values, so that $Y_i - \hat{Y}_i = e_i$, the *residual*. This gives the following two *normal equations* (see Prob. 6.6):

Table 6.1 Corn Produced with Fertilizer Used

Year	n	Y_i	X_i
1971	1	40	6
1972	2	44	10
1973	3	46	12
1974	4	48	14
1975	5	52	16
1976	6	58	18
1977	7	60	22
1978	8	68	24
1979	9	74	26
1980	10	80	32

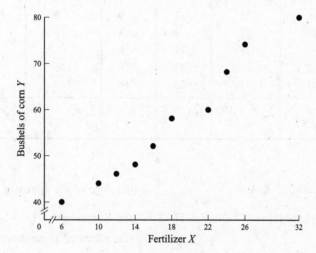

Fig. 6-1

$$\sum Y_i = nb_0 + \hat{b}_1 \sum X_i \tag{6.4}$$

$$\sum X_i Y_i = \hat{b}_0 \sum X_i + \hat{b}_1 \sum X_i^2 \tag{6.5}$$

where n is the number of observations and $\hat{b}_0$ and $\hat{b}_1$ are estimators of the true parameters b_0 and b_1. Solving simultaneously Eqs. (6.4) and (6.5), we get [see Prob. 6.7(a)]

$$\hat{b}_1 = \frac{n \sum X_i Y_i - \sum X_i \sum Y_i}{n \sum X_i^2 - \left(\sum X_i\right)^2} \tag{6.6}$$

The value of $\hat{b}_0$ is then given by [see Prob. 6.7(b)]

$$\hat{b}_0 = \overline{Y} - \hat{b}_1 \overline{X} \tag{6.7}$$

It is often useful to use an equivalent formula for estimating $\hat{b}_1$ [see Prob. 6.10(a)]:

$$\hat{b}_1 = \frac{\sum x_i y_i}{\sum x_i^2} = \frac{\text{cov}(X, Y)}{\sigma_X^2} \tag{6.8}$$

where $x_i = X_i - \overline{X}$, and $y_i = Y_i - \overline{Y}$. The estimated least-squares regression (OLS) equation is then

$$\hat{Y}_i = \hat{b}_0 + \hat{b}_1 X_i \qquad\qquad (6.9)$$

EXAMPLE 2.   Table 6.2 shows the calculations to estimate the regression equation for the corn-fertilizer problem in Table 6.1.   Using Eq. (6.8),

Table 6.2   Corn Produced with Fertilizer Used: Calculations

n	Y_i (Corn)	X_i (Fertilizer)	y_i	x_i	$x_i y_i$	x_i^2
1	40	6	-17	-12	204	144
2	44	10	-13	-8	104	64
3	46	12	-11	-6	66	36
4	48	14	-9	-4	36	16
5	52	16	-5	-2	10	4
6	58	18	1	0	0	0
7	60	22	3	4	12	16
8	68	24	11	6	66	36
9	74	26	17	8	136	64
10	80	32	23	14	322	196
$n = 10$	$\sum Y_i = 570$ $\overline{Y} = 57$	$\sum X_i = 180$ $\overline{X} = 18$	$\sum y_i = 0$	$\sum x_i = 0$	$\sum x_i y_i = 956$	$\sum x_i^2 = 576$

$$\hat{b}_i = \frac{\sum x_i y_i}{\sum x_i^2} = \frac{956}{576} = 1.66 \qquad\qquad \text{(the slope of the estimated regression line)}$$

$$\hat{b}_0 = \overline{Y} - \hat{b}_1 \overline{X} \cong 57 - (1.66)(18) \cong 57 - 29.88 \cong 27.12 \qquad \text{(the } Y \text{ intercept)}$$

$$\hat{Y}_i = 27.12 + 1.66 X_i \qquad\qquad \text{(the estimated regression equation)}$$

Thus, when $X_i = 0$, $\hat{Y} = 27.12 = \hat{b}_0$.   When $X_i = 18 = \overline{X}$, $\hat{Y} = 27.12 + 1.66(18) = 57 = \overline{Y}$.   As a result, the regression line passes through point $\overline{XY}$ (see Fig. 6-2).

6.3   TESTS OF SIGNIFICANCE OF PARAMETER ESTIMATES

In order to test for the statistical significance of the parameter estimates of the regression, the variance of $\hat{b}_0$ and $\hat{b}_1$ is required (see Probs. 6.14 and 6.15):

$$\text{Var } \hat{b}_0 = \sigma_u^2 \frac{\sum X_i^2}{n \sum x_i^2} \qquad\qquad (6.10)$$

$$\text{Var } \hat{b}_1 = \sigma_u^2 \frac{1}{\sum x_i^2} \qquad\qquad (6.11)$$

Since σ_u^2 is unknown, the *residual variance* s^2 is used as an (unbiased) estimate of σ_u^2:

$$s^2 = \hat{\sigma}_u^2 = \frac{\sum e_i^2}{n - k} \qquad\qquad (6.12)$$

where k represents the number of parameter estimates.

Unbiased estimates of the variance of $\hat{b}_0$ and $\hat{b}_1$ are then given by

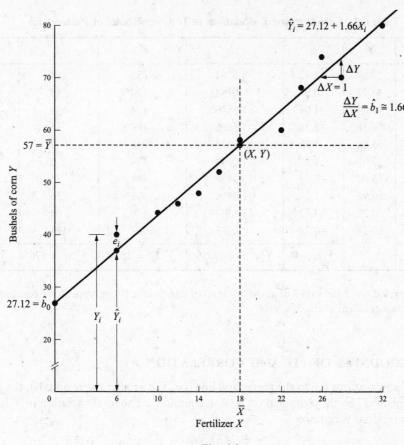

Fig. 6-2

$$s_{\hat{b}_0}^2 = \frac{\sum e_i^2}{n-k} \frac{\sum X_i^2}{n \sum x_i^2} \tag{6.13}$$

$$s_{\hat{b}_1}^2 = \frac{\sum e_i^2}{n-k} \frac{1}{\sum x_i^2} \tag{6.14}$$

so that $s_{\hat{b}_0}$ and $s_{\hat{b}_1}$ are the *standard errors of the estimates*. Since u_i is normally distributed, Y_i and therefore $\hat{b}_0$ and $\hat{b}_1$ are also normally distributed, so that we can use the t distribution with $n-k$ degrees of freedom, to test hypotheses about and construct confidence intervals for $\hat{b}_0$ and $\hat{b}_1$ (see Secs. 4.4 and 5.2).

EXAMPLE 3. Table 6.3 (an extension of Table 6.2) shows the calculations required to test the statistical significance of $\hat{b}_0$ and $\hat{b}_1$. The values of $\hat{Y}_i$ in Table 6.3 are obtained by substituting the values of X_i into the estimated regression equation found in Example 2. (The values of y_i^2 are obtained by squaring y_i from Table 6.2 and are to be used in Sec. 6.4.)

$$s_{\hat{b}_0}^2 = \frac{\sum e_i^2}{n-k} \frac{\sum X_i^2}{n \sum x_i^2} \cong \frac{47.3056}{10-2} \frac{3816}{10(576)} \cong 3.92 \quad \text{and} \quad s_{\hat{b}_0} = \sqrt{3.92} \cong 1.98$$

$$s_{\hat{b}_1}^2 = \frac{\sum e_i^2}{(n-k)\sum x_i^2} \cong \frac{47.3056}{(10-2)576} \cong 0.01 \quad \text{and} \quad s_{\hat{b}_1} \cong \sqrt{0.01} \cong 0.1$$

Therefore $\qquad t_0 = \dfrac{\hat{b}_0 - b_0}{s_{\hat{b}_0}} \cong \dfrac{27.12 - 0}{1.98} \cong 13.7 \quad \text{and} \quad t_1 = \dfrac{\hat{b}_1 - b_1}{s_{\hat{b}_1}} \cong \dfrac{1.66}{0.1} \cong 16.6$

Table 6.3   Corn-Fertilizer Calculations to Test Significance of Parameters

Year	Y_i	X_i	$\hat{Y}_i$	e_i	e_i^2	X_i^2	x_i^2	y_i^2
1	40	6	37.08	2.92	8.5264	36	144	289
2	44	10	43.72	0.28	0.0784	100	64	169
3	46	12	47.04	−1.04	1.0816	144	36	121
4	48	14	50.36	−2.36	5.5696	196	16	81
5	52	16	53.68	−1.68	2.8224	256	4	25
6	58	18	57.00	1.00	1.0000	324	0	1
7	60	22	63.64	−3.64	13.2496	484	16	9
8	68	24	66.96	1.04	1.0816	576	36	121
9	74	26	70.28	3.72	13.8384	676	64	289
10	80	32	80.24	−0.24	0.0576	1024	196	529
$n = 10$				$\sum e_i = 0$	$\sum e_i^2 = 47.3056$	$\sum X_i^2 = 3816$	$\sum x_i^2 = 576$	$\sum y_i^2 = 1634$

Since both t_0 and t_1 exceed $t = 2.306$ with 8 df at the 5% level of significance (from App. 5), we conclude that both b_0 and b_1 are statistically significant at the 5% level.

6.4   TEST OF GOODNESS OF FIT AND CORRELATION

The closer the observations fall to the regression line (i.e., the smaller the residuals), the greater is the variation in Y "explained" by the estimated regression equation.   The total variation in Y is equal to the explained plus the residual variation:

$$\underset{\substack{\text{Total variation in} \\ Y \text{ [or total sum of} \\ \text{squares} \\ \text{(TSS)]}}}{\sum(Y_i - \overline{Y})^2} = \underset{\substack{\text{Explained variation} \\ \text{in } Y \text{ [or regression} \\ \text{sum of squares} \\ \text{(RSS)]}}}{\sum(\hat{Y}_i - \overline{Y})^2} + \underset{\substack{\text{Residual variation} \\ \text{in } Y \text{ [or error sum} \\ \text{of squares} \\ \text{(ESS)]}}}{\sum(Y_i - \hat{Y}_i)^2} \qquad (6.15)$$

Dividing both sides by TSS gives

$$1 = \frac{\text{RSS}}{\text{TSS}} + \frac{\text{ESS}}{\text{TSS}}$$

The *coefficient of determination*, or R^2, is then defined as the proportion of the total variation in Y "explained" by the regression of Y on X:

$$R^2 = \frac{\text{RSS}}{\text{TSS}} = 1 - \frac{\text{ESS}}{\text{TSS}} \qquad (6.16)$$

R^2 can be calculated by

$$R^2 = \frac{\sum \hat{y}^2}{\sum y_i^2} = 1 - \frac{\sum e_i^2}{\sum y_i^2} \qquad (6.17)$$

where

$$\sum \hat{y}_i^2 = \sum(\hat{Y}_i - \overline{Y}_i)^2$$

R^2 ranges in value from 0 (when the estimated regression equation explains none of the variation in Y) to 1 (when all points lie on the regression line).

The *correlation coefficient r* is given by (see Prob. 6.22)

$$r = \sqrt{R^2} = \frac{\text{cov}(X, Y)}{\sigma_X \sigma_Y} = \sqrt{\hat{b}_1 \frac{\sum x_i y_i}{\sum y_i^2}} \qquad (6.18)$$

r ranges in value from -1 (for perfect negative linear correlation) to $+1$ (for perfect positive linear correlation) and does not imply causality or dependence. With qualitative data, the *rank* or (the *Spearman*) *correlation coefficient* r' (see Prob. 6.25) can be used.

EXAMPLE 4. The coefficient of determination for the corn-fertilizer example can be found from Table 6.3:

$$R^2 = 1 - \frac{\sum e_i^2}{\sum y_i^2} \cong 1 - \frac{47.31}{1634} \cong 1 - 0.0290 \cong 0.9710, \text{ or } 97.10\%$$

Thus the regression equation explains about 97% of the total variation in corn output. The remaining 3% is attributed to factors included in the error term. Then $r = \sqrt{R^2} \cong \sqrt{0.9710} \cong 0.9854$, or 98.54%, and is positive because $\hat{b}_1$ is positive. Figure 6-3 shows the total, the explained, and the residual variation of Y.

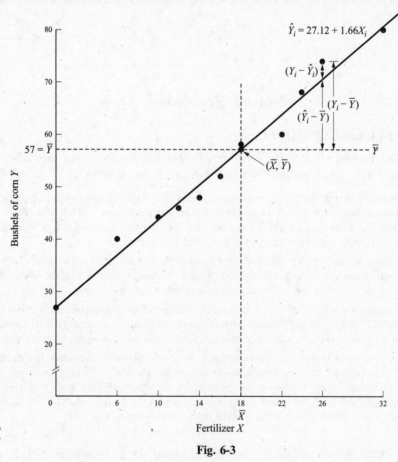

Fig. 6-3

6.5 PROPERTIES OF ORDINARY LEAST-SQUARES ESTIMATORS

Ordinary least-squares (OLS) estimators are *best linear unbiased estimators* (BLUE). Lack of bias means

$$E(\hat{b}) = b$$

so that

$$\text{Bias} = E(\hat{b}) - b$$

Best unbiased or efficient means smallest variance. Thus OLS estimators are the best among all unbiased linear estimators [see Probs. 6.14(*a*) and 6.15(*b*)]. This is known as the *Gauss-Markov theorem* and represents the most important justification for using OLS.

Sometimes, a researcher may want to trade off some bias for a possibly smaller variance and minimize the mean square error, MSE (see Prob. 6.29):

$$\text{MSE}(\hat{b}) = E(\hat{b} - b)^2 = \text{var}(\hat{b}) + (\text{bias } \hat{b})^2$$

An estimator is *consistent* if, as the sample size approaches infinity in the limit, its value approaches the true parameter (i.e., it is asymptotically unbiased) and its distribution collapses on the true parameter (see Prob. 6.30).

EXAMPLE 5.   OLS estimators $\hat{b}_0$ and $\hat{b}_1$ found in Example 2 are unbiased linear estimators of b_0 and b_1 because

$$E(\hat{b}_0) = b_0 \qquad \text{and} \qquad E(\hat{b}_1) = b_1$$

Var $\hat{b}_0$ and var $\hat{b}_1$ found in Example 3 are also lower than for any other linear unbiased estimators.   Therefore $\hat{b}_0$ and $\hat{b}_1$ are BLUE.

Solved Problems

THE TWO-VARIABLE LINEAR MODEL

6.1   What is meant by and what is the function of   (*a*) Simple regression analysis?   (*b*) Linear regression analysis?   (*c*) A scatter diagram?   (*d*) An error term?

   (*a*)   *Simple regression* is used for testing hypotheses about the relationship between a dependent variable Y and an independent or explanatory variable X and for prediction.   This is to be contrasted with *multiple regression* analysis, in which there are not one, but two or more independent or explanatory variables.   Multiple regression analysis is discussed in Chap. 7.

   (*b*)   *Linear regression analysis* assumes that there is an approximate linear relationship between X and Y (i.e., the set of random sample values of X and Y fall on or near a straight line).   This is to be contrasted with *nonlinear regression analysis* (discussed in Sec. 8.1).

   (*c*)   A *scatter diagram* is a figure in which each pair of independent-dependent observations is plotted as a point in the XY plane.   Its purpose is to determine (by inspection) if there exists an approximate linear relationship between the dependent variable Y and the independent or explanatory variable X.

   (*d*)   The *error term* (also known as the *disturbance* or *stochastic term*) measures the deviation of each observed Y value from the true (but unobserved) regression line.   These error terms, designated by u_i and e_i, arise because of   (1) numerous explanatory variables with only slight and irregular effects on Y that are omitted from the exact linear relationship given by Eq. (*6.1*),   (2) possible errors of measurement in Y, and   (3) random human behavior (see Prob. 1.8).

6.2   The data in Table 6.4 reports the aggregate consumption (Y, in billions of U.S. dollars) and disposable income (X, also in billions of U.S. dollars) for a developing economy for the 12 years from 1988 to 1999.   Draw a scatter diagram for the data and determine by inspection if there exists an approximate linear relationship between Y and X.

   From Fig. 6-4 it can be seen that the relationship between consumption expenditures Y and disposable income X is approximately linear, as required by the linear regression model.

6.3   State the general relationship between consumption Y and disposable income X in   (*a*) exact linear form and   (*b*) stochastic form.   (*c*) Why would you expect most observed values of Y not to fall exactly on a straight line?

   (*a*)   The exact or deterministic general relationship between aggregate consumption expenditures Y and aggregate disposable income X can be written as

Table 6.4   Aggregate Consumption (Y) and Disposable Income (X)

Year	n	Y_i	X_i
1988	1	102	114
1989	2	106	118
1990	3	108	126
1991	4	110	130
1992	5	122	136
1993	6	124	140
1994	7	128	148
1995	8	130	156
1996	9	142	160
1997	10	148	164
1998	11	150	170
1999	12	154	178

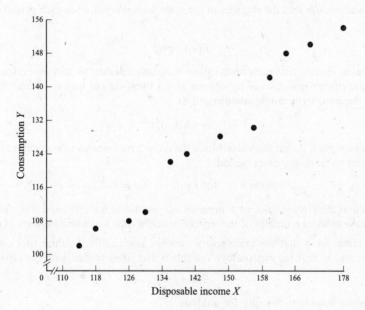

Fig. 6-4

$$Y_i = b_0 + b_1 X_i \tag{6.1}$$

where i refers to each year in time-series analysis (as with the data in Table 6.4) or to each economic unit (such as a family) in cross-sectional analysis.   In Eq. (6.1), b_0 and b_1 are unknown constants called *parameters*.   Parameter b_0 is the constant or Y intercept, while b_1 measures $\Delta Y / \Delta X$, which, in the context of Prob. 6.2, refers to the marginal propensity to consume (MPC) (see Sec. 1.2).   The *specific* linear relationship corresponding to the general linear relationship in Eq. (6.1) is obtained by estimating the values of b_0 and b_1 (represented by $\hat{b}_0$ and $\hat{b}_1$ and read as "b sub zero hat" and "b sub one hat").

(b)   The exact linear relationship in Eq. (6.1) can be made stochastic by adding a random disturbance or error term, u_i, giving

$$Y_i = b_0 + b_1 X_i + u_i \tag{6.2}$$

(c)   Most observed values of Y are not expected to fall precisely on a straight line   (1)   because even though consumption Y is postulated to depend primarily on disposable income X, it also may depend on numerous other omitted variables with only slight and irregular effect on Y (if some of these other variables had instead a significant and regular effect on Y, then they should be included as additional explanatory variables, as in a multiple regression model); (2) because of possible errors in measuring Y; and (3) because of inherent random human behavior, which usually leads to different values of Y for the same value of X under identical circumstances (see Prob. 1.8).

6.4 State each of the five assumptions of the classical regression model (OLS) and give an intuitive explanation of the meaning and need for each of them.

1. The first assumption of the classical linear regression model (OLS) is that the random error term u is normally distributed. As a result, Y and the sampling distribution of the parameters of the regression are also normally distributed, so that tests can be conducted on the significance of the parameters (see Secs. 4.2, 5.2, and 6.3).

2. The second assumption is that the expected value of the error term or its mean equals zero:

$$E(u_i) = 0 \qquad (6.19)$$

Because of this assumption, Eq. (6.1) gives the *average* value of Y. Specifically, since X is assumed fixed, the value of Y in Eq. (6.2) varies above and below its mean as u exceeds or is smaller than 0. Since the average value of u is assumed to be 0, Eq. (6.1) gives the average value of Y.

3. The third assumption is that the variance of the error term is constant in each period and for all values of X:

$$E(u_i)^2 = \sigma_u^2 \qquad (6.20)$$

This assumption ensures that each observation is equally reliable, so that estimates of the regression coefficients are efficient and tests of hypotheses about them are not biased. These first three assumptions about the error term can be summarized as

$$u \sim N(0, \sigma_u^2) \qquad (6.21)$$

4. The fourth assumption is that the value which the error term assumes in one period is uncorrelated or unrelated to its value in any other period:

$$E(u_i u_j) = 0 \qquad \text{for } i \neq j; \qquad i, j = 1, 2, \ldots, n \qquad (6.22)$$

This ensures that the average value of Y depends only on X and not on u, and it is, once again, required in order to have efficient estimates of the regression coefficients and unbiased tests of their significance.

5. The fifth assumption is that the explanatory variable assumes fixed values that can be obtained in repeated samples, so that the explanatory variable is also uncorrelated with the error term:

$$E(X_i u_i) = 0 \qquad (6.23)$$

This assumption is made to simplify the analysis.

THE ORDINARY LEAST-SQUARES METHOD

6.5 (a) What is meant by the *ordinary least-squares* (OLS) *method* of estimating the "best" straight line that fits the sample of XY observations? (b) Why do we take vertical deviations? (c) Why do we not simply take the sum of the deviations *without squaring them*? (d) Why do we not take the sum of the absolute deviations?

(a) The OLS method gives the best straight line that fits the sample of XY observations in the sense that it minimizes the sum of the squared (vertical) deviations of each observed point on the graph from the straight line.

(b) We take vertical deviations because we are trying to explain or predict movements in Y, which is measured along the vertical axis.

(c) We cannot take the sum of the deviations of each of the observed points from the OLS line because deviations that are equal in size but opposite in sign cancel out, so the sum of the deviations equals 0 (see Table 6.2).

(d) Taking the sum of the *absolute* deviations avoids the problem of having the sum of the deviations equal to 0. However, the sum of the *squared* deviations is preferred so as to penalize larger deviations relatively more than smaller deviations.

6.6 Starting from Eq. (6.3) calling for the minimization of the sum of the squared deviations or residuals, derive (a) normal Eq. (6.4) and (b) normal Eq. (6.5). (The reader without knowledge of calculus can skip this problem.)

(a)
$$\sum e_i^2 = \sum (Y_i - \hat{Y}_i)^2 = \sum (Y_i - \hat{b}_0 - \hat{b}_1 X_i)^2$$

Normal Eq. (6.4) is derived by minimizing $\sum e_i^2$ with respect to $\hat{b}_0$:

$$\frac{\partial \sum e_i^2}{\partial \hat{b}_0} = \frac{\partial \sum (Y_i - \hat{b}_0 - \hat{b}_1 X_i)^2}{\partial \hat{b}_0} = 0$$

$$2 \sum (Y_i - \hat{b}_0 - \hat{b}_1 X_i)(-1) = 0$$

$$\sum (Y_i - \hat{b}_0 - \hat{b}_1 X_i) = 0$$

$$\sum Y_i = n\hat{b}_0 + \hat{b}_1 \sum X_i \qquad (6.4)$$

(b) Normal Eq. (6.5) is derived by minimizing $\sum e_i^2$ with respect to $\hat{b}_1$:

$$\frac{\partial \sum e_i^2}{\partial \hat{b}_1} = \frac{\partial \sum (Y_i - \hat{b}_0 - \hat{b}_1 X_i)^2}{\partial \hat{b}_1} = 0$$

$$2 \sum (Y_i - \hat{b}_0 - \hat{b}_1 X_i)(-X_i) = 0$$

$$\sum (Y_i X_i - \hat{b}_0 X_i - \hat{b}_1 X_i^2) = 0$$

$$\sum Y_i X_i = \hat{b}_0 \sum X_i + \hat{b}_1 \sum X_i^2 \qquad (6.5)$$

6.7 Derive (a) Eq. (6.6) to find $\hat{b}_1$ and (b) Eq. (6.7) to find $\hat{b}_0$. [*Hint for part a*: Start by multiplying Eq. (6.5) by n and Eq. (6.4) by $\sum X_i$.]

(a) Multiplying Eq. (6.5) by n and Eq. (6.4) by $\sum X_i$, we get

$$n \sum X_i Y_i = \hat{b}_0 n \sum X_i + \hat{b}_1 n \sum X_i^2 \qquad (6.24)$$

$$\sum X_i \sum Y_i = \hat{b}_0 n \sum X_i + \hat{b}_1 \left(\sum X_i \right)^2 \qquad (6.25)$$

Subtracting Eq. (6.25) from Eq. (6.24), we get

$$n \sum X_i Y_i - \sum X_i \sum Y_i = \hat{b}_1 \left[n \sum X_i^2 - \left(\sum X_i \right)^2 \right] \qquad (6.26)$$

Solving Eq. (6.26) for $\hat{b}_1$, we get

$$\hat{b}_1 = \frac{n \sum X_i Y_i - \sum X_i \sum Y_i}{n \sum X_i^2 - \left(\sum X_i \right)^2} \qquad (6.6)$$

(b) Equation (6.7) is obtained by simply solving Eq. (6.4) for $\hat{b}_0$:

$$\sum Y_i = n\hat{b}_0 + \hat{b}_1 \sum X_i \qquad (6.4)$$

$$\hat{b}_0 = \frac{\sum Y_i}{n} - \hat{b}_1 \frac{\sum X_i}{n}$$

$$= \overline{Y} - \hat{b}_1 \overline{X} \qquad (6.7)$$

6.8 (a) State the difference between b_0 and b_1, on one hand, and $\hat{b}_0$ and $\hat{b}_1$ on the other hand. (b) State the difference between u_i and e_i. (c) Write the equations for the true and estimated *relationships* between X and Y. (*d*) Write the equations for the true and estimated *regression* lines between X and Y.

(*a*) b_0 and b_1 are the parameters of the true but unknown regression line, while $\hat{b}_0$ and $\hat{b}_1$ are the parameters of the estimated regression line.

(*b*) u_i is the random disturbance, error, or stochastic term in the true but unknown relationship between X and Y. However, e_i is the residual between each observed value of Y and its corresponding fitted value $\hat{Y}$ in the estimated relationship.

(*c*) The equations for the true and estimated relationships between X and Y are, respectively,

$$Y_i = b_0 + b_1 X_i + u_i \qquad\qquad (6.2)$$

$$Y_i = \hat{b}_0 + \hat{b}_1 X_i + e_i \qquad\qquad (6.27)$$

(*d*) The equations for the true and estimated regressions between X and Y are, respectively,

$$E(Y_i) = b_0 + b_1 X_i \qquad\qquad (6.28)$$

$$\hat{Y}_i = \hat{b}_0 + \hat{b}_1 X_i \qquad\qquad (6.9)$$

6.9 (*a*) Find the regression equation for the consumption schedule in Table 6.4, using Eq. (*6.6*) to find $\hat{b}_1$. (*b*) Plot the regression line and show the deviations of each Y_i from the corresponding $\hat{Y}_i$.

(*a*) Table 6.5 shows the calculations to find $\hat{b}_1$ and $\hat{b}_0$ for the data in Table 6.4.

$$\hat{b}_1 = \frac{n\sum X_i Y_i - \sum X_i \sum Y_i}{n\sum X_i^2 - \left(\sum X_i\right)^2} = \frac{(12)(225{,}124) - (1740)(1524)}{(12)(257{,}112) - (1740)^2} = \frac{2{,}701{,}488 - 2{,}651{,}760}{3{,}085{,}344 - 3{,}027{,}600}$$

$$= \frac{49{,}728}{57{,}744} \cong 0.86$$

$$\hat{b}_0 = \overline{Y} - \hat{b}_1 \overline{X} \cong 127 - 0.86(145) \cong 127 - 124.70 \cong 2.30$$

Thus the equation for the estimated consumption regression is $\hat{Y}_i = 2.30 + 0.86\hat{X}_i$.

Table 6.5 Aggregate Consumption and Disposable Income: Calculations

n	Y_i	X_i	$X_i Y_i$	X_i^2
1	102	114	11,628	12,996
2	106	118	12,508	13,924
3	108	126	13,608	15,876
4	110	130	14,300	16,900
5	122	136	16,592	18,496
6	124	140	17,360	19,600
7	128	148	18,944	21,904
8	130	156	20,280	24,336
9	142	160	22,720	25,600
10	148	164	24,272	26,896
11	150	170	25,500	28,900
12	154	178	27,412	31,684
$n = 12$	$\sum Y_i = 1524$ $\overline{Y} = 127$	$\sum X_i = 1740$ $\overline{X} = 145$	$\sum X_i Y_i = 225{,}124$	$\sum X_i^2 = 257{,}112$

(*b*) To plot the regression equation, we need to define any two points on the regression line. For example, when $X_i = 114$, $Y_i = 2.30 + 0.86(114) = 100.34$. When $X_i = 178$, $Y_i = 2.30 + 0.86(178) = 155.38$.

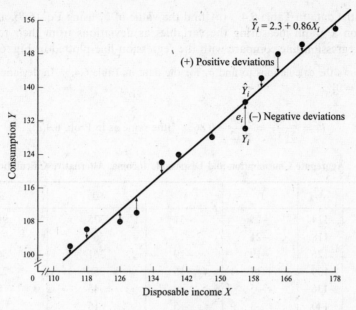

Fig. 6-5

The consumption regression line is plotted in Fig. 6-5, where the positive and negative residuals are also shown.   The regression line represents the best fit to the random sample of consumption–disposable income observations in the sense that it minimizes the sum of the squared (vertical) deviations from the line.

6.10 (a) Starting with Eq. (6.6), derive the equation for $\hat{b}_1$ in deviation form for the case where $\overline{X} = \overline{Y} = 0$.   (b) What is the value of $\hat{b}_0$ when $\overline{X} = \overline{Y} = 0$?

(a) Starting with Eq. (6.6) for $\hat{b}_1$

$$\hat{b}_1 = \frac{n\sum X_i Y_i - \sum X_i \sum Y_i}{n\sum X_i^2 - \left(\sum X_i\right)^2} \tag{6.6}$$

we divide numerator and denominator by n^2 and get

$$\hat{b} = \frac{\sum X_i Y_i/n - \left(\sum X_i/n\right)\left(\sum Y_i/n\right)}{\sum X_i^2/n - \left(\sum X_i/n\right)^2}$$

$$= \frac{\sum X_i Y_i/n - \overline{XY}}{\sum X_i^2/n - \overline{X}^2}$$

$$= \frac{\sum X_i Y_i}{\sum X_i^2} \qquad \text{since } \overline{X} = \overline{Y} = 0 \text{ and canceling the } n \text{ terms}$$

$$= \frac{\sum x_i y_i}{\sum x_i^2} \qquad \text{since } \overline{X} = \overline{Y} = 0 \tag{6.8}$$

(b) Starting with Eq. (6.7) for $\hat{b}_0$, we obtain

$$\hat{b}_0 = \overline{Y} - \hat{b}_1 \overline{X} \tag{6.7}$$

and substituting 0 for $\overline{X}$ and $\overline{Y}$, we get

$$\hat{b}_0 = 0 - \hat{b}_1(0) = 0$$

6.11 With respect to the data in Table 6.4, (a) find the value of $\hat{b}_1$ using Eq. (6.8), and (b) plot the regression line on a graph measuring the variables as deviations from their respective means. How does this regression line compare with the regression line plotted in Fig. 6-5?

(a) Table 6.6 shows the calculations to find $\hat{b}_1$ for the data in Table 6.4. In deviation form (note that $\sum y_i = \sum x_i = 0$):

$$\hat{b}_1 = \frac{\sum x_i y_i}{\sum x_i^2} = \frac{4144}{4812} \cong 0.86 \qquad \text{[the same as in Prob. 6.9(a)]}$$

Table 6.6 Aggregate Consumption and Disposable Income: Alternative Calculations

n	Y_i	X_i	y_i	x_i	$x_i y_i$	x_i^2
1	102	114	−25	−31	775	961
2	106	118	−21	−27	567	729
3	108	126	−19	−19	361	361
4	110	130	−17	−15	255	225
5	122	136	−5	−9	45	81
6	124	140	−3	−5	15	25
7	128	148	1	3	3	9
8	130	156	3	11	33	121
9	142	160	15	15	225	225
10	148	164	21	19	399	361
11	150	170	23	25	575	625
12	154	178	27	33	891	1089
			$\sum y_i = 0$	$\sum x_i = 0$	$\sum x_i y_i = 4144$	$\sum x_i^2 = 4812$

(b) From Prob. 6.10(b) we know that the regression line crosses the origin when plotted on a graph with the axis measuring the variables in deviation form, and from part a of this problem we know that this regression line has the same slope as the regression line in Fig. 6-5. See Fig. 6-6.

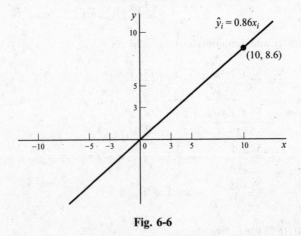

Fig. 6-6

6.12 In the context of Prob. 6.9(a), what is the meaning of: (a) Estimator $\hat{b}_0$? (b) Estimator $\hat{b}_1$? (c) Find the income elasticity of consumption.

(a) Estimator $\hat{b}_0 \cong 2.30$ is the Y intercept, or the value of aggregate consumption, in billions of dollars, when disposable income, also in billions of dollars, is 0. The fact that $\hat{b}_0 > 0$ confirms what was anticipated on theoretical grounds in Example 3 in Chap. 1.

(b) Estimator $\hat{b}_1 = dY/dX \cong 0.86$ is the slope of the estimated regression line.   It measures the marginal propensity to consume (MPC) or the change in consumption per one-unit change in disposable income. Once again, the fact that $0 < \hat{b}_1 < 1$ confirms what was anticipated on theoretical grounds in Example 3 in Chap. 1.

(c) The income elasticity of consumption η measures the percentage change in consumption resulting from a given percentage change in disposable income.   Since the elasticity usually changes at every point in the function, it is measured at the means:

$$\eta = \hat{b}_1 \frac{\overline{X}}{\overline{Y}} \tag{6.29}$$

For the data in Table 6.4

$$\eta = 0.86 \frac{145}{127} \cong 0.98$$

Note that elasticity, as opposed to the slope, is a pure (unit-free) number.

TESTS OF SIGNIFICANCE OF PARAMETER ESTIMATES

6.13 Define   (a) σ_u^2 and s^2,   (b) var $\hat{b}_0$ and var $\hat{b}_1$,   (c) $s_{\hat{b}_0}^2$ and $s_{\hat{b}_1}^2$, and   (d) $s_{\hat{b}_0}$ and $s_{\hat{b}_1}$.

(a) σ_u^2 is the variance of the error term in the true relationship between X_i and Y_i.   However, $s^2 = \sigma_u^2 = \sum e_i^2/(n-k)$ is the residual variance and is an (unbiased) estimate of σ_u^2, which is unknown. k is the number of estimated parameters.   In simple regression analysis, $k = 2$.   Thus $n - k = n - 2$ and refers to the degrees of freedom.

(b) Var $\hat{b}_0 = \sigma_u^2 \sum X_i/n \sum x_i^2$, while var $\hat{b}_1 = \sigma_u^2/\sum x_i^2$.   The variances of $\hat{b}_0$ and $\hat{b}_1$ (or their estimates) are required to test hypotheses about and construct confidence intervals for $\hat{b}_0$ and $\hat{b}_1$.

(c) $$s_{\hat{b}_0}^2 = s^2 \frac{\sum X_i^2}{n \sum x_i^2} = \frac{\sum e_i^2 \sum X_i^2}{(n-k)n \sum x_i^2} \quad \text{and} \quad s_{\hat{b}_1}^2 = \frac{s^2}{\sum x_i^2} = \frac{\sum e_i^2}{(n-k) \sum x_i^2}$$

$s_{\hat{b}_0}^2$ and $s_{\hat{b}_1}^2$ are, respectively, (unbiased) estimates of var $\hat{b}_0$ and var $\hat{b}_1$, which are unknown since σ_u^2 is unknown.

(d) $s_{\hat{b}_0} = \sqrt{s_{\hat{b}_0}^2}$ and $s_{\hat{b}_1} = \sqrt{s_{\hat{b}_1}^2}$. $s_{\hat{b}_0}$ and $s_{\hat{b}_1}$ are, respectively, the standard deviations of $\hat{b}_0$ and $\hat{b}_1$ and are called the *standard errors*.

6.14 Prove that   (a) mean $\hat{b}_1 = b_1$, and   (b) var $\hat{b}_1 = \sigma_u^2/\sum x_i^2$
                     (c) mean $\hat{b}_0 = b_0$, and   (d) var $\hat{b}_0 = \sigma_u^2(\sum X_i^2/n \sum x_i^2)$

(a) $$\hat{b}_1 = \frac{\sum x_i y_i}{\sum x_i^2} = \frac{\sum x_i(Y_i - \overline{Y})}{\sum x_i^2} = \frac{\sum x_i Y_i}{\sum x_i^2} - \frac{\sum x_i \overline{Y}}{\sum x_i^2} = \frac{\sum x_i Y_i}{\sum x_i^2} \quad \text{since } \sum x_i = 0$$
$$= \sum c_i Y_i$$

where $c_i = x_i/\sum x_i^2 =$ constant because of assumption 5 (Sec. 6.1)

$$\hat{b}_1 = \sum c_i Y_i = \sum c_i(b_0 + b_1 X_i + u_i) = b_0 \sum c_i + b_1 \sum c_i X_i + \sum c_i u_i$$
$$= b_1 + \sum c_i u_i = b_1 + \frac{\sum x_i u_i}{\sum x_i^2}$$

since $\sum c_i = \sum x_i/\sum x_i^2 = 0$ (because $\sum x_i = 0$) and

$$\sum c_i X_i = \frac{\sum x_i X_i}{\sum x_i^2} = \frac{\sum (X_i - \overline{X})X_i}{\sum (X_i - \overline{X})^2} = \frac{\sum X_i^2 - \overline{X} \sum X_i}{\sum X_i^2 - 2\overline{X} \sum X_i + n\overline{X}^2} = \frac{\sum X_i^2 - \overline{X} \sum X_i}{\sum X_i^2 - \overline{X} \sum X_i} = 1$$

$$E(\hat{b}_1) = E(b_1) + E\left[\frac{\sum x_i u_i}{\sum x_i^2}\right] = E(b_1) + \frac{1}{\sum x_i^2} E\left(\sum x_i u_i\right) = b_1$$

since b_1 is a constant and $E(\sum x_i u_i) = 0$ because of assumption 5 (Sec. 6.1).

(b) From part a, we obtain

$$b_1 = \frac{\sum x_i Y_i}{\sum x_i^2} = \sum c_i Y_i$$

$$\text{Var } \hat{b}_1 = \text{var}\left(\sum c_i Y_i\right) = \sum c_i^2 \text{ var } Y_i = \sum c_i^2 \sigma_u^2$$

since Y_i varies only because of u_i with X_i assumed fixed.

$$\text{Var } \hat{b}_1 = \sum c_i^2 \sigma_u^2 = \sum\left(\frac{x_i}{\sum x_i^2}\right)^2 \sigma_u^2 = \frac{\sum x_i^2}{\left(\sum x_i^2\right)^2}\sigma_u^2 = \frac{\sigma_u^2}{\sum x_i^2}$$

(c) $$\hat{b}_0 = \overline{Y} - \hat{b}_1 \overline{X} = \frac{\sum Y_i}{n} - \overline{X}\sum c_i Y_i \qquad\qquad \text{(from part } a)$$

$$\hat{b}_0 = \frac{\sum Y_i}{n} - \overline{X}\sum c_i Y_i = \sum\left(\frac{1}{n} - \overline{X}c_i\right)Y_i$$

$$E(b_0) = \sum\left(\frac{1}{n} - \overline{X}c_i\right)E(Y_i) = \sum\left(\frac{1}{n} - \overline{X}c_i\right)(b_0 + b_1 X_i) \qquad \text{[from Eq. (6.1) in Prob. 6.8(d)]}$$

Cross multiplying,

$$E(b_0) = \sum\left(\frac{b_0}{n} - \overline{X}c_i b_0 + \frac{b_1 X_i}{n} - \overline{X}c_i b_1 X_i\right) = b_0 + b_1\overline{X} - b_1\overline{X} = b_0$$

because $\sum c_i = 0$ and $\sum c_i X_i = 1$, from part a.

(d) We saw in part c that

$$\hat{b}_0 = \sum\left(\frac{1}{n} - \overline{X}c_i\right)Y_i$$

$$\text{Var } \hat{b}_0 = \text{var}\left[\sum\left(\frac{1}{n} - \overline{X}c_i\right)Y_i\right] = \sum\left(\frac{1}{n} - \overline{X}c_i\right)^2 \text{var } Y_i = \sigma_u^2\sum\left(\frac{1}{n} - \overline{X}c_i\right)^2$$

$$\text{Var } \hat{b}_0 = \sigma_u^2\sum\left(\frac{1}{n^2} - \frac{2\overline{X}c_i}{n} + \overline{X}^2 c_i^2\right) = \sigma_u^2\left(\frac{1}{n} + \frac{\overline{X}^2}{\sum x_i^2}\right) = \sigma_u^2\frac{\sum x_i^2 + n\overline{X}^2}{n\sum x_i^2}$$

since $\sum c_i = 0$ and $\sum c_i^2 = 1/\sum x_i^2$.

$$\text{Var } \hat{b}_0 = \sigma_u^2\frac{\sum x_i^2 + n\overline{X}^2}{n\sum x_i^2} = \sigma_u^2\frac{\sum X_i^2 - n\overline{X}^2 + n\overline{X}^2}{n\sum x_i^2} = \sigma_u^2\frac{\sum X_i^2}{n\sum x_i^2}$$

since in part a we saw that $\sum x_i^2 = \sum X_i^2 - n\overline{X}^2$.

6.15 For the aggregate consumption-income observations in Table 6.4, find (a) s^2, (b) $s_{\hat{b}_0}^2$ and $s_{\hat{b}_0}$, (c) $s_{\hat{b}_1}^2$ and $s_{\hat{b}_1}$.

(a) The calculations required to find s^2 are shown in Table 6.7, which is an extension of Table 6.6.   The values for $\hat{Y}_i$ in Table 6.7 are obtained by substituting the values for X_i into the regression equation found in Prob. 6.9(a).

$$s^2 = \hat{\sigma}_u^2 = \frac{\sum e_i^2}{n - k} = \frac{115.2752}{12 - 2} = 11.52752 \cong 11.53$$

(b) $$s_{\hat{b}_0}^2 = \frac{\sum e_i^2}{n - k}\frac{\sum X_i^2}{n\sum x_i^2} = \frac{s\sum X_i^2}{n\sum x_i^2} \cong \frac{(11.53)(257{,}112)}{(12)(4812)} \cong 51.34$$

Then $$s_{\hat{b}_0} = \sqrt{s_{\hat{b}_0}^2} \cong \sqrt{51.34} \cong 7.17$$

Table 6.7   Consumption Regression: Calculations to Test Significance of Parameters

Year	Y_i	X_i	$\hat{Y}_i$	e_i	e_i^2	X_i^2	x_i^2
1	102	114	100.34	1.66	2.7556	12,996	961
2	106	118	103.78	2.22	4.9284	13,924	729
3	108	126	110.66	−2.66	7.0756	15,876	361
4	110	130	114.10	−4.10	16.8100	16,900	225
5	122	136	119.26	2.74	7.5076	18,496	81
6	124	140	122.70	1.30	1.6900	19,600	25
7	128	148	129.58	−1.58	2.4964	21,904	9
8	130	156	136.46	−6.46	41.7316	24,336	121
9	142	160	139.90	2.10	4.4100	25,600	225
10	148	164	143.34	4.66	21.7156	26,896	361
11	150	170	148.50	1.50	2.2500	28,900	625
12	154	178	155.38	−1.38	1.9044	31,684	1089
				$\sum e_i = 0$	$\sum e_i^2 = 115.2752$	$\sum X_i^2 = 27,112$	$\sum x_i^2 = 4812$

(c)
$$s_{b_1}^2 = \frac{\sum e_i^2}{(n-k)\sum x_i^2} = \frac{s^2}{\sum x_i^2} \cong \frac{11.53}{4812} \cong 0.0024$$

Then
$$s_{\hat{b}_1} = \sqrt{s_{b_1}^2} \cong \sqrt{0.0024} \cong 0.05$$

6.16 (a) State the null and alternative hypotheses to test the statistical significance of the parameters of the regression equation estimated in Prob. 6.9(a).   (b) What is the form of the sampling distribution of $\hat{b}_0$ and $\hat{b}_1$?   (c) Which distribution must we use to test the statistical significance of b_0 and b_1?   (d) What are the degrees of freedom?

(a) To test for the statistical significance of b_0 and b_1, we set the following null hypothesis, H_0, and alternative hypothesis, H_1 (see Sec. 5.2):

$$H_0:\quad b_0 = 0 \qquad \text{versus} \qquad H_1:\quad b_0 \neq 0$$
$$H_0:\quad b_1 = 0 \qquad \text{versus} \qquad H_1:\quad b_1 \neq 0$$

The hope in regression analysis is to reject H_0 and to accept H_1, that b_0 and $b_1 \neq 0$, with a two-tail test.

(b) Since u_i is assumed to be normally distributed (assumption 1 in Sec. 6.1), Y_i is also normally distributed (since X_i is assumed to be fixed—assumption 5).   As a result, $\hat{b}_0$ and $\hat{b}_1$ also will be normally distributed.

(c) To test the statistical significance of b_0 and b_1, the t distribution (from App. 5) must be used because $\hat{b}_0$ and $\hat{b}_1$ are normally distributed, but var $\hat{b}_0$ and var $\hat{b}_1$ are unknown (since σ_u^2 is unknown) and $n < 30$ (see Sec. 4.4).

(d) The degrees of freedom are $n - k$, where n is the number of observations and k is the number of parameters estimated.   Since in simple regression analysis, two parameters are estimated ($\hat{b}_0$ and $\hat{b}_1$), df $= n - k = n - 2$.

6.17 Test at the 5% level of significance for   (a) b_0 and   (b) b_1 in Prob. 6.9(a).

(a)
$$t_0 = \frac{\hat{b}_0 - b_0}{s_{\hat{b}_0}} \cong \frac{2.30 - 0}{7.17} \cong 0.32$$

Since t_0 is smaller than the tabular value of $t = 2.228$ at the 5% level (two-tail test) and with 10 df (from App. 5), we conclude that b_0 is not statistically significant at the 5% level (i.e., we cannot reject H_0, that $b_0 = 0$).

(b)
$$t_1 = \frac{\hat{b}_1 - b_1}{s_{\hat{b}_1}} \cong \frac{0.86 - 0}{0.05} \cong 17.2$$

So b_1 is statistically significant at the 5% (and 1%) level (i.e., we cannot reject H_1, that $b_1 \neq 0$).

6.18 Construct the 95% confidence interval for   (a) b_0 and   (b) b_1 in Prob. 6.9(a).

(a) The 95% confidence interval for b_0 is given by (Sec. 4.4)

$$b_0 = \hat{b}_0 \pm 2.228 s_{\hat{b}_0} = 2.30 \pm 2.228(7.17) = 2.30 \pm 15.97$$

So b_0 is between -13.67 and 18.27 with 95% confidence.   Note how wide (and meaningless) the 95% confidence interval b_0 is, reflecting the fact that $\hat{b}_0$ is highly insignificant.

(b) The 95% confidence interval for b_1 is given by

$$b_1 = \hat{b}_1 \pm 2.228 s_{\hat{b}_1} = 0.86 \pm 2.228(0.05) = 0.86 \pm 0.11$$

So b_1 is between 0.75 and 0.97 (i.e., $0.75 < b_1 < 0.97$) with 95% confidence.

TEST OF GOODNESS OF FIT AND CORRELATION

6.19 Derive the formula for R^2.

The coefficient of determination R^2 is defined as the proportion of the total variation in Y "explained" by the regression of Y on X.   The total variation in Y or total sum of squares TSS $= \sum (Y_i - \overline{Y})^2 = \sum y_i^2$. The explained variation in Y or regression sum of squares RSS $= \sum (\hat{Y}_i - \overline{Y})^2 = \sum \hat{y}_i^2$.   The residual variation in Y or error sum of squares ESS $= \sum (Y_i - \hat{Y}_i)^2 = \sum e_i^2$.

$$
\begin{aligned}
\text{TSS} &= \text{RSS} &&+ \text{ESS} \\
\sum (Y_i - \overline{Y})^2 &= \sum (\hat{Y}_i - \overline{Y})^2 &&+ \sum (Y_i - \hat{Y}_i)^2 \\
\sum y_i^2 &= \sum \hat{y}_i^2 &&+ \sum e_i^2
\end{aligned}
$$

Dividing both sides by $\sum y_i^2$, we get

$$\frac{\sum y_i^2}{\sum y_i^2} = \frac{\sum \hat{y}_i^2}{\sum y_i^2} + \frac{\sum e_i^2}{\sum y_i^2}$$

$$1 = \frac{\sum \hat{y}_i^2}{\sum y_i^2} + \frac{\sum e_i^2}{\sum y_i^2}$$

Therefore
$$R^2 = \frac{\sum \hat{y}_i^2}{\sum y_i^2} = 1 - \frac{\sum e_i^2}{\sum y_i^2}$$

R^2 is unit-free and $0 \leq R^2 \leq 1$ because $0 \leq \text{ESS} \leq \text{TSS}$. $R^2 = 0$ when, for example, all sample points lie on a horizontal line $Y = \overline{Y}$ or on a circle. $R^2 = 1$ when all sample points lie on the estimated regression line, indicating a perfect fit.

6.20 (a) What does the correlation coefficient measure?   What is its range of values?   (b) What is the relationship between correlation and regression analysis?

(a) The correlation coefficient measures the degree of association between two or more variables.   In the two-variable case, the simple linear correlation coefficient, r, for a set of sample observations is given by

$$r = \sqrt{R^2} = \frac{\sum x_i y_i}{\sqrt{\sum x_i^2}\sqrt{\sum y_i^2}} = \sqrt{\hat{b}_1 \frac{\sum x_i y_i}{\sum y_i^2}}$$

$-1 \le r \le +1$. $r < 0$ means that X and Y move in opposite directions, such as, for example, the quantity demanded of a commodity and its price. $r > 0$ indicates that X and Y change in the same direction, such as the quantity supplied of a commodity and its price. $r = -1$ refers to a perfect negative correlation (i.e., all the sample observations lie on a straight line of negative slope); however, $r = 1$ refers to perfect positive correlation (i.e., all the sample observations lie on a straight line of positive slope). $r = \pm 1$ is seldom found.   The closer r is to ± 1, the greater is the degree of positive or negative linear relationship.   It should be noted that the sign of r is always the same as that of $\hat{b}_1$.   A zero correlation coefficient means that there exists no linear relationship whatsoever between X and Y (i.e., they tend to change with no connection with each other).   For example, if the sample observations fall exactly on a circle, there is a perfect nonlinear relationship but a zero linear relationship, and $r = 0$.

(b) Regression analysis implies (but does not prove) causality between the independent variable X and dependent variable Y.   However, correlation analysis implies no causality or dependence but refers simply to the type and degree of association between two variables.   For example, X and Y may be highly correlated because of another variable that strongly affects both.   Thus correlation analysis is a much less powerful tool than regression analysis and is seldom used by itself in the real world.   In fact, the main use of correlation analysis is to determine the degree of association found in regression analysis.   This is given by the coefficient of determination, which is the square of the correlation coefficient.

6.21   Derive the equation   (a) $r = \sum x_i y_i / (\sqrt{\sum x_i^2} \sqrt{\sum y_i^2})$ (Hint: Start by showing that $\sum x_i y_i$ is a measure of association between X and Y.) and   (b) $r = \sqrt{\hat{b}_1(\sum x_i y_i / \sum y_i^2)}$ [Hint: Start with $r = \sum x_i y_i / (\sqrt{\sum x_i^2} \sqrt{\sum y_i^2})$.]

(a) $\sum x_i y_i$ provides a measure of the association between X and Y because if X and Y both rise or fall, $x_i y_i > 0$, while if X rises and Y falls, or vice versa, $x_i y_i < 0$.   If all or most sample observations involve a rise or fall in both X and Y, $\sum x_i y_i > 0$ and large, implying a large positive correlation.   If all or most sample observations involve opposite changes in X and Y, then $\sum x_i y_i < 0$ and large, implying a large negative correlation.   If, however, some X and Y observations move in the same direction, while others move in opposite directions, $\sum x_i y_i$ will be smaller, indicating a small net positive or negative correlation.   However, measuring the degree of association by $\sum x_i y_i$ has two disadvantages. First, the greater is the number of sample observations, the larger is $\sum x_i y_i$; and second, $\sum x_i y_i$ is expressed in the units of the problem.   These problems can be overcome by dividing $\sum x_i y_i$ by n (the number of sample observations) and by the standard deviation of X and Y ($\sqrt{\sum x_i^2 / n}$ and $\sqrt{\sum y_i^2 / n}$). Then

$$\frac{\sum x_i y_i}{n} = \text{covariance of } X \text{ and } Y \qquad (6.30)$$

and

$$\frac{\sum x_i y_i}{n \sqrt{\dfrac{\sum x_i^2}{n}} \sqrt{\dfrac{\sum y_i^2}{n}}} = \frac{\sum x_i y_i}{\sqrt{\sum x_i^2} \sqrt{\sum y_i^2}} = r$$

(b)

$$r = \frac{\sum x_i y_i}{\sqrt{\sum x_i^2} \sqrt{\sum y_i^2}} = \frac{\sqrt{\sum x_i y_i}}{\sqrt{\sum x_i^2}} \frac{\sqrt{\sum x_i y_i}}{\sqrt{\sum y_i^2}} = \sqrt{\hat{b}_1 \frac{\sum x_i y_i}{\sum y_i^2}}$$

6.22   Find R^2 for the estimated consumption regression of Prob. 6.9 using the equation   (a) $R^2 = \sum \hat{y}_i^2 / \sum y_i^2$ and   (b) $R^2 = 1 - \sum e_i^2 / \sum y_i^2$.

(a) From Prob. 6.19, we know that $\sum y_i^2 = \sum \hat{y}_i^2 + \sum e_i^2$, so $\sum \hat{y}_i^2 = \sum y_i^2 - \sum e_i^2$. Since $\sum y_i^2 = 3684$ (by squaring and adding the y_i terms from Table 6.6) and $\sum e_i^2 = 115.2572$ (from Table 6.7) $\sum \hat{y}_i^2 = 3684 - 115.2572 = 3568.7428$.   Thus

$$R^2 = \frac{\sum \hat{y}_i^2}{\sum y_i^2} = \frac{3568.7428}{3684} \cong 0.9687, \text{ or } 96.87\%$$

(b)   Using $\sum e_i^2 = 115.2572$ and $\sum y_i^2 = 3684$, we get

$$R^2 = 1 - \frac{\sum e_i^2}{\sum y_i^2} = 1 - \frac{115.2572}{3684} \cong 0.9687, \text{ or } 96.87\%$$

(as in part *a*).

6.23   Find r for the estimated consumption regression in Prob. 6.9 using   (*a*) $\sqrt{R^2}$,   (*b*) $r = \sum x_i y_i /$ $(\sqrt{\sum x_i^2} \sqrt{\sum y_i^2})$, and   (*c*) $r = \sqrt{\hat{b}_1 (\sum x_i y_i / \sum y_i^2)}$.

(*a*) $r = \sqrt{R^2} \cong \sqrt{0.9687} \cong 0.9842$ and is positive because $\hat{b}_1 > 0$.

(*b*)   Using $\sum x_i y_i = 4144$ and $\sum x_i^2 = 4812$ from Table 6.6 and $\sum y_i^2 = 3684$ from Prob. 6.22(*a*), we get

$$r = \frac{\sum x_i y_i}{\sqrt{\sum x_i^2} \sqrt{\sum y_i^2}} \cong \frac{4144}{\sqrt{4812}\sqrt{3684}} \cong 0.9841$$

The very small difference between the value of r found here and that found in part *a* results from rounding errors.

(*c*)   Using $\hat{b}_1 \cong 0.86$ found in Prob. 6.9(*a*), we obtain

$$r = \sqrt{\hat{b}_1 \frac{\sum x_i y_i}{\sum y_i^2}} \cong \sqrt{\frac{(0.86)(4144)}{3684}} \cong 0.9836$$

6.24   (*a*) Find the rank or Spearman correlation coefficient between the midterm grade and the IQ ranking of a random sample of 10 students in a large class, as given in Table 6.8, using Eq. (*6.31*). (*b*) When is the rank correlation used?

Table 6.8   Midterm Grade and IQ Ranking

Student	1	2	3	4	5	6	7	8	9	10
Midterm grade	77	78	65	84	84	88	67	92	68	96
IQ ranking	7	6	8	5	4	3	9	1	10	2

(*a*) $$r' = 1 - \frac{6 \sum D^2}{n(n^2 - 1)} \tag{6.31}$$

where D is the difference between ranks of corresponding pairs of the two variables (either in ascending or descending order, with the mean rank assigned to observations of the same value) and n is the number of observations.

The calculations to find r' are given in Table 6.9.

$$r' = 1 - \frac{6 \sum D^2}{n(n^2 - 1)} = 1 - \frac{6(10.50)}{10(99)} = 1 - \frac{63}{990} \cong 0.94$$

(*b*)   Rank correlation is used with qualitative data such as profession, education, or sex, when, because of the absence of numerical values, the coefficient of correlation cannot be found. Rank correlation also is used when precise values for all or some of the variables are not available (so that, once again, the coefficient of correlation cannot be found). Furthermore, with a great number of observations of large values, r' can be found as an estimate of r in order to avoid very time-consuming calculations (however, easy accessibility to computers has practically eliminated this reason for using r').

Table 6.9 Calculations to Find the Coefficient of Rank Correlation

n	Midterm Grade	Ranking on Midterm	IQ Ranking	D	D^2
1	96	1	2	−1	1
2	92	2	1	1	1
3	88	3	3	0	0
4	84	4.5	4	0.5	0.25
5	84	4.5	5	−0.5	0.25
6	78	6	6	0	0
7	77	7	7	0	0
8	68	8	10	−2	4
9	67	9	9	0	0
10	65	10	8	2	4
					$\sum D^2 = 10.50$

PROPERTIES OF ORDINARY LEAST-SQUARES ESTIMATORS

6.25 (*a*) What is meant by an unbiased estimator? How is bias defined? (*b*) Draw a figure showing the sampling distribution of an unbiased and a biased estimator.

(*a*) An estimator is *unbiased* if the mean of its sampling distribution equals the true parameter. The mean of the sampling distribution is the expected value of the estimator. Thus *lack of bias* means that $E(\hat{b}) = b$, where $\hat{b}$ is the estimator of the true parameter, b. *Bias* is then defined as the difference between the expected value of the estimator and the true parameter; that is, bias $= E(\hat{b}) - b$. Note that lack of bias does not mean that $\hat{b} = b$, but that in repeated random sampling, we get, on average, the correct estimate. The hope is that the sample actually obtained is close to the mean of the sampling distribution of the estimator.

(*b*) Figure 6-7*a* shows the sampling distribution of an estimator that is unbiased, and Fig. 6-7*b* shows one that is biased.

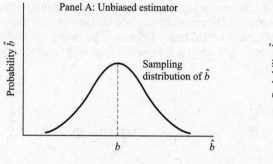

Fig. 6-7

6.26 (*a*) What is meant by the best unbiased or efficient estimator? Why is this important? (*b*) Draw a figure of the sampling distribution of two unbiased estimators, one of which is efficient.

(*a*) The *best unbiased or efficient estimator* refers to the one with the smallest variance among unbiased estimators. It is the unbiased estimator with the most compact or least spread-out distribution. This is very important because the researcher would be more certain that the estimator is closer to the true population parameter being estimated. Another way of saying this is that an efficient estimator has the smallest confidence interval and is more likely to be statistically significant than any other estimator. It should be noted that minimum variance by itself is not very important, unless coupled with the lack of bias.

(b)   Figure 6-8a shows the sampling distribution of an efficient estimator, while Fig. 6-8b shows an inefficient estimator.

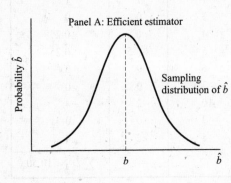
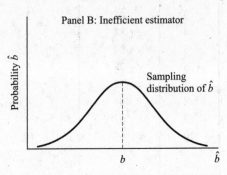

Fig. 6-8

6.27   Why is the OLS estimator so widely used?   Is it superior to all other estimators?

The OLS estimator is widely used because it is BLUE (best linear unbiased estimator). That is, among all unbiased linear estimators, it has the lowest variance. The BLUE properties of the OLS estimator is often referred to as the *Gauss-Markov theorem*. However, *nonlinear* estimators may be superior to the OLS estimator (i.e., they might be unbiased and have lower variance). Since it is often difficult or impossible to find the variance of unbiased nonlinear estimators, however, the OLS estimator remains by far the most widely used. The OLS estimator, being linear, is also easier to use than nonlinear estimators.

6.28   (a) What is meant by the *mean-square error*? Why and when is the rule to minimize the mean-square error useful? (b) Prove that the mean-square error equals the variance plus the square of the bias of the estimator.

(a)
$$\text{MSE}(\hat{b}) = E(\hat{b} - b)^2 = \text{var}\,\hat{b} + (\text{bias}\,\hat{b})^2$$

The rule to minimize the MSE arises when the researcher faces a slightly biased estimator but with a smaller variance than any unbiased estimator. The researcher is then likely to choose the estimator with the lowest MSE. This rule penalizes equally for the larger variance or for the square of the bias of an estimator. However, this is used only when the OLS estimator has an "unacceptably large" variance.

(b)
$$\text{MSE}(\hat{b}) = E(\hat{b} - b)^2$$
$$= E[\hat{b} - E(\hat{b}) + E(\hat{b}) - b]^2$$
$$= E[\hat{b} - E(\hat{b})]^2 + [E(\hat{b}) - b]^2 + 2E\{[\hat{b} - E(\hat{b})][E(\hat{b}) - b]\}$$
$$= \text{var}\,\hat{b} + (\text{bias}\,\hat{b})^2$$

because $E[\hat{b} - E(\hat{b})]^2 = \text{var}\,\hat{b}$, $[E(\hat{b}) - b]^2 = (\text{bias}\,\hat{b})^2$, and $E\{[\hat{b} - E(\hat{b})][E(\hat{b}) - b]\} = 0$ because this expression is equal to $E\{\hat{b}E(\hat{b}) - [E(\hat{b})]^2 - \hat{b}b + bE(\hat{b})\} = [E(\hat{b})]^2 - [E(\hat{b})]^2 - bE(\hat{b}) + bE(\hat{b}) = 0$.

6.29   (a) What is meant by consistency? (b) Draw a figure of the sampling distribution of a consistent estimator.

(a)   Two conditions are required for an estimator to be consistent: (1) as the sample size increases, the estimator must approach more and more the true parameter (this is referred to as *asymptotic unbiasedness*); and   (2) as the sample size approaches infinity in the limit, the sampling distribution of the estimator must collapse or become a straight vertical line with height (probability) of 1 above the value of the true parameter. This large-sample property of consistency is used only in situations when small-sample BLUE or lowest MSE estimators cannot be found.

(b) In Fig. 6-9, $\hat{b}$ is a consistent estimator of b because as n increases, $\hat{b}$ approaches b, and as n approaches infinity in the limit, the sampling distribution of $\hat{b}$ collapses on b.

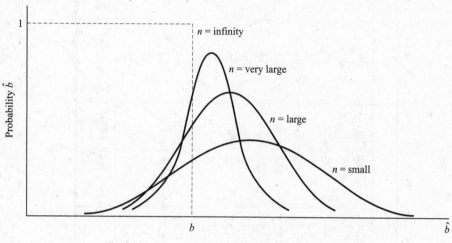

Fig. 6-9

SUMMARY PROBLEM

6.30 Table 6.10 gives the per capita income to the nearest \$100 (Y) and the percentage of the economy represented by agriculture (X) reported by the World Bank World Development Indicators for 1999 for 15 Latin American countries. (a) Estimate the regression equation of Y_i on X_i. (b) Test at the 5% level of significance for the statistical significance of the parameters. (c) Find the coefficient of determination. (d) Report the results obtained in part a in standard summary form.

Table 6.10   Per Capita Income (Y, \$00) and Percentage of the Economy in Agriculture (X)

Country:*	(1)	(2)	(3)	(4)	(5)	(6)	(7)	(8)	(9)	(10)	(11)	(12)	(13)	(14)	(15)
n	1	2	3	4	5	6	7	8	9	10	11	12	13	14	15
Y_i	76	10	44	47	23	19	13	19	8	44	4	31	24	59	37
X_i	6	16	9	8	14	11	12	10	18	5	26	8	8	9	5

Key: (1) Argentina; (2) Bolivia; (3) Brazil; (4) Chile; (5) Colombia; (6) Dominican Republic; (7) Ecuador; (8) El Salvador; (9) Honduras; (10) Mexico; (11) Nicaragua; (12) Panama; (13) Peru; (14) Uruguay; (15) Venezuela.
Source: World Bank World Development Indicators.

(a) The first seven columns of Table 6.11 are used to answer part a. The rest of the table is filled by utilizing the results of part a in order to answer parts b and c of this problem.

$$\hat{b}_1 = \frac{\sum x_i y_i}{\sum x_i^2} = \frac{-1149}{442} \cong -2.60$$

$$\hat{b}_0 = \overline{Y} - \hat{b}_1 \overline{X} = 30.53 - (-2.60)(11) = 59.13$$

$$\hat{Y}_i = 59.13 - 2.60 X_i$$

Table 6.11　Worksheet

n	Y_i	X_i	y_i	x_i	x_iy_i	x_i^2	$\hat{Y}_i$	e_i	e_i^2	X_i^2	Y_i^2
1	76	6	45.47	−5	−227.35	25	43.53	32.47	1054.3009	36	2067.5209
2	10	16	−20.53	5	−102.65	25	17.53	−7.53	56.7009	256	421.4809
3	44	9	13.47	−2	−26.94	4	35.73	8.27	68.3929	81	181.4409
4	47	8	16.47	−3	−49.41	9	38.33	8.67	75.1689	64	271.2609
5	23	14	−7.53	3	−22.59	9	22.73	0.27	0.0729	196	56.7009
6	19	11	−11.53	0	0	0	30.53	−11.53	132.9409	121	132.9409
7	13	12	−17.53	1	−17.53	1	27.93	−14.93	222.9049	144	307.3009
8	19	11	−11.53	−1	11.53	1	33.13	−14.13	199.6569	100	132.9409
9	8	18	−22.53	7	−157.71	49	12.33	−4.33	18.7489	324	507.6009
10	44	5	13.47	−6	−80.82	36	46.13	−2.13	4.5369	25	181.4409
11	4	26	−26.53	15	−397.95	225	−8.47	12.47	155.5009	676	703.8409
12	31	8	0.47	−3	−1.41	9	38.33	−7.33	53.7289	64	0.2209
13	24	8	−6.53	−3	19.59	9	38.33	−14.33	205.3489	64	42.6409
14	59	9	28.47	−2	−56.94	4	35.73	23.27	541.4929	81	810.5409
15	37	5	6.47	−6	−38.82	36	46.13	−9.13	83.3569	25	41.9609
	$\sum Y_i = 458$ $\bar{Y} \approx 30.53$	$\sum X_i = 165$ $\bar{X} = 11.00$			$\sum y_i x_i = -1149$	$\sum x_i^2 = 442$			$\sum e_i^2 = 2872.8535$	$\sum X_i^2 = 2257$	$\sum y_i^2 = 5859.7335$

(b) $$s_{\hat{b}_0}^2 = \frac{\sum e_i^2}{(n-k)} \frac{\sum X_i^2}{n \sum x_i^2} = \frac{(2872.8535)(2257)}{(15-2)(15)(442)} \cong 75.23 \quad \text{and} \quad s_{\hat{b}_0} \cong 8.67$$

$$s_{\hat{b}_1}^2 = \frac{\sum e_i^2}{(n-k) \sum x_i^2} = \frac{(2872.8535)}{(15-2)(442)} \cong 0.050 \quad \text{and} \quad s_{\hat{b}_1} = 0.71$$

$$t_0 = \frac{\hat{b}_0}{s_{\hat{b}_0}^2} = \frac{59.13}{8.67} \cong 6.82$$

$$t_1 = \frac{\hat{b}_1}{s_{\hat{b}_1}^2} = \frac{-2.60}{0.71} \cong -3.66$$

Therefore both $\hat{b}_0$ and $\hat{b}_1$ are statistically significant at the 5% level.

(c) $$R^2 = 1 - \frac{\sum e_i^2}{\sum y_i^2} = 1 - \frac{2872.8535}{5859.7335} \cong 0.51$$

(d) $$\hat{Y}_i = 59.13 - 2.60X_i \qquad R^2 = 0.51$$
$$(6.82) \quad (-3.66)$$

The numbers in parentheses below the estimated parameters are the corresponding t values. An alternative way is to report the standard error of the estimates in parentheses.

Supplementary Problems

THE TWO-VARIABLE LINEAR MODEL

6.31 Draw a scatter diagram for the data in Table 6.12 and determine by inspection if there is an approximate linear relationship between Y_i and X_i.

Ans. The relationship between X and Y in Fig. 6-10 is approximately linear.

Table 6.12 Observations on Variables Y and X

n	Y_i	X_i
1	20	2
2	28	3
3	40	5
4	45	4
5	37	3
6	52	5
7	54	7
8	43	6
9	65	7
10	56	8

6.32 State the assumptions of the classical linear regression (OLS) model in mathematical form.

Ans.

$$u \sim N(0, \sigma_u^2) \tag{6.21}$$

$$E(u_i u_j) = 0 \quad \text{for } i \neq j; \quad i, j = 1, 2, \ldots, n \tag{6.22}$$

$$E(X_i u_i) = 0 \tag{6.23}$$

(See Prob. 6.4.)

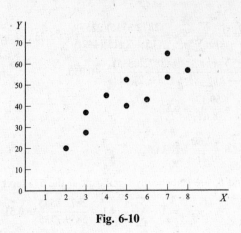

Fig. 6-10

THE ORDINARY LEAST-SQUARES METHOD

6.33   Express mathematically the following statements and formulas: (a) Minimize the sum of the squared deviations of each value of Y from its corresponding fitted value.   (b) Minimize the sum of squared residuals.   (c) The normal equations.   (d) The formulas for estimating $\hat{b}_1$ and $\hat{b}_0$.
Ans.   (a) Min $\sum (Y_i - \hat{Y}_i)^2$ (b) Min $\sum e_i^2$ (c) $\sum Y_i = n\hat{b}_0 + \hat{b}_1 \sum X_i$ and $\sum X_i Y_i = \hat{b}_0 \sum X_i + \hat{b}_1 \sum X_i^2$
(d) $\hat{b}_1 = (n \sum X_i Y_i - \sum X_i \sum Y_i)/[n \sum X_i^2 - (\sum X_i)^2] = \sum x_i y_i / \sum x_i^2$ and $\hat{b}_0 = \overline{Y} - \hat{b}_1 \overline{X}$

6.34   For the data in Table 6.12, find the value of   (a) $\hat{b}_1$ and   (b) $\hat{b}_0$.   (c) Write the equation for the estimated OLS regression line.
Ans.   (a) $\hat{b}_1 \cong 5.94$ (b) $\hat{b}_0 \cong 14.28$ (c) $\hat{Y}_i = 14.28 + 5.94 X_i$

6.35   (a) On a set of axes, plot the data in Table 6.12, plot the estimated OLS regression line in Prob. 6.34, and show the residuals.   (b) Show algebraically that the regression line goes through point $\overline{X}\,\overline{Y}$.
Ans.   (a) See Fig. 6-11   (b) At $X_i = 5 = \overline{X}$, $\hat{Y}_i = 14.28 + 5.94(5) = 43.98 \cong \overline{Y} = 44$ (the slight difference due to rounding)

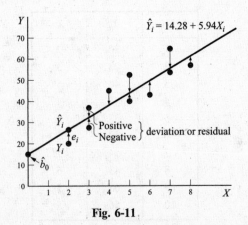

Fig. 6-11

6.36   With reference to the estimated OLS regression line in Prob. 6.34, state   (a) the meaning of $\hat{b}_0$,   (b) the meaning of $\hat{b}_1$, and   (c) the elasticity of Y with respect to X at the means.
Ans.   (a) $\hat{b}_0$ is the Y intercept   (b) $\hat{b}_1$ is the slope of the estimated OLS regression line   (c) $\eta \cong 0.68$

TESTS OF SIGNIFICANCE OF PARAMETER ESTIMATES

6.37   For the data in Table 6.12 in Prob. 6.31, find   (a) s^2 (b) $s_{\hat{b}_0}^2$ and $s_{\hat{b}_0}$, and   (c) $s_{\hat{b}_1}^2$ and $s_{\hat{b}_1}$.
Ans.   (a) $s^2 \cong 46.97$ (b) $s_{\hat{b}_0}^2 \cong 37.31$ and $s_{\hat{b}_0} \cong 6.11$ (c) $s_{\hat{b}_1}^2 \cong 1.31$ and $s_{\hat{b}_1} \cong 1.14$

6.38   Test at the 5% level of significance for   (a) b_0 and   (b) b_1 in Prob. 6.34.
Ans.   (a) b_0 is statistically significant at the 5% level   (b) b_1 is also statistically significant at the 5% level

6.39   Construct the 95% confidence interval for   (a) b_0 and   (b) b_1 in Prob. 6.34.
Ans.   (a) $0.19 < b_0 < 28.37$ (b) $3.31 < b_1 < 8.57$

TEST OF GOODNESS OF FIT AND CORRELATION

6.40   For the estimated OLS regression equation in Prob. 6.34, find   (a) R^2 and   (b) r.
Ans.   (a) $R^2 \cong 0.77$ (b) $r \cong 0.88$

6.41   Find the coefficient of rank correlation for the sample of XY observations in Table 6.12.
Ans. $r' \cong 0.90 \ (\cong r \cong 0.88)$

PROPERTIES OF ORDINARY LEAST-SQUARES ESTIMATORS

6.42   With reference to $\hat{b}_0$ and $\hat{b}_1$ in Prob. 6.34, are they   (a) BLUE?   (b) Asymptotically unbiased?   (c) Consistent?
Ans.   (a) Yes   (b) Yes   (c) Yes

6.43   With reference to $\hat{b}_0$ and $\hat{b}_1$ in Prob. 6.34   (a) What is the MSE?   (b) Do $\hat{b}_0$ and $\hat{b}_1$ minimize the MSE?
Ans.   (a) $\text{MSE}(\hat{b}_0) = \text{var}\,\hat{b}_0$ and $\text{MSE}(\hat{b}_1) = \text{var}\,\hat{b}_1$ (b) Yes

SUMMARY PROBLEM

6.44   Table 6.13 gives data for a random sample of 12 couples on the number of children they have Y_i and the number of children they had stated they wanted at the time of their marriage X_i.   Regress Y_i on X_i and report your results in summary form.

Ans.
$$\hat{Y}_i = 0.22 + 1.14X_i \qquad R^2 = 0.68$$
$$(0.39) \quad (4.56)$$

Table 6.13   Number of Children Had and Wanted

Couple	1	2	3	4	5	6	7	8	9	10	11	12
Y_i	4	3	0	4	4	3	0	4	3	1	3	1
X_i	3	3	0	2	2	3	0	3	2	1	3	2

The numbers in parentheses are t values.   Thus $\hat{b}_1$ is statistically significant at the 5% (and 1%) level of significance, but $\hat{b}_0$ is not.

Multiple Regression Analysis

CHAPTER 7

7.1 THE THREE-VARIABLE LINEAR MODEL

Multiple regression analysis is used for testing hypotheses about the relationship between a dependent variable Y and two or more independent variables X and for prediction. The three-variable linear regression model can be written as

$$Y_i = b_0 + b_1 X_{1i} + b_2 X_{2i} + u_i \tag{7.1}$$

The additional assumption (to those of the simple regression model) is that there is no exact linear relationship between the X values.

Ordinary least-squares (OLS) parameter estimates for Eq. (7.1) can be obtained by minimizing the sum of the squared residuals:

$$\sum e_i^2 = \sum (Y_i - \hat{Y}_i)^2 = \sum (Y_i - \hat{b}_0 - \hat{b}_1 X_{1i} - \hat{b}_2 X_{2i})^2$$

This gives the following three normal equations (see Prob. 7.2):

$$\sum Y_i = n\hat{b}_0 + \hat{b}_1 \sum X_{1i} + \hat{b}_2 \sum X_{2i} \tag{7.2}$$

$$\sum X_{1i} Y_i = \hat{b}_0 \sum X_{1i} + \hat{b}_1 \sum X_{1i}^2 + \hat{b}_2 \sum X_{1i} X_{2i} \tag{7.3}$$

$$\sum X_{2i} Y_i = \hat{b}_0 \sum X_{2i} + \hat{b}_1 \sum X_{1i} X_{2i} + \hat{b}_2 \sum X_{2i}^2 \tag{7.4}$$

which (when expressed in deviation form) can be solved simultaneously for $\hat{b}_1$ and $\hat{b}_2$, giving (see Prob. 7.3)

$$\hat{b}_1 = \frac{\left(\sum x_1 y\right)\left(\sum x_2^2\right) - \left(\sum x_2 y\right)\left(\sum x_1 x_2\right)}{\left(\sum x_1^2\right)\left(\sum x_2^2\right) - \left(\sum x_1 x_2\right)^2} \tag{7.5}$$

$$\hat{b}_2 = \frac{\left(\sum x_2 y\right)\left(\sum x_1^2\right) - \left(\sum x_1 y\right)\left(\sum x_1 x_2\right)}{\left(\sum x_1^2\right)\left(\sum x_2^2\right) - \left(\sum x_1 x_2\right)^2} \tag{7.6}$$

Then
$$\hat{b}_0 = \bar{Y} - \hat{b}_1 \bar{X}_1 - \hat{b}_2 \bar{X}_2 \tag{7.7}$$

Estimator $\hat{b}_1$ measures the change in Y for a unit change in X_1 while holding X_2 constant. $\hat{b}_2$ is analogously defined. Estimators $\hat{b}_1$ and $\hat{b}_2$ are called *partial regression coefficients*. $\hat{b}_0$, $\hat{b}_1$, and $\hat{b}_2$ are BLUE (see Sec. 6.5).

EXAMPLE 1. Table 7.1 extends Table 6.1 and gives the bushels of corn per acre, Y, resulting from the use of various amounts of fertilizer X_1 and insecticides X_2, both in pounds per acre, from 1971 to 1980. Using Eqs. (7.5), (7.6), and (7.7), we get

$$\hat{b}_1 = \frac{(\sum x_1 y)(\sum x_2^2) - (\sum x_2 y)(\sum x_1 x_2)}{(\sum x_1^2)(\sum x_2^2) - (\sum x_1 x_2)^2} = \frac{(956)(504) - (900)(524)}{(576)(504) - (524)^2} \cong 0.65$$

$$\hat{b}_2 = \frac{(\sum x_2 y)(\sum x_1^2) - (\sum x_1 y)(\sum x_1 x_2)}{(\sum x_1^2)(\sum x_2^2) - (\sum x_1 x_2)^2} = \frac{(900)(576) - (956)(524)}{(576)(504) - (524)^2} \cong 1.11$$

$$\hat{b}_0 = \bar{Y} - \hat{b}_1 \bar{X}_1 - \hat{b}_2 \bar{X}_2 \cong 57 - (0.65)(18) - (1.11)(12) \cong 31.98$$

so that $\hat{Y}_i = 31.98 + 0.65 X_{1i} + 1.11 X_{2i}$. To estimate the regression parameters with three or more independent or explanatory variables, see Section 7.6.

7.2   TESTS OF SIGNIFICANCE OF PARAMETER ESTIMATES

In order to test for the statistical significance of the parameter estimates of the multiple regression, the variance of the estimates is required:

$$\text{Var}\,\hat{b}_1 = \sigma_u^2 \frac{\sum x_2^2}{\sum x_1^2 \sum x_2^2 - \left(\sum x_1 x_2\right)^2} \tag{7.8}$$

$$\text{Var}\,\hat{b}_2 = \sigma_u^2 \frac{\sum x_1^2}{\sum x_1^2 \sum x_2^2 - \left(\sum x_1 x_2\right)^2} \tag{7.9}$$

[b_0 is usually not of primary concern; see Prob. 7.7(e)]. Since σ_u^2 is unknown, the residual variance s^2 is used as an unbiased estimate of σ_u^2:

$$s^2 = \hat{\sigma}_u^2 = \frac{\sum e_i^2}{n - k} \tag{6.12}$$

where k = number of parameter estimates.

Unbiased estimates of the variance of $\hat{b}_0$ and $\hat{b}_1$ are then given by

$$s_{\hat{b}_1}^2 = \frac{\sum e_i^2}{n - k} \frac{\sum x_2^2}{\sum x_1^2 \sum x_2^2 - \left(\sum x_1 x_2\right)^2} \tag{7.10}$$

$$s_{\hat{b}_2}^2 = \frac{\sum e_i^2}{n - k} \frac{\sum x_1^2}{\sum x_1^2 \sum x_2^2 - \left(\sum x_1 x_2\right)^2} \tag{7.11}$$

so that $s_{\hat{b}_1}$ and $s_{\hat{b}_2}$ are the standard errors of the estimates. Tests of hypotheses about b_1 and b_2 are conducted as in Sec. 6.3.

EXAMPLE 2. Table 7.2 (an extension of Table 7.1) shows the additional calculations required to test the statistical significance of $\hat{b}_1$ and $\hat{b}_2$. The values for $\hat{Y}_i$ in Table 7.2 are obtained by substituting the values for X_{1i} and X_{2i} into the estimated OLS regression equation found in Example 1. (The values for y_i^2 are obtained by squaring y_i from Table 7.1 and are to be used in Sec. 7.3.) Using the values from Table 7.2 and 7.1, we get

$$s_{\hat{b}_1}^2 = \frac{\sum e_i^2}{n - k} \frac{\sum x_2^2}{\sum x_1^2 \sum x_2^2 - \left(\sum x_1 x_2\right)^2} = \frac{13.6704}{10 - 3} \frac{504}{(576)(504) - (524)^2} \cong 0.06 \quad \text{and} \quad s_{\hat{b}_1} \cong 0.24$$

$$s_{\hat{b}_2}^2 = \frac{\sum e_i^2}{n - k} \frac{\sum x_1^2}{\sum x_1^2 \sum x_2^2 - \left(\sum x_1 x_2\right)^2} = \frac{13.6704}{10 - 3} \frac{576}{(576)(504) - (524)^2} = 0.07 \quad \text{and} \quad s_{\hat{b}_2} \cong 0.27$$

Table 7.1 Corn Produced with Fertilizer and Insecticide Used with Calculations for Parameter Estimation

Year	Y	X_1	X_2	y	x_1	x_2	$x_1 y$	$x_2 y$	$x_1 x_2$	x_1^2	x_2^2
1971	40	6	4	-17	-12	-8	204	136	96	144	64
1972	44	10	4	-13	-8	-8	104	104	64	64	64
1973	46	12	5	-11	-6	-7	66	77	42	36	49
1974	48	14	7	-9	-4	-5	36	45	20	16	25
1975	52	16	9	-5	-2	-3	10	15	6	4	9
1976	58	18	12	1	0	0	0	0	0	0	0
1977	60	22	14	3	4	2	12	6	8	16	4
1978	68	24	20	11	6	8	66	88	48	36	64
1979	74	26	21	17	8	9	136	153	72	64	81
1980	80	32	24	23	14	12	322	276	168	196	144
$n = 10$	$\sum Y = 570$ $\bar{Y} = 57$	$\sum X_1 = 180$ $\bar{X}_1 = 18$	$\sum X_2 = 120$ $\bar{X}_2 = 12$	$\sum y = 0$	$\sum x_1 = 0$	$\sum x_2 = 0$	$\sum x_1 y = 956$	$\sum x_2 y = 900$	$\sum x_1 x_2 = 524$	$\sum x_1^2 = 576$	$\sum x_2^2 = 504$

Table 7.2. Corn-Fertilizer-Insecticide Calculations to Test Significance of Parameters

Year	Y	X_1	X_2	$\hat{Y}$	e	e^2	y^2
1971	40	6	4	40.32	−0.32	0.1024	289
1972	44	10	4	42.92	1.08	1.1664	169
1973	46	12	5	45.33	0.67	0.4489	121
1974	48	14	7	48.85	−0.85	0.7225	81
1975	52	16	9	52.37	−0.37	0.1369	25
1976	58	18	12	57.00	1.00	1.0000	1
1977	60	22	14	61.82	−1.82	3.3124	9
1978	68	24	20	69.78	−1.78	3.1684	121
1979	74	26	21	72.19	1.81	3.2761	289
1980	80	32	24	79.42	0.58	0.3364	529
$n = 10$					$\sum e = 0$	$\sum e^2 = 13.6704$	$\sum y^2 = 1634$

Therefore, $t_1 = \hat{b}_1/s_{\hat{b}_1} \cong 0.65/0.24 \cong 2.70$, and $t_2 = \hat{b}_2/s_{\hat{b}_2} = 1.11/0.27 \cong 4.11$. Since both t_1 and t_2 exceed $t = 2.365$ with 7 df at the 5% level of significance (from App. 5), both b_1 and b_2 are statistically significant at the 5% level.

7.3 THE COEFFICIENT OF MULTIPLE DETERMINATION

The *coefficient of multiple determination* R^2 is defined as the proportion of the total variation in Y "explained" by the multiple regression of Y on X_1 and X_2, and (as shown in Sec. 6.4) it can be calculated by (see Prob. 7.14)

$$R^2 = \frac{\sum \hat{y}_i^2}{\sum y_i^2} = 1 - \frac{\sum e_i^2}{\sum y_i^2} = \frac{\hat{b}_1 \sum yx_1 + \hat{b}_2 \sum yx_2}{\sum y^2}$$

Since the inclusion of additional independent or explanatory variables is likely to increase the RSS $= \sum \hat{y}_i^2$ for the same TSS $= \sum y_i^2$ (see Sec. 6.4), R^2 increases. To factor in the reduction in the degrees of freedom as additional independent or explanatory variables are added, the *adjusted R^2 or $\bar{R}^2$*, is computed (see Prob. 7.16):

$$\bar{R}^2 = 1 - (1 - R^2)\frac{n-1}{n-k} \tag{7.12}$$

where n is the number of observations, and k the number of parameters estimated.

EXAMPLE 3. R^2 for the corn-fertilizer-insecticide example can be found from Table 7.2:

$$R^2 = 1 - \frac{\sum e_i^2}{\sum y_i^2} = 1 - \frac{13.6704}{1634} \cong 1 - 0.0084 = 0.9916, \text{ or } 99.16\%$$

This compares with an R^2 of 97.10% in the simple regression, with fertilizer as the only independent or explanatory variable.

$$\bar{R}^2 = 1 - (1 - R^2)\frac{n-1}{n-k} = 1 - (1 - 0.9916)\frac{10-1}{10-3} = 1 - 0.0084(1.2857) = 0.9892, \text{ or } 98.92\%$$

7.4 TEST OF THE OVERALL SIGNIFICANCE OF THE REGRESSION

The overall significance of the regression can be tested with the ratio of the explained to the unexplained *variance*. This follows an F distribution (see Sec. 5.5) with $k-1$ and $n-k$ degrees of freedom, where n is number of observations and k is number of parameters estimated:

$$F_{k-1,n-k} = \frac{\sum \hat{y}_i^2/(k-1)}{\sum e_i^2/(n-k)} = \frac{R^2/(k-1)}{(1-R^2)/(n-k)} \tag{7.13}$$

If the calculated F ratio exceeds the tabular value of F at the specified level of significance and degrees of freedom (from App. 7), the hypothesis is accepted that the regression parameters are not all equal to zero and that R^2 is significantly different from zero.

In addition, the F ratio can be used to test any linear restriction of regression parameters by using the form

$$F_{p,n-k} - \frac{\left(\dfrac{\sum e_{Ri}^2 - \sum e_i^2}{p}\right)}{\left(\dfrac{\sum e_i^2}{n-k}\right)}$$

where p is the number of restriction being tested, $\sum e_{Ri}^2$ indicates the sum of squared residuals for the restricted regression where the restrictions are assumed to be true, and $\sum e_i^2$ indicates the sum of squared residuals for the unrestricted regression (i.e., the usual residuals). The null hypothesis is that the p restrictions are true, in which case the residuals from the restricted and unrestricted models should be identical, and F would take the value of zero. If the restrictions are not true, the unrestricted model will have lower errors, increasing the value of F. If F exceeds the tabular value, the null hypothesis is rejected. This test will be used extensively in Sec. 11.6.

EXAMPLE 4. To test the overall significance of the regression estimated in Example 1 at the 5% level, we can use $R^2 = 0.9916$ (from Example 3), so that

$$F_{2,7} = \frac{0.9916/2}{(1-0.9916)/7} \cong 413.17$$

Since the calculated value of F exceeds the tabular value of $F = 4.74$ at the 5% level of significance and with df $= 2$ and 7 (from App. 7), the hypothesis is accepted that b_1 and b_2 are not both zero and that R^2 is significantly different from zero.

7.5 PARTIAL-CORRELATION COEFFICIENTS

The *partial-correlation coefficient* measures the net correlation between the dependent variable and one independent variable after excluding the common influence of (i.e., holding constant) the other independent variables in the model. For example, $r_{YX_1 \cdot X_2}$ is the partial correlation between Y and X_1, after removing the influence of X_2 from both Y and X_1 [see Prob. 7.23(a)]:

$$r_{YX_1 \cdot X_2} = \frac{r_{YX_1} - r_{YX_2} r_{X_1 X_2}}{\sqrt{1 - r_{X_1 X_2}^2}\sqrt{1 - r_{YX_2}^2}} \tag{7.14}$$

$$r_{YX_2 \cdot X_1} = \frac{r_{YX_2} - r_{YX_1} r_{X_1 X_2}}{\sqrt{1 - r_{X_1 X_2}^2}\sqrt{1 - r_{YX_1}^2}} \tag{7.15}$$

where $r_{YX_1} =$ simple-correlation coefficient between Y and X_1, and r_{YX_2} and $r_{X_1 X_2}$ are analogously defined. Partial-correlation coefficients range in value from -1 to $+1$ (as do simple-correlation coefficients), have the sign of the corresponding estimated parameter, and are used to determine the relative importance of the different explanatory variables in a multiple regression.

EXAMPLE 5. Substituting the values from Tables 7.1 and 7.2 into Eq. (*6.18*) for the simple-correlation coefficient, we get

$$r_{YX_1} = \frac{\sum x_1 y}{\sqrt{\sum x_1^2}\sqrt{\sum y^2}} = \frac{956}{\sqrt{576}\sqrt{1634}} \cong 0.9854$$

$$r_{YX_2} = \frac{\sum x_2 y}{\sqrt{\sum x_2^2}\sqrt{\sum y^2}} = \frac{900}{\sqrt{504}\sqrt{1634}} \cong 0.9917$$

$$r_{X_1 X_2} = \frac{\sum x_2 x_1}{\sqrt{\sum x_2^2}\sqrt{\sum x_1^2}} = \frac{524}{\sqrt{504}\sqrt{576}} \cong 0.9725$$

Thus

$$r_{YX_1 \cdot X_2} = \frac{r_{YX_1} - r_{YX_2} r_{X_1 X_2}}{\sqrt{1 - r_{X_1 X_2}^2}\sqrt{1 - r_{YX_2}^2}} = \frac{0.9854 - (0.9917)(0.9725)}{\sqrt{1 - 0.9725^2}\sqrt{1 - 0.9917^2}}$$

$$\cong 0.7023, \text{ or } 70.23\%$$

and

$$r_{YX_2 \cdot X_1} = \frac{r_{YX_2} - r_{YX_1} r_{X_1 X_2}}{\sqrt{1 - r_{X_1 X_2}^2}\sqrt{1 - r_{YX_1}^2}} = \frac{0.9917 - (0.9854)(0.9725)}{\sqrt{1 - 0.9725^2}\sqrt{1 - 0.9854^2}} \cong 0.8434, \text{ or } 84.34\%$$

Therefore, X_2 is more important than X_1 in explaining the variation Y.

EXAMPLE 6. The overall results of the corn-fertilizer-insecticide example can be summarized as

$$\hat{Y} = 31.98 + 0.65X_1 + 1.11X_2$$

$$t \text{ values } (2.70) \quad (4.11)$$

$$R^2 = 0.992 \qquad \bar{R}^2 = 0.989 \qquad F_{2,7} = 413.17$$

$$r_{YX_1 \cdot X_2} = 0.70 \qquad r_{YX_2 \cdot X_1} = 0.84$$

Even though results are usually obtained from the computer (see Chap. 12), it is crucial to work through a problem "by hand," as we have done, in order to clearly understand the procedure.

7.6 MATRIX NOTATION

Calculations increase substantially as the number of independent variables increase. Matrix notation can aid in solving larger regressions algebraically. The following solution works with any number of independent variables, and is therefore extremely flexible. Students not familiar with linear algebra may skip this section with no loss of continuity.

The regression from Sec. 1 can be written with matrices as

$$Y = Xb + u$$

where

$$Y = \begin{bmatrix} Y_1 \\ Y_2 \\ \vdots \\ Y_n \end{bmatrix} \qquad X = \begin{bmatrix} 1 & X_{11} & X_{21} \\ 1 & X_{12} & X_{22} \\ \vdots & \vdots & \vdots \\ 1 & X_{1n} & X_{2n} \end{bmatrix} \qquad b = \begin{bmatrix} b_0 \\ b_1 \\ b_2 \end{bmatrix} \qquad u = \begin{bmatrix} u_1 \\ u_2 \\ \vdots \\ u_n \end{bmatrix}$$

$$\hat{b} = \begin{bmatrix} \hat{b}_0 \\ \hat{b}_1 \\ \hat{b}_2 \end{bmatrix} = (X'X)^{-1} X'Y$$

$$s_{\hat{b}}^2 = \begin{pmatrix} s_{\hat{b}_0}^2 & \text{cov}(b_0, b_1) & \text{cov}(b_0, b_2) \\ \text{cov}(b_0, b_1) & s_{\hat{b}_1}^2 & \text{cov}(b_1, b_2) \\ \text{cov}(b_0, b_2) & \text{cov}(b_1, b_2) & s_{\hat{b}_2}^2 \end{pmatrix} = \frac{e'e}{(n-k)} (X'X)^{-1} \quad \text{(symmetrical, so lower and upper triangle are identical)}$$

EXAMPLE 7. Recalculation of corn-fertilizer-insecticide example with matrices

$$\hat{b} = \begin{bmatrix} 1 & 1 & 1 & 1 & 1 & 1 & 1 & 1 & 1 & 1 \\ 6 & 10 & 12 & 14 & 16 & 18 & 22 & 24 & 26 & 32 \\ 4 & 4 & 5 & 7 & 9 & 12 & 14 & 20 & 21 & 24 \end{bmatrix} \begin{bmatrix} 1 & 6 & 4 \\ 1 & 10 & 4 \\ 1 & 12 & 5 \\ 1 & 14 & 7 \\ 1 & 16 & 9 \\ 1 & 18 & 12 \\ 1 & 22 & 14 \\ 1 & 24 & 20 \\ 1 & 26 & 21 \\ 1 & 37 & 24 \end{bmatrix}^{-1}$$

$$\times \begin{bmatrix} 1 & 1 & 1 & 1 & 1 & 1 & 1 & 1 & 1 & 1 \\ 6 & 10 & 12 & 14 & 16 & 18 & 22 & 24 & 26 & 32 \\ 4 & 4 & 5 & 7 & 9 & 12 & 14 & 20 & 21 & 24 \end{bmatrix} \begin{bmatrix} 40 \\ 44 \\ 46 \\ 48 \\ 52 \\ 58 \\ 60 \\ 68 \\ 74 \\ 80 \end{bmatrix}$$

$$\hat{b} = \begin{bmatrix} 1.36 & -0.18 & 0.16 \\ -0.18 & 0.03 & -0.03 \\ 0.16 & -0.03 & 0.04 \end{bmatrix} \begin{bmatrix} 570 \\ 11{,}216 \\ 7740 \end{bmatrix} = \begin{bmatrix} 31.98 \\ 0.65 \\ 1.11 \end{bmatrix}$$

therefore, $\hat{b}_0 = 31.98$, $\hat{b}_1 = 0.65$, and $\hat{b}_2 = 1.11$.

$$e = Y - X\hat{b} = \begin{bmatrix} 40 \\ 44 \\ 46 \\ 48 \\ 52 \\ 58 \\ 60 \\ 68 \\ 74 \\ 80 \end{bmatrix} - \begin{bmatrix} 1 & 6 & 4 \\ 1 & 10 & 4 \\ 1 & 12 & 5 \\ 1 & 14 & 7 \\ 1 & 16 & 9 \\ 1 & 18 & 12 \\ 1 & 22 & 14 \\ 1 & 24 & 20 \\ 1 & 26 & 21 \\ 1 & 32 & 24 \end{bmatrix} \begin{bmatrix} 31.98 \\ 0.65 \\ 1.11 \end{bmatrix} = \begin{bmatrix} -0.32 \\ 1.08 \\ 0.67 \\ -0.85 \\ -0.37 \\ 1.00 \\ -1.82 \\ -1.78 \\ 1.81 \\ 0.58 \end{bmatrix}$$

$$s_b^2 = \frac{13.6704}{(10-3)} \begin{bmatrix} 1.36 & -0.18 & 0.16 \\ -0.18 & 0.03 & -0.03 \\ 0.16 & -0.03 & 0.04 \end{bmatrix} = \begin{bmatrix} 2.66 & -0.35 & 0.31 \\ -0.34 & 0.06 & -0.07 \\ 0.31 & -0.07 & 0.07 \end{bmatrix}$$

therefore $s_{\hat{b}_0}^2 = 2.66$, $s_{\hat{b}_1}^2 = 0.06$, and $s_{\hat{b}_2}^2 = 0.07$.

Solved Problems

THE THREE-VARIABLE LINEAR MODEL

7.1 (a) Write the equation of the multiple regression linear model for the case of 2 and k independent or explanatory variables. (b) State the assumptions of the multiple regression linear model.

(a) For the case of 2 independent or explanatory variables, we have

$$Y_i = b_0 + b_1 X_{1i} + b_2 X_{2i} + u_i \tag{7.1}$$

For the case of k independent or explanatory variables, we have

$$Y_i = b_0 + b_1 X_{1i} + b_2 X_{2i} + \cdots + b_k X_{ki} + u_i$$

where X_{2i} represents, for example, the ith observation on independent variable X_2.

(b) The first five assumptions of the multiple regression linear model are exactly the same as those of the simple OLS regression model (see Prob. 6.4). That is, the first three assumptions can be summarized as $u_i \sim N(0, \sigma_u^2)$. The fourth assumption is $E(u_i u_j) = 0$ for $i \neq j$; and the fifth assumption is $E(X_i u_i) = 0$. The only additional assumption required for the multiple OLS regression linear model is that there is no exact linear relationship between the Xs. If two or more explanatory variables are perfectly linearly correlated, it will be impossible to calculate OLS estimates of the parameters because the system of normal equations will contain two or more equations that are not independent. If two or more explanatory variables are highly but not perfectly linearly correlated, then OLS parameter estimates can be calculated, but the effect of each of the highly linearly correlated variables on the explanatory variable cannot be isolated (see Sec. 9.1).

7.2 With the OLS procedure in the case of two independent or explanatory variables, derive (a) normal Eq. (7.2),   (b) normal Eq. (7.3), and   (c) normal Eq. (7.4).   (The reader without knowledge of calculus can skip this problem.)

(a) Normal Eq. (7.2) is derived by minimizing $\sum e_i^2$ with respect to $\hat{b}_0$:

$$\frac{\partial e_i^2}{\partial \hat{b}_0} = \frac{\partial \sum (Y_i - \hat{b}_0 - \hat{b}_1 X_{1i} - \hat{b}_2 X_{2i})^2}{\partial \hat{b}_0} = 0$$

$$-2 \sum (Y_i - \hat{b}_0 - \hat{b}_1 X_{1i} - \hat{b}_2 X_{2i}) = 0 \tag{7.2}$$

$$\sum Y_i = n\hat{b}_0 + \hat{b}_1 \sum X_{1i} + \hat{b}_2 \sum X_{2i}$$

(b) Normal Eq. (7.3) is derived by minimizing $\sum e_i^2$ with respect to $\hat{b}_1$:

$$\frac{\partial \sum e_i^2}{\partial \hat{b}_1} = \frac{\partial \sum (Y_i - \hat{b}_0 - \hat{b}_1 X_{1i} - \hat{b}_2 X_{2i})^2}{\partial \hat{b}_1} = 0$$

$$-2 \sum X_{1i}(Y_i - \hat{b}_0 - \hat{b}_1 X_{1i} - \hat{b}_2 X_{2i}) = 0 \tag{7.3}$$

$$\sum X_{1i} Y_i = \hat{b}_0 \sum X_{1i} + \hat{b}_1 \sum x_{1i}^2 + \hat{b}_2 \sum X_{1i} X_{2i}$$

(c) Normal Eq. (7.4) is derived by minimizing $\sum e_i^2$ with respect to $\hat{b}_2$:

$$\frac{\partial \sum e_i^2}{\partial \hat{b}_2} = \frac{\partial \sum (Y_i - \hat{b}_0 - \hat{b}_1 X_{1i} - \hat{b}_2 X_{2i})^2}{\partial \hat{b}_2} = 0$$

$$-2 \sum X_{2i}(Y_i - \hat{b}_0 - \hat{b}_1 X_{1i} - \hat{b}_2 X_{2i}) = 0 \tag{7.4}$$

$$\sum X_{2i} Y_i = \hat{b}_0 \sum X_{2i} + \hat{b}_1 \sum X_{1i} X_{2i} + \hat{b}_2 \sum X_{2i}^2$$

7.3 For the two independent or explanatory variable multiple linear regression model, (a) derive the normal equations in deviation form.   (*Hint:* Start by deriving the expression for $\hat{y}_i$; the reader without knowledge of calculus can skip this part of this problem.)   (b) How are Eqs. (7.5), (7.6), and (7.7) derived for $\hat{b}_1$, $\hat{b}_2$, and $\hat{b}_0$?

(a)
$$\hat{Y}_i = \hat{b}_0 + \hat{b}_1 X_{1i} + \hat{b}_2 X_{2i}$$
$$\bar{Y} = \hat{b}_0 + \hat{b}_1 \bar{X}_1 + \hat{b}_2 \bar{X}_2$$

Subtracting, we get

$$\hat{y}_i = \hat{Y}_i - \bar{Y} = \hat{b}_1 x_{1i} + \hat{b}_2 x_{2i}$$

Therefore, $e_i = y_i - \hat{y}_i = y_i - \hat{b}_1 x_{1i} - \hat{b}_2 x_{2i}$

$$\sum e_i^2 = \sum (y_i - \hat{y}_i)^2 = \sum (y_i - \hat{b}_1 x_{1i} - \hat{b}_2 x_{2i})^2$$

$$\frac{\partial \sum e_i^2}{\partial \hat{b}_1} = \frac{\partial \sum (y_i - \hat{b}_1 x_{1i} - \hat{b}_2 x_{2i})^2}{\partial \hat{b}_1} = 0$$

$$-2 \sum x_{1i}(y_i - \hat{b}_1 x_{1i} - \hat{b}_2 x_{2i}) = 0$$

$$\sum x_{1i} y_i = \hat{b}_1 \sum x_{1i}^2 + \hat{b}_2 \sum x_{1i} x_{2i} \qquad (7.16)$$

$$\frac{\partial \sum e_i^2}{\partial \hat{b}_2} = \frac{\partial \sum (y_i - \hat{b}_1 x_{1i} - \hat{b}_2 x_{2i})^2}{\partial \hat{b}_2} = 0$$

$$-2 \sum x_{2i}(y_i - \hat{b}_1 x_{1i} - \hat{b}_2 x_{2i}) = 0$$

$$\sum x_{2i} y_i = \hat{b}_1 \sum x_{1i} x_{2i} + \hat{b}_2 \sum x_{2i}^2 \qquad (7.17)$$

(b)   Equations (7.5) and (7.6) to calculate $\hat{b}_1$ and $\hat{b}_2$, respectively, are obtained by solving Eqs. (7.16) and (7.17) simultaneously.   It is always possible to calculate $\hat{b}_1$ and $\hat{b}_2$, except if there is an exact linear relationship between X_1 and X_2 or if the number of observations on each variable of the model is 3 or fewer.   Parameter $\hat{b}_0$ can then be calculated by substituting into Eq. (7.7) the values of $\hat{b}_1$ and $\hat{b}_2$ [calculated with Eqs. (7.5) and (7.6)] and $\bar{Y}$, $\bar{X}_1$, and $\bar{X}_2$ (calculated from the given values of the problem).

7.4   With reference to multiple regression analysis with two independent or explanatory variables, indicate the meaning of   (a) $\hat{b}_0$,   (b) $\hat{b}_1$,   (c) $\hat{b}_2$.   (d) Are $\hat{b}_0, \hat{b}_1$, and $\hat{b}_2$ BLUE?

(a)   Parameter b_0 is the constant term or intercept of the regression and gives the estimated value of Y_i, when $X_{1i} = X_{2i} = 0$.

(b)   Parameter b_1 measures the change in Y for each one-unit change in X_1 while holding X_2 constant. Slope parameter b_1 is a partial regression coefficient because it corresponds to the partial derivative of Y with respect to X_1, or $\partial Y / \partial X_1$.

(c)   Parameter b_2 measures the change in Y for each one-unit change in X_2 while holding X_1 constant. Slope parameter b_2 is the second partial regression coefficient because it corresponds to the partial derivative of Y with respect to X_2, or $\partial Y / \partial X_2$.

(d)   Since $\hat{b}_0$, $\hat{b}_1$, and $\hat{b}_2$ are obtained by the OLS method, they are also best linear unbiased estimators (BLUE; see Sec. 6.5).   That is, $E(\hat{b}_0) = b_0$, $E(\hat{b}_1) = b_1$, and $E(\hat{b}_2) = b_2$, and $s_{\hat{b}_0}$, $s_{\hat{b}_1}$, and $s_{\hat{b}_2}$ are lower than for any other unbiased linear estimator.   Proof of these properties is very cumbersome without the use of matrix algebra, so they are not provided here.

7.5   Table 7.3 gives the real per capita income in thousands of U.S. dollars Y with the percentage of the labor force in agriculture X_1 and the average years of schooling of the population over 25 years of age X_2 for 15 developed countries in 1981.   (a) Find the least-squares regression equation of Y on X_1 and X_2.   (b) Interpret the results of part a.

(a)   Table 7.4 shows the calculations required to estimate the parameters of the OLS regression equation of Y on X_1 and X_2.

Table 7.3   Per Capita Income, Labor Force in Agriculture, and Years of Schooling

n	1	2	3	4	5	6	7	8	9	10	11	12	13	14	15
Y	6	8	8	7	7	12	9	8	9	10	10	11	9	10	11
X_1	9	10	8	7	10	4	5	5	6	8	7	4	9	5	8
X_2	8	13	11	10	12	16	10	10	12	14	12	16	14	10	12

$$\hat{b}_1 = \frac{\left(\sum x_1 y\right)\left(\sum x_2^2\right) - \left(\sum x_2 y\right)\left(\sum x_1 x_2\right)}{\left(\sum x_1^2\right)\left(\sum x_2^2\right) - \left(\sum x_1 x_2\right)^2} = \frac{(-28)(74) - (38)(-12)}{(60)(74) - (-12)^2}$$

$$= \frac{-2072 + 456}{4440 - 144} \cong -0.38$$

$$\hat{b}_2 = \frac{\left(\sum x_2 y\right)\left(\sum x_1^2\right) - \left(\sum x_1 y\right)\left(\sum x_1 x_2\right)}{\left(\sum x_1^2\right)\left(\sum x_2^2\right) - \left(\sum x_1 x_2\right)^2} = \frac{(38)(60) - (-28)(-12)}{(60)(74) - (-12)^2}$$

$$= \frac{2280 - 336}{4440 - 144} \cong 0.45$$

$$\hat{b}_0 = \bar{Y} - \hat{b}_1 \bar{X}_1 - \hat{b}_2 \bar{X}_2 \cong 9 - (-0.38)(7) - (0.45)(12) = 9 + 2.66 - 5.40 \cong 6.26$$

Thus the estimated OLS regression equation of Y on X_1 and X_2 is

$$\hat{Y}_i = 6.26 - 0.38 X_{1i} + 0.45 X_{2i}$$

(b)   The estimated OLS regression equation indicates that the level of real per capita income Y is inversely related to the percentage of the labor force in agriculture X_1 but directly related to the years of schooling of the population over 25 years (as might have been anticipated). Specifically, $\hat{b}_1$ indicates that a 1 percentage point decline in the labor force in agriculture is associated with an increase in per capita income of 380 U.S. dollars while holding X_2 constant.   However, an increase of 1 year of schooling for the population over 25 years of age is associated with an increase in per capita income of 450 U.S. dollars, while holding X_1 constant.   When $X_{1i} = X_{2i} = 0$, $\hat{Y}_i = \hat{b}_0 = 6.26$.

7.6   Table 7.5 extends Table 6.11 and gives the per capita GDP (gross domestic product) to the nearest \$100 (Y) and the percentage of the economy represented by agriculture (X_1), and the male literacy rate (X_2) reported by the World Bank World Development Indicators for 1999 for 15 Latin American countries.   (a) Find the least-squares regression equation of Y on X_1 and X_2. (b) Interpret the results of part a and compare them with those of Prob. 6.30.

(a)   Table 7.6 shows the calculations required to estimate the parameters of the OLS regression equation of Y on X_1 and X_2.

$$\hat{b}_1 = \frac{\left(\sum x_1 y\right)\left(\sum x_2^2\right) - \left(\sum x_2 y\right)\left(\sum x_1 x_2\right)}{\left(\sum x_1^2\right)\left(\sum x_2^2\right) - \left(\sum x_1 x_2\right)^2} = \frac{(-1149)(1093.7335) - (1637.7335)(-543)}{(442)(1093.7335) - (-543)^2} \cong -1.95$$

$$\hat{b}_2 = \frac{\left(\sum x_2 y\right)\left(\sum x_1^2\right) - \left(\sum x_1 y\right)\left(\sum x_1 x_2\right)}{\left(\sum x_1^2\right)\left(\sum x_2^2\right) - \left(\sum x_1 x_2\right)^2} = \frac{(1637.7335)(442) - (-1149)(-543)}{(442)(1093.7335) - (-543)^2} \cong 0.53$$

$$\hat{b}_0 = \bar{Y} - \hat{b}_1 \bar{X}_1 - \hat{b}_2 \bar{X}_2 = 30.53 - (-1.95)(11) - (0.53)(88.53) = 5.06$$

Thus the estimated OLS regression equation of Y on X_1 and X_2 is

$$\hat{Y} = 5.06 - 1.95 X_1 + 0.53 X_2$$

(b)   The estimated OLS equation indicates that the level of per capita income Y is inversely related to the percentage of the economy represented by agriculture X_1 but directly related to the literacy rate of the male population (as might have been anticipated).   Specifically, $\hat{b}_1$ indicates that a 1 point decline in the percentage of the economy represented by agriculture is associated with an increase in per capita

Table 7.4 Worksheet for Estimating the Parameters for the Data in Table 7.3

n	Y	X_1	X_2	y	x_1	x_2	x_1y	x_2y	x_1x_2	x_1^2	x_2^2
1	6	9	8	−3	2	−4	−6	12	−8	4	16
2	8	10	13	−1	3	1	−3	−1	3	9	1
3	8	8	11	−1	1	−1	−1	1	−1	1	1
4	7	7	10	−2	0	−2	0	4	0	0	4
5	7	10	12	−2	3	0	−6	0	0	9	0
6	12	4	16	3	−3	4	−9	12	−12	9	16
7	9	5	10	0	−2	−2	0	0	4	4	4
8	8	5	10	−1	−2	−2	2	2	4	4	4
9	9	6	12	0	−1	0	0	0	0	1	0
10	10	8	14	1	1	2	1	2	2	1	4
11	10	7	12	1	0	0	0	0	0	0	0
12	11	4	16	2	−3	4	−6	8	−12	9	16
13	9	9	14	0	2	2	0	0	4	4	4
14	10	5	10	1	−2	−2	−2	−2	4	4	4
15	11	8	12	2	1	0	2	0	0	1	0
$n=15$	$\sum Y = 135$ $\bar{Y}=9$	$\sum X_1 = 105$ $\bar{X}_1 = 7$	$\sum X_2 = 180$ $\bar{X}_2 = 12$	$\sum y = 0$	$\sum x_1 = 0$	$\sum x_2 = 0$	$\sum x_1 y = -28$	$\sum x_2 y = 38$	$\sum x_1 x_2 = -12$	$\sum x_1^2 = 60$	$\sum x_2^2 = 74$

Table 7.5 Per Capita Income, Agricultural Proportion, and Literacy

Country:*	(1)	(2)	(3)	(4)	(5)	(6)	(7)	(8)	(9)	(10)	(11)	(12)	(13)	(14)	(15)
n	1	2	3	4	5	6	7	8	9	10	11	12	13	14	15
Y_i	76	10	44	47	23	19	13	19	8	44	4	31	24	59	37
X_1	6	16	9	8	14	11	12	10	18	5	26	8	8	9	5
X_2	97	92	85	96	91	83	93	81	74	93	67	92	94	97	93

*Key: (1) Argentina; (2) Bolivia; (3) Brazil; (4) Chile; (5) Colombia; (6) Dominican Republic; (7) Ecuador; (8) El Salvador; (9) Honduras; (10) Mexico; (11) Nicaragua; (12) Panama; (13) Peru; (14) Uruguay; (15) Venezuela.
Source: World Bank World Development Indicators.

income of 195 U.S. dollars while holding X_2 constant. However, an increase in the male literacy rate of 1 point is associated with an increase in per capita income of 53 U.S. dollars, while holding X_1 constant. When $X_{1i} = X_{2i} = 0$, $\hat{Y}_i = \hat{b}_0 = 5.06$. If X_2 is found to be statistically significant [see Prob. 7.12(b)] and should, therefore, be included in the regression, $\hat{b}_1 = -2.60$ found in Prob. 6.30 is not a reliable estimate of b_1.

TESTS OF SIGNIFICANCE OF PARAMETER ESTIMATES

7.7 Define (a) σ_u^2 and s^2, (b) var $\hat{b}_1$ and var $\hat{b}_2$, (c) $s_{\hat{b}_1}^2$ and $s_{\hat{b}_2}^2$, (d) $s_{\hat{b}_1}$ and $s_{\hat{b}_2}$. (e) Why is b_0 usually not of primary concern?

(a) σ_u^2 is the variance of the error term in the true relationship between X_{1i}, X_{2i}, and Y_i. However, $s^2 = \hat{\sigma}_u^2 = \sum e_i^2/(n-k)$ is the residual variance and is an unbiased estimate of σ_u^2, which is unknown. k is the number of estimated parameters. In the two independent or explanatory variable multiple regression, $k = 3$. Thus $n - k = n - 3 = \text{df}$.

(b)
$$\text{Var } \hat{b}_1 = \sigma_u^2 \frac{\sum x_2^2}{\sum x_1^2 \sum x_2^2 - \left(\sum x_1 x_2\right)^2}$$

while
$$\text{Var } \hat{b}_2 = \sigma_u^2 \frac{\sum x_1^2}{\sum x_1^2 \sum x_2^2 - \left(\sum x_1 x_2\right)^2}$$

The variances of $\hat{b}_1$ and $\hat{b}_2$ (or their estimates) are required to test hypotheses about and construct confidence intervals for b_1 and b_2.

(c)
$$s_{\hat{b}_1}^2 = s^2 \frac{\sum x_2^2}{\sum x_1^2 \sum x_2^2 - \left(\sum x_1 x_2\right)^2} = \frac{\sum e_i^2}{n-k} \frac{\sum x_2^2}{\sum x_1^2 \sum x_2^2 - \left(\sum x_1 x_2\right)^2}$$

$$s_{\hat{b}_2}^2 = s^2 \frac{\sum x_1^2}{\sum x_1^2 \sum x_2^2 - \left(\sum x_1 x_2\right)^2} = \frac{\sum e_i^2}{n-k} \frac{\sum x_1^2}{\sum x_1^2 \sum x_2^2 - \left(\sum x_1 x_2\right)^2}$$

$s_{\hat{b}_1}^2$ and $s_{\hat{b}_2}^2$ are, respectively, unbiased estimates of var $\hat{b}_1$ and var $\hat{b}_2$, which are unknown because σ_u^2 is unknown.

(d) $s_{\hat{b}_1} = \sqrt{s_{\hat{b}_1}^2}$ and $s_{\hat{b}_2} = \sqrt{s_{\hat{b}_2}^2}$. $s_{\hat{b}_1}$ and $s_{\hat{b}_2}$ are, respectively, the standard deviations of $\hat{b}_1$ and $\hat{b}_2$ and are called the *standard errors*.

(e) Unless sufficient observations near $X_{1i} = X_{2i} = 0$ are available, intercept parameter b_0 is usually not of primary concern and a test of its statistical significance can be omitted. Equation (7.18) for var $\hat{b}_0$ is very cumbersome and also for that reason is seldom given and used:

$$\text{Var } \hat{b}_0 = \sigma_u^2 \cdot \frac{\sum X_1^2 \sum X_2^2 - \left(\sum X_1 X_2\right)^2}{n\left[\sum X_1^2 X_2^2 - \left(\sum X_1 X_2\right)^2\right] - \sum X_1\left(\sum X_1 \sum X_2^2 - \sum X_2 \sum X_1 X_2\right)} \qquad (7.18)$$
$$+ \sum X_2\left(\sum X_1 \sum X_1 X_2 - \sum X_2 \sum X_1^2\right)$$

Table 7.6 Worksheet

n	Y	X_1	X_2	y	x_1	x_2	x_1y	x_2y	x_1x_2	x_1^2	x_2^2
1	76	6	97	45.47	−5	8.47	−227.35	385.1309	−42.35	25	71.7409
2	10	16	92	−20.53	5	3.47	−102.65	−71.2391	17.35	25	12.0409
3	44	9	85	13.47	−2	−3.53	−26.94	−47.5491	7.06	4	12.4609
4	47	8	96	16.47	−3	7.47	−49.41	123.0309	−22.41	9	55.8009
5	23	14	91	−7.53	3	2.47	−22.59	−18.5991	7.41	9	6.1009
6	19	11	83	−11.53	0	−5.53	0	63.7609	0	0	30.5809
7	13	12	93	−17.53	1	4.47	−17.53	−78.3591	4.47	1	19.9809
8	19	10	81	−11.53	−1	−7.53	11.53	86.8209	7.53	1	56.7009
9	8	18	74	−22.53	7	−14.53	−157.71	327.3609	−101.71	49	211.1209
10	44	5	93	13.47	−6	4.47	−80.82	60.2109	−26.82	36	19.9809
11	4	26	67	−26.53	15	−21.53	−397.95	571.1909	−322.95	225	463.5409
12	31	8	92	0.47	−3	3.47	−1.41	1.6309	−10.41	9	12.0409
13	24	8	94	−6.53	−3	5.47	19.59	−35.7191	−16.41	9	29.9209
14	59	9	97	28.47	−2	8.47	−56.94	241.1409	−16.94	4	71.7409
15	37	5	93	6.47	−6	4.47	−38.82	28.9209	−26.82	36	19.9809
	$\sum Y_1 = 458$ $\bar{Y} \cong 30.53$	$\sum X_1 = 165$ $\bar{X}_1 = 11.00$	$\sum X_2 = 1328$ $\bar{X}_2 \cong 88.53$				$\sum x_1y = -1149$	$\sum x_2y = 1637.7335$	$\sum x_1x_2 = -543$	$\sum x_1^2 = 442$	$\sum x_2^2 = 1093.7335$

However, $s_{\hat{b}_0}$ is sometimes given in the computer printout, so tests of the statistical significance of b_0 can be conducted easily.

7.8   For the data in Table 7.3, find   (a) s^2,   (b) $s_{\hat{b}_1}^2$ and $s_{\hat{b}_1}$, and   (c) $s_{\hat{b}_2}^2$ and $s_{\hat{b}_2}$.

(a)   The calculations required to find s^2 are shown in Table 7.7, which is an extension of Table 7.4.   The values of $\hat{Y}_i$ are obtained by substituting the values of X_{1i} and X_{2i} into the estimated OLS regression equation found in Prob. 7.5(a):

$$s^2 = \hat{\sigma}_u^2 = \frac{\sum e_i^2}{n-k} = \frac{12.2730}{15-3} \cong 1.02$$

Table 7.7   Per Capita Income Regression: Calculation to Test Significance of Parameters

Country	Y	X_1	X_2	$\hat{Y}$	e	e^2
1	6	9	8	6.44	−0.44	0.1936
2	8	10	13	8.31	−0.31	0.0961
3	8	8	11	8.17	−0.17	0.0289
4	7	7	10	8.10	−1.10	1.2100
5	7	10	12	7.86	−0.86	0.7396
6	12	4	16	11.94	0.06	0.0036
7	9	5	10	8.86	0.14	0.0196
8	8	5	10	8.86	−0.86	0.7396
9	9	6	12	9.38	−0.38	0.1444
10	10	8	14	9.52	0.48	0.2304
11	10	7	12	9.00	1.00	1.0000
12	11	4	16	11.94	−0.94	0.8836
13	9	9	14	9.14	−0.14	0.0196
14	10	5	10	8.86	1.14	1.2996
15	11	8	12	8.62	2.38	5.6644
$n = 15$					$\sum e = 0$	$\sum e^2 = 12.2730$

(b)   Using the value of s^2 found in part a and the values in Table 7.4, we get

$$s_{\hat{b}_1}^2 = s^2 \frac{\sum x_2^2}{\sum x_1^2 \sum x_2^2 - \left(\sum x_1 x_2\right)^2} \cong 1.02 \frac{74}{(60)(74) - (-12)^2} \cong 0.02$$
$$s_{\hat{b}_1} \cong \sqrt{0.02} \cong 0.14$$

(c) $$s_{\hat{b}_2}^2 = s^2 \frac{\sum x_1^2}{\sum x_1^2 \sum x_2^2 - \left(\sum x_1 x_2\right)^2} \cong 1.02 \frac{60}{(60)(74) - (-12)^2} \cong 0.01$$
$$s_{\hat{b}_2} \cong \sqrt{0.01} \cong 0.10$$

7.9   Test at the 5% level of significance for   (a) b_1 and   (b) b_2 in Prob. 7.5(a).

(a) $$t_1 = \frac{\hat{b}_1 - b_1}{s_{\hat{b}_1}} = \frac{-0.38 - 0}{0.14} \cong 2.71$$

Since the absolute value of t_1 exceeds the tabular value of $t = 2.179$ (from App. 5) at the 5% level (two-tail test) and $n - k = 15 - 3 = 12\,\mathrm{df}$, we conclude that b_1 is statistically significant at the 5% level (i.e., we cannot reject H_1, that $b_1 \neq 0$):

(b)
$$t_2 = \frac{\hat{b}_2 - b_2}{s_{\hat{b}_2}} \cong \frac{0.45 - 0}{0.10} = 4.50$$

So b_2 is statistically significant at the 5% (and 1%) level (i.e., H_1, that $b_2 \neq 0$ cannot be rejected).

7.10 Construct the 95% confidence interval for   (a) b_1 and   (b) b_2 in Prob. 7.5(a).

(a) the 95% confidence interval for b_1 is given by
$$b_1 = \hat{b}_1 \pm 2.179 s_{\hat{b}_1} = -0.38 \pm 2.179(0.14) = -0.38 \pm 0.31$$

So b_1 is between -0.69 and -0.07 (i.e., $-0.69 \leq b_1 \leq -0.07$) with 95% confidence.

(b) The 95% confidence interval for b_2 is given by
$$b_2 = \hat{b}_2 \pm 2.179 s_{\hat{b}_2} = 0.45 \pm 2.179(0.10) = 0.45 \pm 0.22$$

So b_2 is between 0.23 and 0.67 (i.e., $0.23 \leq b_2 \leq 0.67$) with 95% confidence.

7.11 For the data in Table 7.5, find   (a) s^2,   (b) $s_{b_1}^2$ and s_{b_1}, and   (c) $s_{b_2}^2$ and s_{b_2}.

(a) The calculations required to find s^2 are shown in Table 7.8, which is an extension of Table 7.6. The values of $\hat{Y}_i$ are obtained by substituting the values of X_{1i} and X_{2i} into the estimated OLS regression equation found in Prob. 7.6(a):
$$s^2 = \hat{\sigma}_u^2 = \frac{\sum e_i^2}{n - k} = \frac{2752.9517}{15 - 3} \cong 229.41$$

(b) Using the value of s^2 found in part a and the values in Table 7.6, we get

Table 7.8   Per Capita GDP Regression: Calculation to Test Significance of Parameters

n	Y	X_1	X_2	$\hat{Y}$	e	e^2
1	76	6	97	44.77	31.23	975.3129
2	10	16	92	22.62	-12.62	159.2644
3	44	9	85	32.56	11.44	130.8736
4	47	8	96	40.34	6.66	44.3556
5	23	14	91	25.99	-2.99	8.9401
6	19	11	83	27.60	-8.60	73.9600
7	13	12	93	30.95	-17.95	322.2025
8	19	10	81	28.49	-9.49	90.0601
9	8	18	74	9.18	-1.18	1.3924
10	44	5	93	44.60	-0.60	0.3600
11	4	26	67	-10.13	14.13	199.6569
12	31	8	92	38.22	-7.22	52.1284
13	24	8	94	39.28	-15.28	233.4784
14	59	9	97	38.92	20.08	403.2064
15	37	5	93	44.60	-7.60	57.7600
						$\sum e^2 = 2752.9517$

$$s_{\hat{b}_1}^2 = s^2 \frac{\sum x_2^2}{\sum x_1^2 \sum x_2^2 - (\sum x_1 \sum x_2)^2} \cong 229.41 \frac{1093.7335}{(442)(1093.7335) - (-543)^2} \cong 1.33$$

$$s_{\hat{b}_1} = \sqrt{s_{\hat{b}_1}^2} = \sqrt{1.33} \cong 1.15$$

(c)　　　$$s_{\hat{b}_2}^2 = s^2 \frac{\sum x_1^2}{\sum x_1^2 \sum x_2^2 - (\sum x_1 \sum x_2)^2} \cong 229.41 \frac{442}{(442)(1093.7335) - (-543)^2} \cong 0.54$$

$$s_{\hat{b}_1} = \sqrt{s_{\hat{b}_2}^2} = \sqrt{0.54} \cong 0.73$$

7.12　Test at the 5% level of significance for　(a) b_1 and　(b) b_2 in Prob. 7.6(a).

(a)　　　　　　　　　　　　　　$$t_1 = \frac{\hat{b}_1 - b_1}{s_{\hat{b}_1}} = \frac{-1.95 - 0}{1.15} \cong -1.69$$

Since the absolute value of t_1 does not exceed the tabular value of $t = 2.179$ (from App. 5) at the 5% level (two-tail test) and $n - k = 15 - 3 = 12$ df, we conclude that b_1 is not statistically significant at the 5% level (i.e., we cannot reject H_0, that $b_1 = 0$).

(b)　　　　　　　　　　　　　　$$t_2 = \frac{\hat{b}_2 - b_2}{s_{\hat{b}_2}} = \frac{0.53 - 0}{0.73} \cong 0.73$$

b_2 is also not statistically significant at the 5% level (i.e., H_0, that $b_2 = 0$ cannot be rejected).

7.13　Construct the 95% confidence interval for　(a) b_1 and　(b) b_2 in Prob. 7.6(a).

(a)　The 95% confidence interval for b_1 is given by

$$b_1 = \hat{b}_1 \pm 2.179 s_{\hat{b}_1} = -1.95 \pm 2.179(1.15) = -1.95 \pm 2.51$$

So b_1 is between -4.46 and 0.56 (i.e., $-4.46 \leq b_1 \leq 0.56$) with 95% confidence.　Since the confidence interval contains 0, we can see that b_1 is not statistically significant.

(b)　The 95% confidence interval for b_2 is given by

$$b_2 = \hat{b}_2 \pm 2.179 s_{\hat{b}_2} = 0.53 \pm 2.179(0.73) = 0.53 \pm 1.59$$

So b_1 is between -1.06 and 2.12 (i.e., $-1.06 \leq b_1 \leq 2.12$) with 95% confidence. Again, the confidence interval contains 0, and we can see that b_2 is not statistically significant.

THE COEFFICIENT OF MULTIPLE DETERMINATION

7.14　Starting with $R^2 = 1 - \sum e_i^2 / \sum y_i^2$, derive $R^2 = (\hat{b}_1 \sum yx_1 + \hat{b}_2 \sum yx_2)/\sum y_i^2$　(*Hint*: Start by showing that $\sum e_i^2 = \sum y_i^2 - \hat{b}_1 \sum yx_1 - \hat{b}_2 \sum yx_2$. The reader without knowledge of calculus can skip this problem.)

$$\sum e_i^2 = \sum e_i(y_i - \hat{y}_i) = \sum e_i(y_i - \hat{b}_1 x_{1i} - \hat{b}_2 x_{2i}) = \sum e_i y_i - \hat{b}_1 \sum e_i x_{1i} - \hat{b}_2 \sum e_i x_{2i}$$

But in the OLS process

$$\frac{\partial \sum e_i^2}{\partial \sum \hat{b}_1} = -\sum e_i x_{1i} = 0 \quad \text{and} \quad \sum e_i x_{1i} = 0$$

$$\frac{\partial \sum e_i^2}{\partial \sum \hat{b}_2} = -\sum e_i x_{2i} = 0 \quad \text{and} \quad \sum e_i x_{2i} = 0$$

Therefore

$$\sum e_i^2 = \sum e_i y_i = \sum (y_i - \hat{y}_i) y_i = \sum y_i (y_i - \hat{b}_1 x_{1i} - \hat{b}_2 x_{2i})$$
$$= \sum y_i^2 - \hat{b}_1 \sum y_i x_{1i} - \hat{b}_2 \sum y_i x_{2i}$$

Substituting into the equation for R^2, we obtain

$$R^2 = 1 - \frac{\sum e_i^2}{\sum y_i^2} = 1 - \frac{\sum y_i^2 - \hat{b}_1 \sum y_i x_{1i} - \hat{b}_2 \sum y_i x_{2i}}{\sum y_i^2} = \frac{\hat{b}_1 \sum y_i x_{1i} + \hat{b}_2 \sum y_i x_{2i}}{\sum y_i^2}$$

or omitting the i for simplicity, we get (as in Sec. 7.3)

$$R^2 = \frac{\hat{b}_1 \sum yx_1 + \hat{b}_2 \sum yx_2}{\sum y^2}$$

7.15 Find R^2 for the OLS regression equation estimated in Prob. 7.5(a), using   (a) $R^2 = \sum \hat{y}_i^2 / \sum y_i^2$, (b) $R^2 = 1 - \sum e_i^2 / \sum y_i^2$, and   (c) $R^2 = (\hat{b}_1 \sum yx_1 + \hat{b}_2 \sum yx_2) / \sum y_i^2$.

(a) From Prob. 6.20, we know that

$$\sum y_i^2 = \sum \hat{y}_i^2 + \sum e_i^2 \qquad \text{so that} \qquad \sum \hat{y}_i^2 = \sum y_i^2 - \sum e_i^2$$

Since $\sum y_i^2 = 40$ (by squaring and adding the y_i values from Table 7.4) and $\sum e_i^2 = 12.2730$ (from Table 7.7), $\sum \hat{y}_i^2 = 40 - 12.2730 = 27.7270$. Thus $R^2 = \sum \hat{y}_i^2 / \sum y_i^2 = 27.7270/40 \cong 0.6932$, or 69.32%.

(b) Using $\sum e_i^2 = 12.2730$ and $\sum y_i^2 = 40$, we get $R^2 = 1 - \sum e_i^2 / \sum y_i^2 = 1 - 12.2730/40 \cong 0.6932$, or 69.32%, the same as in part a.

(c) Using $\hat{b}_1 = -0.38$ and $\hat{b}_2 = 0.45$ [found in Prob. 7.5(a)], $\sum yx_1 = -28$ and $\sum yx_2 = 38$ (from Table 7.4), and $\sum y_i^2 = 40$, we get

$$R^2 = \frac{\hat{b}_1 \sum yx_1 + \hat{b}_2 \sum yx_2}{\sum y^2} = \frac{(-0.38)(-28) + (0.45)(38)}{40} \cong \frac{27.74}{40} = 0.6935, \text{ or } 69.35\%$$

This value of R^2 differs slightly from that found in parts a and b because of rounding errors.

7.16 (a) From $R^2 = 1 - (\sum e_i^2 / \sum y_i^2)$, derive $\bar{R}^2$. (b) What is the range of values for $\bar{R}^2$? (*Hint for part a*: Start from the similarity between $\sum e_i^2$ and var e and $\sum y_i^2$ and var Y.)

(a) The difficulty with (the unadjusted) R^2 is that it does not take into consideration the degrees of freedom. However, var $e = s^2 = \sum e_i^2 / (n-k)$, where $n - k = \text{df}$, and var $Y = \sum (Y_i - \bar{Y})^2 / (n-1)$, where $n - 1 = \text{df}$. Therefore, $\sum e_i^2 = s^2(n-k)$ and $\sum (Y_i - \bar{Y})^2 = \sum y_i^2 = \text{var } Y(n-1)$, so that

$$R^2 = 1 - \frac{\sum e_i^2}{\sum y_i^2} = 1 - \frac{s^2(n-k)}{\text{var } Y(n-1)}$$

Thus $1 - R^2 = (s^2/\text{var } Y)(n-k)/(n-1)$. But $1 - \bar{R}^2 = s^2/\text{var } Y$, so that

$$1 - R^2 = (1 - \bar{R}^2) \frac{(n-k)}{(n-1)}$$

Solving for $\bar{R}^2$, we get

$$\bar{R}^2 = 1 - (1 - R^2) \frac{(n-1)}{(n-k)} \qquad\qquad (7.12)$$

(b) When $k = 1$, $(n-1)/(n-k) = 1$ and $R^2 = \bar{R}^2$. When $k > 1$, $(n-1)/(n-k) > 1$ and $R^2 > \bar{R}^2$. When n is large, for a given k, $(n-1)/(n-k)$ is close to unity and $\bar{R}^2$ and R^2 will not differ much. When n is small and k is large in relation to n, $\bar{R}^2$ will be much smaller than R^2 and $\bar{R}^2$ can even be negative (even though $0 \leq R^2 \leq 1$).

7.17 (a) Find $\bar{R}^2$ for the OLS regression equation estimated in Prob. 7.5(a). (b) How does $\bar{R}^2$ computed in part a compare with R^2 from Prob. 7.15(a) in R^2 from Prob. 6.31(c)?

(a) Using $R^2 = 0.6932$ found in Prob. 7.15(b), we get

$$\bar{R}^2 = 1 - (1 - R^2)\frac{n - 1}{n - k} = 1 - (1 - 0.6932)\frac{15 - 1}{15 - 3} \cong 0.6410$$

(b) $R^2 = 0.33$ in the simple regression, with only the percentage of the labor force in agriculture, X_1, as an independent or explanatory variable [see Prob. 6.31(c)]. $R^2 = 0.69$ by adding the years of schooling for the population over 25 years of age, X_2, as the second independent or explanatory variable. However, when consideration is taken of the fact that the addition of X_2 reduces the degrees of freedom by 1 (from $n - k = 15 - 2 = 13$ in the simple regression of Y on X_1, to $n - k = 15 - 3 = 12$ in the multiple regression of Y on X_1 and X_2), $\bar{R}^2$ is reduced to 0.64. The fact that b_2 was found to be statistically significant [in Prob. 7.9(b)] and $R^2 = \bar{R}^2 = 0.33$ in the simple regression of Y on X_1 and rises to $\bar{R}^2 = 0.64$ in the multiple regression of Y on X_1 and X_2 justifies the retention of X_2 as an additional independent or explanatory variable in the regression equation.

7.18 (a) How can $\sum e_i^2$ (required to conduct tests of significance) be found without first finding $\hat{Y}_i$?
(b) Find $\sum e_i^2$ for the data in Table 7.3 without finding $\hat{Y}_i$ (Table 7.7).

(a) Using the estimated values of $\hat{b}_1$ and $\hat{b}_2$ and $\sum yx_1$, $\sum yx_2$, and $\sum y^2$, we first get

$$R^2 = \frac{\hat{b}_1 \sum yx_1 + \hat{b}_2 \sum yx_2}{\sum y^2}$$

Then $R^2 = 1 - (\sum e_i^2 / \sum y_i^2)$, so that $\sum e_i^2 = (1 - R^2)\sum y_i^2$. This method of finding $\sum e_i^2$ involves much fewer calculations than using $\hat{Y}_i$ (the only additional calculation besides those required to estimate $\hat{b}_1$ and $\hat{b}_2$ is $\sum y_i^2$).

(b) From the value of $R^2 = 0.6935$ found in Prob. 7.15(c) [which utilizes only the estimated values of $\hat{b}_1$ and $\hat{b}_2$ found in Prob. 7.5(a) and the values calculated in Table 7.4] and $\sum y_i^2 = 40$ from Prob. 7.15(a), we get

$$\sum e_i^2 = (1 - R^2)\sum y_i^2 = (1 - 0.6935)(40) = 12.26$$

This compares with $\sum e_i^2 = 12.2730$ found in Table 7.7. (The small difference in the value of $\sum e_i^2$ found by these two methods is obviously due to rounding errors.) Note, however, that finding $\sum e_i^2$ as done above eliminates entirely the need for Table 7.7.

TEST OF THE OVERALL SIGNIFICANCE OF THE REGRESSION

7.19 (a) State the null and alternative hypotheses in testing the overall significance of the regression.
(b) How is the overall significance of the regression tested? What is its rationale? (c) Give the formula for the explained and unexplained or residual variance.

(a) Testing the overall significance of the regression refers to testing the hypothesis that *none* of the independent variables helps to explain the variation of the dependent variable about its mean. Formally, the null hypothesis is

$$H_0: \quad b_1 = b_2 = \cdots = b_k = 0$$

against the alternative hypothesis:

$$H_1: \quad \text{not all } b_i \text{ values are 0}$$

(b) The overall significance of the regression is tested by calculating the F ratio of the explained to the unexplained or residual variance. A "high" value for the F statistic suggests a significant relationship between the dependent and independent variables, leading to the rejection of the null hypothesis that the coefficients of all explanatory variables are jointly zero.

(c) Explained variance $= \sum (\hat{Y}_i - \bar{Y})^2 / (k - 1) = \text{RSS}/(k - 1) = \sum \hat{y}_i^2 / (k - 1)$, where k is number of estimated parameters (see Sec. 6.4).   Unexplained variance $= \sum (Y_i - \hat{Y}_i)^2 / (n - k) = \text{ESS}/(n - k) = \sum e_i^2 / (n - k)$.

7.20 (a) Give the formula for the calculated F ratio or statistic for the case of a simple regression and for a regression with $n = 15, k = 3$. (b) Can the calculated F statistic be "large" and yet none of the estimated parameters be statistically significant?

(a)
$$F_{1,n-2} = \frac{\sum \hat{y}_i^2/1}{\sum e_i^2/(n-2)}$$

where the subscripts on F denote the number of degrees of freedom in the numerator and denominator, respectively. In this simple regression case, $F_{1,n-2} = t_{n-2}^2$ for the same level of significance. For a multiple regression with $n = 15$ and $k = 3$, $F_{2,12} = (\sum \hat{y}_i^2/2)/(\sum e_i^2/12)$.

(b) It is possible for the calculated F statistic to be "large" and yet none of the estimated parameters to be statistically significant. This might occur when the independent variables are highly correlated with each other (see Sec. 9.2). The F test is often of limited usefulness because it is likely to reject the null hypothesis, regardless of whether the model explains "a great deal" of the variation of Y.

7.21 (a) Prove that $[\sum \hat{y}_i^2/(k-1)]/[\sum e_i^2/(n-k)] = [R^2/(k-1)]/[(1-R^2)/(n-k)]$. (b) In view of the result of part a, what is an alternative way to state the hypothesis for testing the overall significance of the regression?

(a)
$$\frac{\sum \hat{y}_i^2/(k-1)}{\sum e_i^2/(n-k)} = \frac{\sum \hat{y}_i^2}{\sum e_i^2}\frac{n-k}{k-1} = \frac{\sum \hat{y}_i^2/\sum y_i^2}{\sum e_i^2/\sum y_i^2}\frac{n-k}{k-1} = \frac{R^2}{(1-R^2)}\frac{n-k}{k-1} = \frac{R^2/(k-1)}{(1-R^2)/(n-k)}$$

(b) The F ratio, as a test of significance of the explanatory power of all independent variables jointly, is roughly equivalent to testing the significance of the R^2 statistic. If the alternative hypothesis is accepted, we would expect R^2, and therefore F, to be "high."

7.22 Test at the 5% level the overall significance of the OLS regression estimated in Prob. 7.5(a) by using (a) $\sum \hat{y}_i^2/(k-1)]/[\sum e_i^2/(n-k)]$ and (b) $[R^2/(k-1)]/[(1-R^2)/(n-k)]$.

(a) Using $\sum \hat{y}_i^2 = 27.727$ from Prob. 7.15(a) and $\sum e_i^2 = 12.2730$ from Table 7.7, we get

$$F_{2,12} = \frac{27.727/2}{12.273/12} \cong 13.59$$

Since the calculated value of F ratio exceeds the tabular value of $F = 3.88$ at the 5% level of significance and 2 and 12 degrees of freedom (see App. 7), the alternative hypothesis that not all b_i's are zero is accepted at the 5% level.

(b) Using $R^2 = 0.6932$ from Prob. 7.15(b), we get

$$F_{2,12} = \frac{R^2/(k-1)}{(1-R^2)/(n-k)} = \frac{0.6932/2}{(1-0.6932)/12} \cong 13.54$$

and we accept the hypothesis that R^2 is significantly different from zero at the 5% level.

PARTIAL-CORRELATION COEFFICIENTS

7.23 (a) How can the influence of X_2 be removed from both Y and X_1 in finding $r_{YX_1 \cdot X_2}$? (b) What is the range of values for partial-correlation coefficients? (c) What is the sign of partial-correlation coefficients? (d) What is the use of partial correlation coefficients?

(a) In order to remove the influence of X_2 on Y, we regress Y on X_2 and find the residual $e_1 = Y^*$. To remove the influence of X_2 on X_1, we regress X_1 on X_2 and find the residual $e_2 = X_1^*$. Y^* and X_1^* then represent the variations in Y and X_1, respectively, left unexplained after removing the influence of X_2 from both Y and X_1. Therefore, the partial correlation coefficient is merely the simple correlation coefficient between the residuals Y^* and X_1^* (that is, $r_{YX_1 \cdot X_2} = r_{Y^* X_1^*}$).

(b) Partial correlation coefficients range in value from -1 to $+1$ (just as in the case of simple correlation coefficients). For example, $r_{YX_1 \cdot X_2} = -1$ refers to the case where there is an exact or perfect negative linear relationship between Y and X_1 after removing the common influence of X_2 from both Y and X_1.

However, $r_{YX_1 \cdot X_2} = 1$ indicates a perfect positive linear *net* relationship between Y and X_1. And $r_{YX_1 \cdot X_2} = 0$ indicates no linear relationship between Y and X_1 when the common influence of X_2 has been removed from both Y and X_1. As a result, X_1 can be omitted from the regression.

(c) The sign of partial correlation coefficients is the same as that of the corresponding estimated parameter. For example, for the estimated regression equation $\hat{Y} = \hat{b}_0 + \hat{b}_1 X_1 + \hat{b}_2 X_2$, $r_{YX_1 \cdot X_2}$ has the same sign as $\hat{b}_1$ and $r_{YX_2 \cdot X_1}$ has the same sign as $\hat{b}_2$.

(d) Partial correlation coefficients are used in multiple regression analysis to determine the relative importance of each explanatory variable in the model. The independent variable with the highest partial correlation coefficient with respect to the dependent variable contributes most to the explanatory power of the model and is entered first in a *stepwise* multiple regression analysis. It should be noted, however, that partial correlation coefficients give an ordinal, not a cardinal, measure of net correlation, and the sum of the partial correlation coefficients between the dependent and all the independent variables in the model need not add up to 1.

7.24 For the regression estimated in Prob. 7.5(a), find (a) $r_{YX_1 \cdot X_2}$ and (b) $r_{YX_2 \cdot X_1}$. (c) Does X_1 or X_2 contribute more to the explanatory power of the model?

(a) To find $r_{YX_1 \cdot X_2}$, we need to find first r_{YX_1}, r_{YX_2}, and $r_{X_1 X_2}$. Using the values from Table 7.4, we get

$$r_{YX_1} = \frac{\sum x_1 y}{\sqrt{\sum x_1^2}\sqrt{\sum y^2}} = \frac{-28}{\sqrt{60}\sqrt{40}} \cong -0.5715$$

$$r_{YX_2} = \frac{\sum x_2 y}{\sqrt{\sum x_2^2}\sqrt{\sum y^2}} = \frac{38}{\sqrt{74}\sqrt{40}} \cong 0.6984$$

$$r_{X_1 X_2} = \frac{\sum x_2 x_1}{\sqrt{\sum x_2^2}\sqrt{\sum x_1^2}} = \frac{-12}{\sqrt{74}\sqrt{60}} \cong -0.1801$$

Then
$$r_{YX_1 \cdot X_2} = \frac{r_{YX_1} - r_{YX_2} r_{X_1 X_2}}{\sqrt{1 - r_{X_1 X_2}^2}\sqrt{1 - r_{YX_2}^2}} = \frac{(-0.5715) - (0.6984)(-0.1801)}{\sqrt{1 - (-0.1801)^2}\sqrt{1 - 0.6984^2}} \cong -0.6331$$

(b) Using the values of r_{YX_1}, r_{YX_2}, and $r_{X_1 X_2}$ calculated in part a, we get

$$r_{YX_2 \cdot X_1} = \frac{r_{YX_2} - r_{YX_1} r_{X_1 X_2}}{\sqrt{1 - r_{X_1 X_2}^2}\sqrt{1 - r_{YX_1}^2}} = \frac{(0.6984) - (-0.5715)(-0.1801)}{\sqrt{1 - (-0.1801)^2}\sqrt{1 - (-0.5715)^2}} \cong 0.8072$$

(c) Since $r_{YX_2 \cdot X_1}$ exceeds the absolute value of $r_{YX_1 \cdot X_2}$, we conclude that X_2 contributes more than X_1 to the explanatory power of the model.

MATRIX NOTATION

7.25 (a) Why is matrix notation used? (b) What are the advantages? (c) What are the disadvantages?

(a) Matrix notation is a mathematical way to represent a system of several linear equation in an organized fashion. Since, by our assumptions (Chap. 6), the standard regression is linear and contains multiple observations of the same linear equation, linear algebra lends itself well to econometrics.

(b) One advantage of matrix notation is conciseness in the notation since one does not have to write summations and ellipses. Also, the matrix solution works for any number of independent variables (from 0 to k).

(c) The main disadvantage of matrix notation is that it requires a more advanced knowledge of linear algebra and matrix mathematics.

7.26 Derive the OLS solution using matrix notation.

In matrices, the regression is written

$$Y = Xb + u$$

We want to minimize the sum of squared errors, or in matrix notation

$$\text{Min } u'u \text{ or Min } (Y - X\hat{b})'(Y - X\hat{b})$$

Taking the first derivative and setting it equal to zero:

$$2X'(Y - X\hat{b}) = 0$$

Expanding terms and simplifying

$$X'Y - X'X\hat{b} = 0$$

Solving for $\hat{b}$

$$X'X\hat{b} = X'Y$$
$$(X'X)^{-1}X'X\hat{b} = (X'X)^{-1}X'Y$$

Since any matrix times its inverse is equal to the identity matrix I

$$I\hat{b} = (X'X)^{-1}X'Y$$

Since any matrix multiplied by I is equal to itself

$$\hat{b} = (X'X)^{-1}X'Y$$

7.27 For the regression in Prob. 7.6, identify the matrices (a) X and (b) Y.

$$(a) \qquad X = \begin{bmatrix} 1 & 6 & 97 \\ 1 & 16 & 92 \\ 1 & 9 & 85 \\ 1 & 8 & 96 \\ 1 & 14 & 91 \\ 1 & 11 & 83 \\ 1 & 12 & 93 \\ 1 & 10 & 81 \\ 1 & 18 & 74 \\ 1 & 5 & 93 \\ 1 & 26 & 67 \\ 1 & 8 & 92 \\ 1 & 8 & 94 \\ 1 & 9 & 97 \\ 1 & 5 & 93 \end{bmatrix}$$

$$Y = \begin{bmatrix} 76 \\ 10 \\ 44 \\ 47 \\ 23 \\ 19 \\ 13 \\ 19 \\ 8 \\ 44 \\ 4 \\ 31 \\ 24 \\ 59 \\ 37 \end{bmatrix}$$

(b)

7.28   For the regression in Prob. 7.6, identify the matrices   (a) $X'X$ and   (b) $(X'X)^{-1}$.

(a)
$$X'X = \begin{bmatrix} 15 & 165 & 1328 \\ 165 & 2257 & 14{,}065 \\ 1328 & 14{,}065 & 118{,}666 \end{bmatrix}$$

(b)
$$(X'X)^{-1} = \begin{bmatrix} 24.7479 & -0.3187 & -0.2392 \\ -0.3187 & 0.0058 & 0.0029 \\ -0.2392 & 0.0029 & 0.0023 \end{bmatrix}$$

SUMMARY PROBLEM

7.29   Table 7.9 gives the hypothetical quantity demanded of a commodity, Y, it price, X_1, and consumers' income, X_2, from 1985 to 1999.   (a) Fit an OLS regression to these observations. (b) Test at the 5% level for the statistical significance of the slope parameters.   (c) Find the unadjusted and adjusted coefficient of multiple correlation.   (d) Test for the overall significance of the regression.   (e) Find the partial correlation coefficients and indicate which independent variable contributes more to the explanatory power of the model.   (f) Find the coefficient of price elasticity of demand η_P and income elasticity of demand η_M at the means.   (g) Report all the results in summary and round off all calculations to four decimal places.

(a)   Table 7.10 gives the calculations required to fit the linear regression.

$$\hat{b}_1 = \frac{(\sum x_1 y)(\sum x_2^2) - (\sum x_2 y)(\sum x_1 x_2)}{(\sum x_1^2)(\sum x_2^2) - (\sum x_1 x_2)^2} = \frac{(-505)(2{,}800{,}000) - (107{,}500)(-11{,}900)}{(60)(2{,}800{,}000) - (-11{,}900)^2} \cong -5.1061$$

$$\hat{b}_2 = \frac{(\sum x_2 y)(\sum x_1^2) - (\sum x_1 y)(\sum x_1 x_2)}{(\sum x_1^2)(\sum x_2^2) - (\sum x_1 x_2)^2} = \frac{(107{,}500)(60) - (-505)(-11{,}900)}{(60)(2{,}800{,}000) - (-11{,}900)^2} \cong 0.0167$$

$$\hat{b}_0 = \bar{Y} - \hat{b}_1 \bar{X}_1 - \hat{b}_2 \bar{X}_2 = 70 - (-5.1061)(6) - (0.0167)(1100) \cong 82.2666$$

$$\hat{Y} = 82.2666 - 5.1061 X_1 + 0.0167 X_2$$

(b)   We can find $\sum e_i^2$ by first calculating R^2 from Table 7.10:

Table 7.9　Quantity Demanded of a Commodity, Price, and Consumers Income, 1985–1999

Year	Y	X_1	X_2
1985	40	9	400
1986	45	8	500
1987	50	9	600
1988	55	8	700
1989	60	7	800
1990	70	6	900
1991	65	6	1000
1992	65	8	1100
1993	75	5	1200
1994	75	5	1300
1995	80	5	1400
1996	100	3	1500
1997	90	4	1600
1998	95	3	1700
1999	85	4	1800

$$R^2 = \frac{\hat{b}_1 \sum yx_1 + \hat{b}_2 \sum yx_2}{\sum y^2} = \frac{(-5.1061)(-505) + (0.0167)(107{,}500)}{4600} \cong 0.9508$$

But　　$$R^2 = 1 - \frac{\sum e^2}{\sum y^2}$$

so $\sum e^2 = (1 - R^2) \sum y^2 = (1 - 0.9508)4600 \cong 226.32$

$$s_{b_1}^2 = \frac{\sum e_i^2}{n-k} \frac{\sum x_2^2}{\sum x_1^2 \sum x_2^2 - \left(\sum x_1 x_2\right)^2}$$

$$= \frac{226.32}{15-3} \frac{2{,}800{,}000}{(60)(2{,}800{,}000) - (-11{,}900)^2} \cong 2.0011 \quad \text{and} \quad s_{\hat{b}_1} \cong 1.4146$$

$$s_{b_2}^2 = \frac{\sum e_i^2}{n-k} \frac{\sum x_1}{\sum x_1^2 \sum x_2^2 - \left(\sum x_1 x_2\right)^2}$$

$$= \frac{226.32}{15-3} \frac{60}{(60)(2{,}800{,}000) - (-11{,}900)^2} \cong 0.00004 \quad \text{and} \quad s_{\hat{b}_2} \cong 0.0065$$

$$t_1 = \frac{\hat{b}_1}{s_{\hat{b}_1}} = \frac{-5.1061}{1.4146} \cong -3.6096 \quad \text{and} \quad t_2 = \frac{\hat{b}_2}{s_{\hat{b}_2}} = \frac{0.0167}{0.0065} \cong 2.5692$$

Therefore, both $\hat{b}_1$ and $\hat{b}_2$ are statistically significant at the 5% level.

(c)　$R^2 = 0.9508$ (found in part b).　Therefore

$$\bar{R}^2 = 1 - (1 - R^2)\frac{n-1}{n-k} = 1 - (1 - 0.9508)\frac{15-1}{15-3} \cong 0.9426$$

(d)　　　　$$F_{k-1, n-k} = \frac{R^2/(k-1)}{(1-R^2)/(n-k)} = \frac{0.9508/(3-1)}{(1-0.9508)/(15-3)} \cong 115.9512$$

Therefore, R^2 is significantly different from 0 at the 5% level.

(e)　To find $r_{YX_1 \cdot X_2}$ and $r_{YX_2 \cdot X_1}$, we must first find (from Table 7.10)

Table 7.10　Quantity Demanded Regression: Calculations

Year	Y	X_1	X_2	y	x_1	x_2	yx_1	yx_2	x_1x_2	x_1^2	x_2^2	y^2
1985	40	9	400	−30	3	−700	−90	21,000	−2100	9	490,000	900
1986	45	8	500	−25	2	−600	−50	15,000	−1200	4	360,000	625
1987	50	9	600	−20	3	−500	−60	10,000	−1500	9	250,000	400
1988	55	8	700	−15	2	−400	−30	6000	−800	4	160,000	225
1989	60	7	800	−10	1	−300	−10	3000	−300	1	90,000	100
1990	70	6	900	0	0	−200	0	0	0	0	40,000	0
1991	65	6	1000	−5	0	−100	0	500	0	0	10,000	25
1992	65	8	1100	−5	2	0	−10	0	0	4	0	25
1993	75	5	1200	5	−1	100	−5	500	−100	1	10,000	25
1994	75	5	1300	5	−1	200	−5	1000	−200	1	40,000	25
1995	80	5	1400	10	−1	300	−10	3000	−300	1	90,000	100
1996	100	3	1500	30	−3	400	−90	12,000	−1200	9	160,000	900
1997	90	4	1600	20	−2	500	−40	10,000	−1000	4	250,000	400
1998	95	3	1700	25	−3	600	−75	15,000	−1800	9	360,000	625
1999	85	4	1800	15	−2	700	−30	10,500	−1400	4	490,000	225
$n = 15$	$\sum Y = 1050$ $\bar{Y} = 70$	$\sum X_1 = 90$ $\bar{X}_1 = 6$	$\sum X_2 = 16,500$ $\bar{X}_2 = 1100$	$\sum y = 0$	$\sum x_1 = 0$	$\sum x_2 = 0$	$\sum yx_1 = -505$	$\sum yx_2 = 107,500$	$\sum x_1x_2 = -11,900$	$\sum x_1^2 = 60$	$\sum x_2^2 = 2,800,000$	$\sum y^2 = 4600$

$$r_{YX_1} = \frac{\sum x_1 y}{\sqrt{\sum x_1^2}\sqrt{\sum y^2}} = \frac{-505}{\sqrt{60}\sqrt{4600}} \cong -0.9613$$

$$r_{YX_2} = \frac{\sum x_2 y}{\sqrt{\sum x_2^2}\sqrt{\sum y^2}} = \frac{107,000}{\sqrt{2,800,000}\sqrt{4600}} \cong 0.9472$$

$$r_{X_1 X_2} = \frac{\sum x_2 x_1}{\sqrt{\sum x_2^2}\sqrt{\sum x_1^2}} = \frac{-11,900}{\sqrt{2,800,000}\sqrt{60}} \cong -0.9181$$

$$r_{YX_1 \cdot X_2} = \frac{r_{YX_1} - r_{YX_2} r_{X_1 X_2}}{\sqrt{1 - r_{X_1 X_2}^2}\sqrt{1 - r_{YX_2}^2}} = \frac{(-0.9613) - (0.9472)(-0.9181)}{\sqrt{1 - (-0.9181)^2}\sqrt{1 - (-0.9472)^2}} \cong -0.7213$$

$$r_{YX_2 \cdot X_1} = \frac{r_{YX_2} - r_{YX_1} r_{X_1 X_2}}{\sqrt{1 - r_{X_1 X_2}^2}\sqrt{1 - r_{YX_1}^2}} = \frac{(0.9472) - (-0.9613)(-0.9181)}{\sqrt{1 - (-0.9181)^2}\sqrt{1 - (-0.9613)^2}} \cong 0.5919$$

Thus X_1 contributes more than X_2 to the explanatory power of the model.

(f) $$\eta_P = \hat{b}_1 \frac{\bar{X}_1}{\bar{Y}} = -5.1061 \frac{6}{70} \cong -0.4377$$

$$\eta_M = \hat{b}_2 \frac{\bar{X}_2}{\bar{Y}} = 0.0167 \frac{1100}{70} \cong 0.2624$$

(g) $\hat{Y}_1 = 82.2666 - 5.1061 X_1 + 0.0167 X_2$ $R^2 = 0.9508$ $\bar{R}^2 = 0.9426$ $F_{2,12} = 115.9512$

     t values (-3.6096) (2.5692)

$$r_{YX_1 \cdot X_2} = -0.7213 \qquad r_{YX_2 \cdot X_1} = 0.5919$$

$$\eta_P = -0.4377 \qquad \eta_M = 0.2624$$

Supplementary Problems

THE THREE-VARIABLE LINEAR MODEL

7.30   Table 7.11 extends Table 6.12 and gives observations on Y, X_1, and X_2. Find the OLS regression equation of Y on X_1 and X_2.
     Ans. $\hat{Y}_i = 4.76 + 5.29 X_{1i} + 2.13 X_{2i}$

Table 7.11   Observations on Y, X_1, and X_2

n	1	2	3	4	5	6	7	8	9	10
Y	20	28	40	45	37	52	54	43	65	56
X_1	2	3	5	4	3	5	7	6	7	8
X_2	5	6	6	5	5	7	6	6	7	7

7.31   With reference to the estimated OLS regression equation of Y on X_1 and X_2 in Prob. 7.30 interpret (*a*) $\hat{b}_0$, (*b*) $\hat{b}_1$, and (*c*) $\hat{b}_2$.
     Ans.   (*a*) $\hat{b}_0 = 4.76$ is the constant or Y intercept; $\hat{Y}_i = \hat{b}_0 = 4.76$, when $X_{1i} = X_{2i} = 0$ (*b*) $b_1 = 5.29$, indicating that a one-unit increase in X_1 (while holding X_2 constant) results in an increase in $\hat{Y}_i$ of 5.29 units   (*c*) $\hat{b}_2 = 2.13$, indicating that a one-unit increase in X_2 (while holding X_1 constant) results in an increase in $\hat{Y}_i$ of 2.13 units

TESTS OF SIGNIFICANCE OF PARAMETER ESTIMATES

7.32 With reference to the data in Table 7.11, find (a) s^2, (b) $s^2_{\hat{b}_1}$ and $s_{\hat{b}_1}$, and (c) $s^2_{\hat{b}_2}$ and $s_{\hat{b}_2}$.
        Ans. (a) $s^2 = 50$ (b) $s^2_{\hat{b}_1} \cong 3.16$ and $s_{\hat{b}_1} \cong 1.78$ (c) $s^2_{\hat{b}_2} \cong 18.95$ and $s_{\hat{b}_2} \cong 4.35$

7.33 Test at the 5% level of significance for (a) b_1 and (b) b_2 in Prob. 7.30.
        Ans. (a) b_1 is statistically significant at the 5% level (b) b_2 is not statistically significant at the 5% level

7.34 Construct the 95% confidence interval for (a) b_1 and (b) b_2 in Prob. 7.30.
        Ans. (a) $1.08 \leq b_1 \leq 9.50$ (b) $-8.16 \leq b_2 \leq 12.42$

THE COEFFICIENT OF MULTIPLE DETERMINATION

7.35 For the estimated OLS regression found in Prob. 7.30, find (a) R^2 and (b) $\bar{R}^2$. (c) Should X_2 be included in the regression?
        Ans. (a) $R^2 \cong 0.79$ [using $R^2 = 1 - (\sum e_i^2 / \sum y_i^2)$] (b) $\bar{R}^2 \cong 0.73$ (c) Since b_2 was not found to be statistically significant [in Prob. 7.33(b)] and $\bar{R}^2$ fell from $R^2 = \bar{R}^2 = 0.77$ with only X_1 as an independent variable [see Prob. 6.40(a)] to $\bar{R}^2 = 0.73$ (above), X_2 should not be included in the regression.

7.36 For $R^2 = 0.60, n = 10$, and $k = 1$, find $\bar{R}^2$.
        Ans. $\bar{R}^2 = 0.60$

7.37 For $R^2 = 0.60, n = 10$, and $k = 2$, find $\bar{R}^2$.
        Ans. $\bar{R}^2 = 0.55$

7.38 For $R^2 = 0.60$ and $k = 2$ (as in Prob. 7.37) but $n = 100$, find $\bar{R}^2$.
        Ans. $\bar{R}^2 = 0.596$

7.39 For $R^2 = 0.40, n = 10$, and $k = 5$, find $\bar{R}^2$.
        Ans. $\bar{R}^2 = -0.08$ (but is interpreted as being equal to 0)

TEST OF THE OVERALL SIGNIFICANCE OF THE REGRESSION

7.40 For the estimated OLS regression in Prob. 7.30, find (a) the explained variance, (b) the unexplained or residual variance, and (c) the F ratio or statistic.
        Ans. (a) $\sum \hat{y}^2 / (k-1) \cong 649$ (b) $\sum e^2 / (n-k) = 50$ (c) $F_{2,7} = 12.98$

7.41 Test the overall significance of the OLS regression estimated in Prob. 7.30 at (a) the 5% level and (b) at the 1% level.
        Ans. (a) Since the calculated F ratio (12.98) exceeds the tabular or theoretical value of F (4.74) at $\alpha = 0.05$ and df $= 2$ and 7, we accept the hypothesis that the estimated OLS regression parameters are jointly significant at the 5% level. (b) Since the tabular value of F is 9.55 at $\alpha = 0.01$, the alternative hypothesis is accepted at the 1% level of significance also.

PARTIAL CORRELATION COEFFICIENTS

7.42 For the estimated OLS regression in Prob. 7.30, find (a) $r_{YX_1 \cdot X_2}$ and (b) $r_{YX_2 \cdot X_1}$. (c) Which independent variable contributes more to the explanatory power of the model?
        Ans. (a) $r_{YX_1 \cdot X_2} = 0.74$ (b) $r_{YX_2 \cdot X_1} = 0.18$ (c) X_1

MATRIX NOTATION

7.43 (a) What is the first column of the X matrix? (b) Where is the variance of $\hat{b}_1$ in the $s^2_{\hat{b}}$ matrix?
        Ans. (a) a column of 1s (b) second row, second column

SUMMARY PROBLEM

7.44   Table 7.12 extends Table 6.13 and gives data for a random sample of 12 couples on the number of children they had, Y, the number of children they stated that they wanted at the time of their marriage, X_1, and the years of education of the wife, X_2. (a) Find the OLS regression equation of Y on X_1 and X_2. (b) Calculate t values and test at the 5% level for the statistical significance of the slope parameters. (c) Find the unadjusted and adjusted coefficient of multiple correlation. (d) Test for the overall significance of the regression. (e) Find the partial correlation coefficients and indicate which independent variable contributes more to the explanatory power of the model. Carry out all calculations to two decimal places.

Table 7.12   Number of Children Had and Wanted and Education of Wife

Couple	1	2	3	4	5	6	7	8	9	10	11	12
Y	4	3	0	4	4	3	0	4	3	1	3	1
X_1	3	3	0	2	2	3	0	3	2	1	3	2
X_2	12	14	18	10	10	14	18	12	15	16	14	15

Ans.   (a) $\hat{Y} = 6.90 + 0.53X_1 - 0.39X_2$ (b) Since $t_1 = 3.12$ and $t_2 = -5.57$, both $\hat{b}_1$ and $\hat{b}_2$ are statistically significant at the 5% level.   (c) $R^2 = 0.92$ and $\bar{R}^2 = 0.90$ (d) Since $F_{2,9} = 51.31$, R^2 is statistically significant at the 5% level.   (e) $r_{YX_1 \cdot X_2} = 0.71$ and $r_{YX_2 \cdot X_1} = -0.87$; thus X_2 contributes more than X_1 to the explanatory power of the model.

CHAPTER 8

Further Techniques and Applications in Regression Analysis

8.1 FUNCTIONAL FORM

Theory or the scatter of points frequently suggests nonlinear relationships. It is possible to transform some nonlinear functions into linear ones so that the OLS method can still be used. Some of the most common of these and their transformations are shown in Table 8.1. Applying the OLS method to the transformed linear functions gives unbiased slope estimates. In Eq. (8.1), b_1 is the elasticity of Y with respect to X.

Table 8.1 Functional Forms and their Transformations

Function	Transformation	Form	Equation
$Y = b_0 X^{b_1} e^u$	$Y^* = b_0^* + b_1 X^* + u$	Double log	(8.1)
$\ln Y = b_0 + b_1 X + u$	$Y^* = b_0 + b_1 X + u$	Semilog	(8.2)
$Y = b_0 + (b_1/X) + u$	$Y = b_0 + b_1 Z + u$	Reciprocal	(8.3)
$Y = b_0 + b_1 X + b_2 X^2 + u$	$Y = b_0 + b_1 X + B_2 W + u$	Polynomial	(8.4)

where $Y^* = \ln Y$, $b_0^* = \ln b_0$, $X^* = \ln X$, $u = \ln e^u$, $Z = 1/X$, $W = X^2$
$\ln$ = the natural logarithm to the base $e \cong 2.718$

EXAMPLE 1. Suppose that we postulate a demand function of the form

$$Y = b_0 X_1^{b_1} X_2^{b_2} e^u$$

where Y = quantity demanded of a commodity
X_1 = its price
X_2 = consumers' income

181

Utilizing the data in Table 7.9 and applying the OLS method to this demand function transformed into double-log linear form, we get

$$\ln Y = 1.96 - 0.26 \ln X_1 + 0.39 \ln X_2 \qquad R^2 = 0.97$$
$$\qquad\qquad (-3.54) \qquad (6.64)$$

where -0.26 and 0.39 are, respectively, unbiased estimates of the price and income elasticity of demand (see Prob. 8.2).   The fit here seems better than for the linear form [see Prob. 7.29(g)].

8.2   DUMMY VARIABLES

Qualitative explanatory variables (such as wartime vs. peacetime, periods of strike vs. nonstrike, male vs. females, etc.) can be introduced into regression analysis by assigning the value of 1 for one classification (e.g., wartime) and 0 for the other (e.g., peacetime).   These are called *dummy variables* and are treated as any other variable.   Dummy variables can be used to capture changes (shifts) in the intercept [Eq. (8.5)], changes in slope [Eq. (8.6)], and changes in both intercept and slope [Eq. (8.7)]:

$$Y = b_0 + b_1 X + b_2 D + u \qquad\qquad\qquad (8.5)$$
$$Y = b_0 + b_1 X + b_2 XD + u \qquad\qquad\qquad (8.6)$$
$$Y = b_0 + b_1 X + b_2 D + b_3 XD + u \qquad\qquad (8.7)$$

where D is 1 for one classification and 0 otherwise and X is the usual quantitative explanatory variable. Dummy variables also can be used to capture differences among more than two classifications, such as seasons and regions [Eq. (8.8)]:

$$Y = b_0 + b_1 X + b_2 D_1 + b_3 D_2 + b_4 D_3 + u \qquad\qquad (8.8)$$

where b_0 is the intercept for the first season or region and D_1, D_2, and D_3 refer, respectively, to season or region 2, 3, and 4.   Note that for any number of classifications k, $k - 1$ dummies are required (see Probs. 8.9, 8.26, and 8.27).   For qualitative dependent variables, see Sec. 8.5.

EXAMPLE 2.   Table 8.2 gives gross private domestic investment Y and gross national product X, both in billions of current dollars, for the United States from 1939 to 1954.   Using $D = 1$ for the war years (1942–1945) and $D = 0$ for the peace years, we get

$$\hat{Y} = -2.58 - 0.16X - 20.81D \qquad R^2 = 0.94$$
$$\qquad\quad (10.79) \quad (-6.82)$$

D is statistically significant at the 5% level.   Thus $\hat{b}_0 = -2.58$ for peacetime and -23.39 for wartime, while $\hat{b}_1 = 0.16$ is the common slope coefficient.   (For tests of a difference in slope, as well as differences in intercept and slope, see Probs. 8.7 and 8.8.)

Table 8.2   Gross Private Domestic Investment and Gross National Product (in Billions of Dollars); United States, 1939–1954

Year	1939	1940	1941	1942	1943	1944	1945	1946	1947	1948	1949	1950	1951	1952	1953	1954
Y	9.3	13.1	17.9	9.9	5.8	7.2	10.6	30.7	34.0	45.9	35.3	53.8	59.2	52.1	53.3	52.7
X	90.8	100.0	124.9	158.3	192.0	210.5	212.3	209.3	232.8	259.1	258.0	286.2	330.2	347.2	366.1	366.3

Source: Economic Report of the President, U.S. Government Printing Office, Washington, DC, 1980, p. 203.

8.3   DISTRIBUTED LAG MODELS

It is often the case that the current value of the dependent variable is a function of or depends on the weighted sum of present t and past values of the independent variable (and the error term), with generally different weights assigned to various time periods:

$$Y_t = a + b_0 X_t + b_1 X_{t-1} + b_2 X_{t-2} + \cdots + u_t \qquad (8.9)$$

Estimating the *distributed lag model* [Eq. (*8.9*)] presents two difficulties: (1) the data on one observation or time period are lost for each lagged value of X; and (2) the Xs are likely to be related to each other, so that it may be difficult or impossible to isolate the effect of each X on Y.

These difficulties can be eliminated by deriving from Eq. (8.9) the *Koyck lag model* [Eq. (*8.10*)], which assumes that the weights decline geometrically (see Prob. 8.11):

$$Y_t = a(1 - \lambda) + b_0 X_t + \lambda Y_{t-1} + v_t \qquad (8.10)$$

where $0 < \lambda < 1$ and $v_t = u_t - \lambda u_{t-1}$. However, Eq. (*8.10*) violates two assumptions of the OLS model and results in biased and inconsistent estimators that require adjustment (see Sec. 9.3).

Alternatively, the *Almon lag model* can be used. This allows for a more flexible lag structure to be approximated empirically by a polynomial of degree at least one more than the number of turning points in the function (see Prob. 8.13). Assuming a three-period lag [Eq. (*8.11*)] taking the form of a second-degree polynomial [Eq. (*8.12*)], we can derive Eq. (*8.13*) (see Prob. 8.14):

$$Y_t = a + b_0 X_t + b_1 X_{t-1} + b_2 X_{t-2} + b_3 X_{t-3} + u_t \qquad (8.11)$$

where
$$b_i = c_0 + c_1 i + c_2 i^2 \qquad (8.12)$$

so that
$$Y_t = a + c_0 Z_{1t} + c_1 Z_{2t} + c_2 Z_{3t} + v_t \qquad (8.13)$$

where
$$Z_{1t} = \sum_{i=0}^{3} X_{t-i} \qquad Z_{2t} = \sum_{i=1}^{3} i X_{t-i} \quad \text{and} \quad Z_{3t} = \sum_{i=1}^{3} i^2 X_{t-i}$$

The values of the $\hat{b}_i$ terms in Eq. (*8.11*) are obtained by substituting the estimated values of c_0, c_1, and c_2 from Eq. (*8.13*) into Eq. (*8.12*) (see Prob. 8.15).

EXAMPLE 3. Table 8.3 gives the level of imports Y and the gross domestic product X, both in billions of 1996 dollars, for the United States from 1980 to 1999. Fitting the Koyck model, we get

$$\hat{Y}_t = -329.99 + 0.57 X_t + 0.03 Y_{t-1} \qquad R^2 = 0.99$$
$$\qquad\quad (3.15) \quad\ (2.95)$$

where $\hat{\lambda} = 0.57$ and $\hat{\alpha}(1 - 0.57) = -329.99$, so that $\hat{\alpha} = -767.42$.

Table 8.3 Imports and Gross Domestic Product (in Billions of 1996 Dollars): United States, 1980–1999

Year	1980	1981	1982	1983	1984	1985	1986	1987	1988	1989
Imports	585.6	584.2	537.4	547.8	663.3	656.4	703.8	766.1	827.3	866.2
GDP	19,603.6	20,083.7	19,677.5	20,529.4	22,020.5	22,868.1	23,649.6	24,453.0	25,473.2	26,367.3
Year	1990	1991	1992	1993	1994	1995	1996	1997	1998	1999
Imports	877.3	819.2	829.6	873.8	988.4	1102.9	1159.0	1269.2	1321.9	1446.5
GDP	26,831.6	26,705.7	27,520.4	28,250.6	29,390.9	30,175.3	31,252.5	32,638.0	34,062.6	35,503.1

Source: St. Louis Federal Reserve (Bureau of Economic Analysis).

8.4 FORECASTING

Forecasting refers to the estimation of the value of the dependent variable Y_F given the actual or projected value of the independent variable X_F. The forecast-error variance σ_F^2 is given by

$$\sigma_F^2 = \sigma_u^2 \left[1 + \frac{1}{n} + \frac{(X_F - \bar{X})^2}{\sum(X_i - \bar{X})^2} \right] \tag{8.14}$$

where n is the number of observations and σ_u^2 is the variance of u. Since σ_u^2 is seldom known, we use s^2 as an unbiased estimate of σ_u^2, so that the estimated forecast-error variance, s_F^2, is

$$s_F^2 = s^2 \left[1 + \frac{1}{n} + \frac{(X_F - \bar{X})^2}{\sum(X_i - \bar{X})^2} \right] \tag{8.15}$$

The 95% confidence interval for the forecast Y_F is

$$\hat{Y}_F \pm t_{0.025} s_F$$

where $\hat{Y}_F = \hat{b}_0 + \hat{b}_1 X_F$ and t refers to the t distribution with $n - 2$ degrees of freedom.

EXAMPLE 4. Returning to the corn-fertilizer example in Chap. 6, recall that $\hat{Y}_i = 27.12 + 1.66\,X_i$, $n = 10$, $\bar{X} = 18$, $\sum(X_i - \bar{X})^2 = 576$ (from Example 6.2), and $s^2 = \sum e_i^2/(n-2) \cong 47.31/8 \cong 5.91$ (from Example 6.3). Projecting for 1981 an amount of fertilizer used per acre of $X_F = 35$, we get

$$s_F^2 = 5.91 \left[1 + \frac{1}{10} + \frac{(35 - 18)^2}{576} \right] \cong 9.46 \quad \text{and} \quad s_F \cong 3.08$$

$$\hat{Y}_F = 27.12 + 1.66(35) = 45.38$$

Then the 95% confidence or forecast interval for Y_F in 1981 is $45.38 \pm (2.31)(3.08)$, or between 38.27 and 52.49. (See Prob. 8.19 for forecasting in multiple regression analysis.)

8.5 BINARY CHOICE MODELS

If the dependent variable is a dummy variable, an OLS regression is not appropriate. An OLS regression could yield incongruous predictions greater than 1 or less than 0. Also, the regression would violate the assumption of no heteroscedasticity because of the discrete nature of the dependent variable. To estimate the model, we first set up an underlying model

$$Y_i^* = b_0 + b_1 X_i + u_i$$

Here, Y^* is considered an underlying propensity for the dummy variable to take the value of 1 and is a continuous variable so that

$$Y_i = \begin{cases} 1 & \text{if} \quad Y_i^* \geq 0 (u_i \geq -b_0 - b_1 X_i) \\ 0 & \text{if} \quad Y_i^* < 0 (u_i < -b_0 - b_1 X_i) \end{cases}$$

The *maximum-likelihood* estimate of the coefficients is calculated by setting up the *log-likelihood function*

$$\ln L = \Sigma_1 [\ln(P(u_i \geq -b_0 - b_1 X_i \mid Y_i = 1))] + \Sigma_0 [\ln(P(u_i < -b_0 - b_1 X_i \mid Y_i = 0))]$$

where Σ_1 and Σ_0 indicate sum of all probabilities for those data points where $Y_i = 1$ and 0, respectively, and $\hat{b}_0$ and $\hat{b}_1$ are chosen to maximize the log-likelihood function. If the standard normal distribution is used to find the probabilities, it is a *probit model*; if the logistic distribution is used, it is a *logit model*. Since these functions are nonlinear, estimation by computer is usually required (see Chap. 12).

EXAMPLE 5. We estimate the relationship between the openness of a country Y and a country's per capita income in dollars X in 1992. We hypothesize that higher per capita income should be associated with free trade, and test this at the 5% significance level. Data are given in Table 8.4. The variable Y takes the value of 1 for free trade, 0 otherwise.

Since the dependent variable is a binary variable, we set up the indicator function

$$Y^* = b_0 + b_1(X) + u$$

Table 8.4   Openness of Trade, and GDP per Capita: International Data for 1992

Country	Burundi	Chad	Congo	Egypt	Hong Kong
Y	0	0	0	0	1
X	569	408	2240	1869	16,471
Country	India	Indonesia	Ivory Coast	Kenya	Malaysia
Y	0	1	0	0	1
X	1282	2102	1104	914	5746
Country	Morocco	Nigeria	Rwanda	Singapore	South Africa
Y	1	0	0	1	1
X	2173	978	762	12,653	3068
Country	Tunisia	Uganda	Uruguay	Venezuela	Zimbabwe
Y	1	1	1	1	0
X	3075	547	5185	7082	1162

Source: Per capita GDP, World Bank World Development Indicators. Openness, Sachs-Warner Dates.

If $Y^* \geq 0$, $Y = 1$ (open).   If $Y^* < 0$, $Y = 0$ (not open).

Probit estimation gives the following results:

$$\hat{Y}^* = -1.9942 + 0.0010(X) \qquad s_{\hat{b}_0} = 0.8247, s_{\hat{b}_1} = 0.0005, \ln L = -6.8647$$

To test significance, we can use the usual t test, but since probit uses the standard normal distribution, the z tables can be used:

$t_{\hat{b}_0} = \hat{b}_0/s_{\hat{b}_0} = -1.9942/0.8247 = -2.42 < -1.96$ (from App. 3); therefore significant at the 5% level.

$t_{\hat{b}_1} = \hat{b}_1/s_{\hat{b}_1} = 0.0010/0.0005 = 2 > 1.96$; therefore significant at the 5% level.

8.6   INTERPRETATION OF BINARY CHOICE MODELS

The interpretation of b_1 changes in a binary choice model. b_1 is the effect of X on Y^*.   The *marginal effect* of X on $P(Y = 1)$ is easier to interpret and is given by

$$f(b_0 + b_1 \bar{X}) \cdot b_1$$

where Probit: $f(x) = \phi(x) = \dfrac{1}{\sqrt{2\pi}} e^{-(x^2/2)}$

Logit: $f(x) = \lambda(x) = \dfrac{e^x}{(1 + e^x)^2}$

To test the fit of the model (analogous to R^2), the maximized log-likelihood value ($\ln L$) can be compared to the maximized log likelihood in a model with only a constant ($\ln L_0$) in the *likelihood ratio index*

$$\text{LRI} = 1 - \frac{\ln L}{\ln L_0}$$

Another measure of goodness of fit is to compare predicted values of Y to actual values.   Customarily, if $P(u_i \geq -\hat{b}_0 - \hat{b}_1 X_i) > 0.5$, then $\hat{Y}_i = 1$.

EXAMPLE 6.   Continuing with the interpretation from Sec. 8.5.   The marginal effect of X (GDP/cap) on the probability of a country to be open is

$$\phi(b_0 + b_1\overline{X}) \cdot b_1 = \phi(-1.9942 + 0.0010(3469.5))(0.0010) = 0.0001$$

This can also be interpreted as the marginal effect of X on the expected value of Y.

$$\text{LRI} = 1 - \frac{\ln L}{\ln L_0} = 1 - (-6.8647)/(-13.8629) = 0.50 \qquad (\ln L_0 = -13.8629)$$

Predicted probabilities are given in Table 8.5. The model predicts 18 out of 20 countries correctly, or 90%. (*Note*: If values for X when $Y = 1$ are all greater or all less than values when $Y = 0$, the binary choice model cannot be estimated.)

Table 8.5   Predicted Probabilities for the Probit Model

Country	Burundi	Chad	Congo	Egypt	Hong Kong
$P(Y = 1)$	0.08	0.06	0.60	0.45	> 0.99
Country	India	Indonesia	Ivory Coast	Kenya	Malaysia
$P(Y = 1)$	0.24	0.54	0.19	0.14	> 0.99
Country	Morocco	Nigeria	Rwanda	Singapore	South Africa
$P(Y = 1)$	0.57	0.15	0.11	> 0.99	0.86
Country	Tunisia	Uganda	Uruguay	Venezuela	Zimbabwe
$P(Y = 1)$	0.86	0.07	> 0.99	> 0.99	0.20

Predicted

Actual		$Y = 0$	$Y = 1$
	$Y = 0$	9	1
	$Y = 1$	1	9

Solved Problems

FUNCTIONAL FORM

8.1   (*a*) How is the form of the functional relationship decided?   (*b*) What are some of the most useful transformations into linear functions?   (*c*) Are the estimated parameters obtained from the application of the OLS method to transformed linear functions unbiased estimates of the true population parameters?

(*a*)   Economic theory can sometimes suggest the functional form of an economic relationship.   For example, microeconomic theory postulates an average (short-run) cost curve that is U-shaped and an average fixed-cost curve that constantly falls and approaches the quantity axis asymptotically as total fixed costs are spread over more and more units produced.   The scatter of points also suggest the appropriate functional form in a two-variable relationship.   When neither theory nor scatter of points is of help, the linear function is usually tried first because of its simplicity.

(*b*)   Some of the most useful and common transformations of nonlinear into linear functions are the double logarithm or double log, the semilog, the reciprocal, and the polynomial functions (see Table 8.1).   One of the advantages of the double-log form is that the slope parameters represent elasticities (see Prob.

8.2). The semilog function is appropriate when the dependent variable grows at about a constant rate over time, as in the case of the labor force and population (see Prob. 8.4). The reciprocal and polynomial functions are appropriate to estimate average-cost and total-cost curves (see Prob. 8.5).

(c) The estimation of a transformed double-log function by the OLS method results in unbiased slope estimators. However, $\hat{b}_0 = $ antilog $\hat{b}_0^*$ is a biased but consistent estimator of b_0. The fact that $\hat{b}_0$ is biased is not of much consequence because the constant is usually not of primary interest [see Prob. 7.7(e)]. In the other transformed functions in Table 8.1, $\hat{b}_0$ also is unbiased. The double-log linear model is appropriate when $\ln Y$ plotted against $\ln X$ lies approximately on a straight line.

8.2 Prove that in the double-log demand function of the form

$$Q = b_0 P^{b_1} Y^{b_2} e^u$$

where Q is the quantity demanded, P is the price, and Y is the income,   (a) b_1 is the price elasticity of demand, or η_P, and   (b) b_2 is the income elasticity of demand, or η_Y. (The reader without knowledge of calculus can skip this problem.)

(a) The definition of price elasticity of demand is

$$\eta_P = \frac{dQ}{dP} \cdot \frac{P}{Q}$$

The derivative of the Q function with respect to P is

$$\frac{dQ}{dP} = b_1(b_0 P^{b_1-1} Y^{b_2} e^u) = b_1(b_0 P^{b_1} Y^{b_2} e^u)P^{-1} = b_1 \cdot \frac{Q}{P}$$

Substituting the value of dQ/dP into the formula for η_P, we get

$$\eta_P = \frac{dQ}{dP} \cdot \frac{P}{Q} = b_1 \cdot \frac{Q}{P} \cdot \frac{P}{Q} = b_1$$

(b) The definition of income elasticity of demand is

$$\eta_Y = \frac{dQ}{dY} \cdot \frac{Y}{Q}$$

The derivative of the Q function with respect to Y is

$$\frac{dQ}{dY} = b_2(b_0 P^{b_1} Y^{b_2-1} e^u) = b_2(b_0 P^{b_1} Y^{b_2} e^u)Y^{-1} = b_2 \cdot \frac{Q}{Y}$$

Substituting the value of dQ/dY into the formula for η_Y, we get

$$\eta_Y = \frac{dQ}{dY} \cdot \frac{Y}{Q} = b_2 \cdot \frac{Q}{Y} \cdot \frac{Y}{Q} = b_2$$

8.3 Table 8.6 gives the output in tons Q, the labor input in hours L, and capital input in machine-hours K, of 14 firms in an industry. Fit the data to the Cobb-Douglas production function

$$Q = b_0 L^{b_1} K^{b_2} e^u$$

Table 8.6   Output and Labor and Capital Inputs of 14 Firms in an Industry

Firm	1	2	3	4	5	6	7	8	9	10	11	12	13	14
Q	240	400	110	530	590	470	450	160	290	490	350	550	560	430
L	1480	1660	1150	1790	1880	1860	1940	1240	1240	1850	1570	1700	2000	1850
K	410	450	380	430	480	450	490	395	430	460	435	470	480	440

The data are first transformed into natural log form, as shown in Table 8.7, and then the OLS method is applied to the transformed variables as explained in Sec. 6.2 (the computer does all of this).   The results are

$$\ln Q = -23.23 + 1.43 \ln L + 3.05 \ln K \qquad R^2 = 0.88$$
$$\qquad\qquad\quad (2.55) \qquad (2.23)$$

Table 8.7   Output and Labor and Capital Input in Original and Log Form

Firm	Q	L	K	$\ln Q$	$\ln L$	$\ln K$
1	240	1480	410	5.48064	7.29980	6.01616
2	400	1660	450	5.99146	7.41457	6.10925
3	110	1150	380	4.70048	7.04752	5.94017
4	530	1790	430	6.27288	7.48997	6.06379
5	590	1880	480	6.38012	7.53903	6.17379
6	470	1860	450	6.15273	7.52833	6.10925
7	450	1940	490	6.10925	7.57044	6.19441
8	160	1240	395	5.07517	7.12287	5.97889
9	290	1240	430	5.66988	7.12287	6.06379
10	490	1850	460	6.19441	7.52294	6.13123
11	350	1570	435	5.85793	7.35883	6.07535
12	550	1700	470	6.30992	7.43838	6.15273
13	560	2000	480	6.32794	7.60090	6.17379
14	430	1850	440	6.06379	7.52294	6.08677

The estimated coefficients 1.43 and 3.05 refer, respectively, to the output elasticity of L and K.   Since $1.43 + 3.05 = 4.48 > 1$, there are increasing returns to scale in this industry (e.g., increasing the inputs of both L and K by 10% causes output to increase by 44.8%).

8.4   Table 8.8 gives the number of nonfarm persons employed N (in millions) in the United States from 1980 to 1999.   Fit an OLS regression line to the data in Table 8.8.

Table 8.8.   Millions of Persons Employed in the United States from 1980 to 1999

Year	1980	1981	1982	1983	1984	1985	1986	1987	1988	1989
N	90.4	91.2	89.5	90.2	94.4	97.4	99.3	102.0	105.2	107.9
Year	1990	1991	1992	1993	1994	1995	1996	1997	1998	1999
N	109.4	108.2	108.6	110.7	114.2	117.2	119.6	122.7	125.9	128.8

Source: Bureau of Labor Statistics.

Since employment tends to grow at about a constant rate over time T, we fit a semilog function of the form of Eq. (8.2) to the transformed data in Table 8.9.   The result is

$$\ln N = 4.46 + 0.02T \qquad R^2 = 0.99$$
$$(26.77)$$

Table 8.9   Millions Employed in the
United States, 1980–1999:
Original and Transformed Data

Year	N	$\ln N$	T
1980	90.4	4.5042	1
1981	91.2	4.5131	2
1982	89.5	4.4942	3
1983	90.2	4.5020	4
1984	94.4	4.5475	5
1985	97.4	4.5788	6
1986	99.3	4.5981	7
1987	102.0	4.6250	8
1988	105.2	4.6559	9
1989	107.9	4.6812	10
1990	109.4	4.6950	11
1991	108.2	4.6840	12
1992	108.6	4.6877	13
1993	110.7	4.7068	14
1994	114.2	4.7380	15
1995	117.2	4.7639	16
1996	119.6	4.7842	17
1997	122.7	4.8097	18
1998	125.9	4.8355	19
1999	128.8	4.8583	20

8.5   Fit a short-run average-cost curve to the data in Table 8.10, which gives average cost AC and output Q for a firm over a 12-week period.

Table 8.10   Average Cost and Output of a Firm over a 12-Week Period

Week	1	2	3	4	5	6	7	8	9	10	11	12
AC	82	86	100	100	95	85	110	88	86	108	87	87
Q	149	121	190	100	109	138	209	170	158	201	130	181

Since microeconomic theory postulates U-shaped short-run cost curves, we fit

$$AC = b_0 - b_1 Q + b_2 W + u \qquad \text{where } W = Q^2$$

The result is

$$\widehat{AC} = 244.86 - 2.20Q + 0.01Q^2 \qquad R^2 = 0.94$$
$$(-9.84) \quad (10.42)$$

DUMMY VARIABLES

8.6   (a) Write an equation for peacetime and one for wartime for Eqs. (8.5) to (8.7), if C = consumption, Y_d = disposable income, and $D = 1$ for war years and $D = 0$ for peace years.   (b) Draw a figure for Eqs. (8.5) to (8.7) showing a consumption function for peace years and one for war years.   (a) What are the advantages of estimating Eqs. (8.5) to (8.7) as opposed to estimating two regressions, one for peace years and one for war years, in each case?

(a)   Letting a equations refer to peacetime and b equations refer to wartime, we get

$$C = b_0 + b_1 Y_d + u \qquad (8.5a)$$
$$C' = (b_0 + b_2) + b_1 Y_d + u \qquad (8.5b)$$
$$C = b_0 + b_1 Y_d + u \qquad (8.6a)$$
$$C' = b_0 + (b_1 + b_2) Y_d + u \qquad (8.6b)$$
$$C = b_0 + b_1 Y_d + u \qquad (8.7a)$$
$$C' = (b_0 + b_2) + (b_1 + b_3) Y_d + u \qquad (8.7b)$$

Note that all peacetime equations are identical because $D = 0$.   During wartime, consumption is less than in peacetime because of controls, reduced availability of goods and services, and moral suasion. Thus b_2 and b_3 (the coefficients of D) are expected to be negative for war years, so that the equations for war years have a lower intercept and/or slope than the peacetime equations.

(b)   See Fig. 8-1.

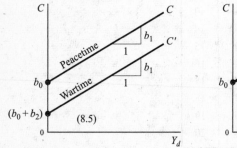 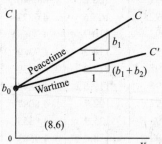 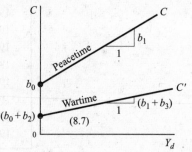

Fig. 8-1

(c)   The advantages of estimating Eqs. (8.5) to (8.7) as opposed to estimating a separate regression in each case, one for peacetime and one for wartime, are   (1) the degrees of freedom are greater,   (2) a variety of hypotheses can easily be tested to see if the differences in constants and/or slopes are statistically significant, and   (3) computer time is saved.

8.7   Table 8.11 gives the quantity of milk (in thousands of quarts) supplied by a firm per month Q at various prices P over a 14-month period.   The firm faced a strike in some of its plants during the fifth, sixth, and seventh months.   Run a regression of Q on P (a) testing only for a shift in the intercept during periods of strike and nonstrike and   (b) testing for a shift in the intercept and slope.

Table 8.11   Quantity Supplied of Milk (in Thousands of Quarts) at Various Prices

Month	1	2	3	4	5	6	7	8	9	10	11	12	13	14
Q	98	100	103	105	80	87	94	113	116	118	121	123	126	128
P	0.79	0.80	0.82	0.82	0.93	0.95	0.96	0.88	0.88	0.90	0.93	0.94	0.96	0.97

(a)   Letting $D = 1$ during the months of strike and $D = 0$ otherwise, we get

$$\hat{Q} = -32.47 + 165.97P - 37.64D \qquad R^2 = 0.98$$
$$(15.65) \quad (-23.59)$$

Since D is statistically significant at better than the 1% level, the intercept is $b_0 = -32.47$ during the period of no strike, and it equals $b_0 + b_2 = -32.47 - 37.64 = -70.11$ during the strike period.

(b)
$$\hat{Q} = -29.74 + 162.86P - 309.62D + 287.14PD \qquad R^2 = 0.99$$
$$\phantom{\hat{Q} = -29.74 + }(27.16) \quad (-5.67) \quad (4.98)$$

D and pD are statistically significant at better than the 1% level.   The intercept and slope are, respectively, -29.74 and 162.86 during the period of no strike.   During the strike period, the intercept is $\hat{b}_0 + \hat{b}_2 = -29.74 - 309.26 = -339$, while the slope is $\hat{b}_1 + \hat{b}_3 = 162.86 + 287.14 = 450$ (since the firm, presumably, is able to step up the increase in output in its nonstriking plants).

8.8   Table 8.12 gives the consumption expenditures C, the disposable income Y_d, and the sex of the head of the household S of 12 random families.   (a) Regress C on Y_d.   (b) Test for a different intercept for families with a male or a female as head of the household.   (c) Test for a different slope or MPC (marginal propensity to consume) for families with a male or a female as head of the household.   (d) Test for both different intercept and slope.   (e) Which is the "best" result?

Table 8.12   Consumption, Disposable Income, and Sex of Head of Household of 12 Random Families

Family	1	2	3	4	5	6	7	8	9	10	11	12
C	18,535	11,350	12,130	15,210	8680	16,760	13,480	9680	17,840	11,180	14,320	19,860
Y_d	22,550	14,035	13,040	17,500	9430	20,635	16,470	10,720	22,350	12,200	16,810	23,000
S	M	M	F	M	F	M	M	F	M	F	F	M

(a)
$$\hat{C} = 1663.60 + 0.75Y_d \qquad R^2 = 0.978$$
$$\phantom{\hat{C} = 16}(2.73) \quad (21.12)$$

(b)   Letting $D = 1$ for families headed by a female and $D = 0$ otherwise, we get

$$\hat{C} = 186.12 + 0.82Y_d + 832.09D \qquad R^2 = 0.984$$
$$\phantom{\hat{C} = 186.12 + }(16.56) \quad (1.82)$$

(c)
$$\hat{C} = 709.18 + 0.79Y_d + 0.05Y_dD \qquad R^2 = 0.983$$
$$\phantom{\hat{C} = 709.18 + }(18.11) \quad (1.51)$$

(d)
$$\hat{C} = -184.70 + 0.83Y_d + 1757.99D - 0.06Y_dD \qquad R^2 = 0.985$$
$$\phantom{\hat{C} = -184.70 + }(13.65) \quad (1.03) \quad (-0.57)$$

(e)   Since neither D nor Y_dD is statistically significant at the 5% level in parts b, c, and d, there is no difference in the consumption patterns of households headed by males or females.   Thus the best results are those given in part a.

8.9   Table 8.13 gives the retail sales (in billions of 1996 dollars) of the United States from the first quarter of 1995 to the fourth quarter of 1999.   (a) Prepare a table showing sales, a time trend, and dummy variables to take into account seasonal effects.   (b) Using the data from the table in part a, run a regression of sales on inventories and the seasonal dummies and interpret the results.

Table 8.13   Retail Sales in the United States (in Billions of 1996 $)

Sales	540.5	608.5	606.6	648.3	568.4	632.8	626.0	674.6	587.0	640.2
quarter	I	II	III	IV	I	II	III	IV	I	II
year	1995				1996				1997	

Sales	645.9	686.9	597.0	675.3	663.6	723.3	639.5	716.5	721.9	779.9
quarter	III	IV	I	II	III	IV	I	II	III	IV
year	1997		1998				1999			

Source: St. Louis Federal Reserve (U.S. Department of Commerce, Census Bureau).

(a)   Taking the first quarter as the base, and letting $D_1 = 1$ for the second quarter and 0 otherwise, $D_2 = 1$ for the third quarter and 0 otherwise, and $D_3 = 1$ for the fourth quarter and 0 otherwise, we get Table 8.14.

Table 8.14   Sales, Time Trend, and Seasonal Dummies

Year	Quarter	Sales	Time Trend	D_1	D_2	D_3
1995	I	540.5	1	0	0	0
1995	II	608.5	2	1	0	0
1995	III	606.6	3	0	1	0
1995	IV	648.3	4	0	0	1
1996	I	568.4	5	0	0	0
1996	II	632.8	6	1	0	0
1996	III	626.0	7	0	1	0
1996	IV	674.6	8	0	0	1
1997	I	587.0	9	0	0	0
1997	II	640.2	10	1	0	0
1997	III	645.9	11	0	1	0
1997	IV	686.9	12	0	0	1
1998	I	597.0	13	0	0	0
1998	II	675.3	14	1	0	0
1998	III	663.6	15	0	1	0
1998	IV	723.3	16	0	0	1
1999	I	639.5	17	0	0	0
1999	II	716.5	18	1	0	0
1999	III	721.9	19	0	1	0
1999	IV	779.9	20	0	0	1

(b)   Using the data from Table 8.14 to regress sales, S, on the time trend, T, D_1, D_2, D_3, we get

$$\hat{S} = 526.56 + 6.66\,T + 61.52 D_1 + 53.01\,D_2 + 96.15\,D_3 \qquad R_2 = 0.98$$
$$\text{(13.78)   (7.95)   (6.81)   (12.23)}$$

Since all dummy variables are statistically significant at the 5% level, we obtain

$$\hat{S} = 526.56 + 6.66\,T \qquad \text{in quarter I}$$
$$\hat{S} = 588.08 + 6.66\,T \qquad \text{in quarter II}$$

$$\hat{S} = 579.57 + 6.66\ T \qquad \text{in quarter III}$$

$$\hat{S} = 622.71 + 6.66\ T \qquad \text{in quarter IV}$$

These results remain unchanged when four dummies are used, one for each of the four seasons, *but* the constant from the regression equation is dropped. Using the four seasonal dummies and the constant together would make it impossible to estimate the OLS regression (see Sec. 9.2).

DISTRIBUTED LAG MODELS

8.10 (*a*) What is meant by a *distributed lag model*? (*b*) Write the equation for a general distributed lag model with an infinite number of lags and for one with k lags. (*c*) What practical difficulties arise in estimating a distributed lag model with k lags?

(*a*) Often the effect of a policy variable may be distributed over a series of time periods (i.e., the dependent variable may be "sluggish" to respond to a policy change), requiring a series of lagged explanatory variables to account for the full adjustment process through time. A *distributed lag model* is one in which the current value of the dependent variable Y_t depends on the weighted sum of present and past values of the independent variables (X_t, X_{t-1}, X_{t-2}, etc.) and the error term, with generally different weights assigned to various time periods (usually declining successively for earlier time periods).

(*b*)
$$Y_t = a + b_0 X_t + b_1 X_{t-1} + b_2 X_{t-2} + \cdots + u_t \tag{8.9}$$
$$Y_t = a + b_0 X_t + b_1 X_{t-1} + b_2 X_{t-2} + \cdots + b_k X_{t-k} + u_t \tag{8.9a}$$

Note that in Eqs. (*8.9*) and (*8.9a*), a is constant, while b_0 is the coefficient of X_t. This has been done in order to simplify the algebraic manipulation in Prob. 8.11(*a*).

(*c*) In the estimation of a distributed lag model, the inclusion of each lagged term uses up one degree of freedom. When the number of independent lagged terms k is small, the model can be estimated with OLS, as done in Chap. 7. However, with k large (in relation to the length of the time series), an inadequate number of degrees of freedom may be left to estimate the model or to be confident in the estimated parameters. Moreover, the lagged explanatory variables in a distributed lag model are likely to be strongly correlated, so it may be difficult to adequately separate their independent effects on the dependent variable [see Prob. 7.3(*b*)].

8.11 (*a*) Derive the Koyck distributed lag model. (*b*) What problems arise in the estimation of this model? (*Hint for part a*: Start with the general distributed lag model and assume that the weights decline geometrically, with λ referring to a constant larger than 0 and smaller than 1; then lag the relationship by one period, multiply by λ, and subtract it from the original relationship.)

(*a*) Starting with Eq. (*8.9*), it is assumed that all the usual assumptions of OLS are satisfied (see Prob. 7.1):
$$Y_t = a + b_0 X_t + b_1 X_{t-1} + b_2 X_{t-2} + \cdots + u_t \tag{8.9}$$

Geometrically declining weights and $0 < \lambda < 1$ gives
$$b_i = \lambda^i b_0 \qquad i = 1, 2, \ldots \tag{8.16}$$

Substituting Eq. (*8.16*) into Eq. (*8.9*), we obtain
$$Y_t = a + b_0 X_t + \lambda b_0 X_{t-1} + \lambda^2 b_0 X_{t-2} + \cdots + u_t$$

Lagging by one period, we have
$$Y_{t-1} = a + b_0 X_{t-1} + \lambda b_0 X_{t-2} + \lambda^2 b_0 X_{t-3} + \cdots + u_{t-1}$$

Multiplying by λ yields
$$\lambda Y_{t-1} = \lambda a + \lambda b_0 X_{t-1} + \lambda^2 b_0 X_{t-2} + \cdots + \lambda u_{t-1}$$

and subtracting from Eq. (*8.9*) yields

$$Y_t - \lambda Y_{t-1} = a - \lambda a + b_0 X_t + \lambda b_0 X_{t-1} - \lambda b_0 X_{t-1}$$
$$+ \lambda^2 b_0 X_{t-2} - \lambda^2 b_0 X_{t-2} + \cdots + u_t - \lambda u_{t-1}$$
$$Y_t - \lambda Y_{t-1} = a(1 - \lambda) + b_0 X_t + u_t - \lambda u_{t-1}$$
$$Y_t = a(1 - \lambda) + b_0 X_t + \lambda Y_{t-1} + v_t \qquad (8.10)$$

where $v_t = u_t - \lambda u_{t-1}$. Note that in Eq. (8.10) the number of regressors has been reduced to only two, with only one X.

(b) Two serious problems arise in the estimation of a Koyck distributed lag model. First, if u_t in Eq. (8.9) satisfies all the OLS assumptions (see Prob. 6.4), then $v_t = u_t - \lambda u_{t-1}$ in Eq. (8.10) does not. Specifically, $E(v_t v_{t-1}) \neq 0$ because v_t and v_{t-1} are both defined with u_{t-1} in common (i.e., $v_t = u_t - \lambda u_{t-1}$ and $v_{t-1} = u_{t-1} - \lambda u_{t-2}$). In addition, $E(v_t Y_{t-1}) \neq 0$. Violations of these OLS assumptions result in biased and inconsistent estimators for the Koyck lag model [Eq. (8.10)], requiring elaborate correction procedures (some of which are discussed in Sec. 9.3). The second serious problem is that the Koyck model rigidly assumes geometrically declining weights. This may seldom be the case in the real world, thus requiring a more flexible lag scheme (see Prob. 8.13).

8.12 Table 8.15 gives the level of inventories Y and sales X (in billions of dollars) in U.S. manufacturing from 1981 to 1999. (a) Fit the Koyck model to the data in Table 8.15. (b) What is the value of $\hat{\lambda}$ and $\hat{a}$?

Table 8.15   Inventories and Sales in U.S. Manufacturing, 1981–1999 (in Billions of Dollars)

Year	1981	1982	1983	1984	1985	1986	1987	1988	1989	1990
Y	546	574	590	650	664	663	710	767	815	841
X	345	344	396	417	428	445	473	522	533	542
Year	1991	1992	1993	1994	1995	1996	1997	1998	1999	
Y	835	843	870	935	996	1014	1062	1100	1151	
X	542	585	609	672	701	730	769	797	872	

Source: St. Louis Federal Reserve (U.S. Department of Commerce, Census Bureau).

(a) $$\hat{Y}_t = 88{,}426.14 + 0.60 \, X_t + 0.50 \, Y_{t-1} \qquad R^2 = 0.99$$
$$(4.49) \quad (4.22)$$

(b) $\hat{\lambda} = 0.50$ and $\hat{a}(1 - 0.50) = 88{,}426.14$,   so $\hat{a} = 176{,}852.28$

8.13 (a) What is the lag structure in the Almon lag model? (b) What are the advantages and disadvantages of the Almon lag model with respect to the Koyck model?

(a) While the Koyck lag model assumes geometrically declining weights, the Almon lag model allows for any lag structure, to be approximated empirically by a polynomial of degree at least one more than the number of turning points in the function. For example, a lag structure of the form of an inverted U (i.e., with $b_1 > b_0$) can be approximated by a polynomial of at least the second degree. This may arise, as in the case of an investment function, when because of delays in recognition and in making decisions, the level of investment in the current period is more responsive to demand conditions in a few earlier periods than in the current period.

(b) The Almon lag model has at least two important advantages with respect to the Koyck lag model. First (and as pointed out earlier), the Almon model has a flexible lag structure as opposed to the rigid lag structure of the Koyck model. Second, since the Almon lag model does not replace the lagged independent variables (the Xs) with the lagged dependent variable, it does not violate any of the OLS assumptions (as does the Koyck model). One disadvantage of the Almon model is that the number of coefficients to be estimated is not reduced by as much as in the Koyck model. Another disadvantage is that in actual empirical work, neither the period nor the form of the lag may be suggested by theory or be known a priori.

8.14 Derive the Almon transformation for (a) a three-period lag taking the form of a second-degree polynomial and (b) a four-period lag taking the form of a third-degree polynomial.

(a) Starting with Eqs. (8.11) and (8.12)

$$Y_t = a + b_0 X_t + b_1 X_{t-1} + b_2 X_{t-2} + b_3 X_{t-3} + u_t \qquad (8.11)$$

$$b_i = c_0 + c_1 i + c_2 i^2 \qquad \text{with } i = 0, 1, 2, 3 \qquad (8.12)$$

and substituting Eq. (8.12) into Eq. (8.11), we get

$$Y_t = a + c_0 X_t + (c_0 + c_1 + c_2)X_{t-1} + (c_0 + 2c_1 + 4c_2)X_{t-2} + (c_0 + 3c_1 + 9c_2)X_{t-3} + u_t$$

Rearranging the terms in the last expression:

$$Y_t = a + c_0 \left(\sum_{i=0}^{3} X_{t-i} \right) + c_1 \left(\sum_{i=1}^{3} i X_{t-i} \right) + c_2 \left(\sum_{i=1}^{3} i^2 X_{t-i} \right) + u_t$$

and letting $Z_{1t} = \sum_{i=0}^{3} X_{t-i}$, $Z_{2t} = \sum_{i=1}^{3} i X_{t-i}$, and $Z_{3t} = \sum_{i=1}^{3} i^2 X_{t-i}$, we get

$$Y_t = a + c_0 Z_{1t} + c_1 Z_{2t} + c_2 Z_{3t} + u_t \qquad (8.13)$$

(b) With a four-period lag taking the form of a third-degree polynomial, we have

$$Y_t = a + b_0 X_t + b_1 X_{t-1} + b_2 X_{t-2} + b_3 X_{t-3} + b_4 X_{t-4} + u_t$$

$$b_i = c_0 + c_1 i + c_2 i^2 + c_3 i^3 \qquad \text{with } i = 0, 1, 2, 3, 4$$

Substituting the second into the first, we get

$$Y_t = a + c_0 X_t + (c_0 + c_1 + c_2 + c_3)X_{t-1} + (c_0 + 2c_1 + 4c_2 + 8c_3)X_{t-2}$$
$$+ (c_0 + 3c_1 + 9c_2 + 27c_3)X_{t-3} + (c_0 + 4c_1 + 16c_2 + 64c_3)X_{t-4} + u_t$$

Rearranging the terms in the last expression, we have

$$Y_t = a + c_0 \left(\sum_{i=0}^{4} X_{t-i} \right) + c_1 \left(\sum_{i=1}^{4} i X_{t-i} \right) + c_2 \left(\sum_{i=1}^{4} i^2 X_{t-i} \right) + c_3 \left(\sum_{i=1}^{4} i^3 X_{t-i} \right) + u_t$$

and letting the terms in parentheses equal, respectively, Z_{1t}, Z_{2t}, Z_{3t}, and Z_{4t}, we get

$$Y_t = a + c_0 Z_{1t} + c_1 Z_{2t} + c_2 Z_{3t} + c_3 Z_{4t} + u_t$$

8.15 Using the data from Table 8.15 and assuming a three-period lag taking the form of a second-degree polynomial, (a) Prepare a table with the original variables and the calculated Z values to be used to estimate the Almon lag model. (b) Regress the level of inventories, Y, on the Z values in the table in part a, i.e., estimate regression Eq. (8.13). (c) Find the $\hat{b}$ values and write out estimated Eq. (8.11).

(a) The Z values given in Table 8.16 are calculated as follows:

Table 8.16 Inventories, Sales, and Z Values in U.S. Manufacturing, 1981–1999 (in Billions of Dollars)

Year	Y	X	Z_1	Z_2	Z_3
1981	546	345	—	—	—
1982	574	344	—	—	—
1983	590	396	—	—	—
1984	650	417	1502	2119	4877
1985	664	428	1585	2241	5097
1986	663	445	1686	2450	5660
1987	710	473	1763	2552	5910
1988	767	522	1868	2647	6105
1989	815	533	1973	2803	6419
1990	841	542	2070	2996	6878
1991	835	542	2139	3174	7372
1992	843	585	2202	3225	7507
1993	870	609	2278	3295	7631
1994	935	672	2408	3405	7827
1995	996	701	2567	3645	8373
1996	1014	730	2712	3872	8870
1997	1062	769	2872	4148	9582
1998	1100	797	2997	4332	9998
1999	1151	872	3168	4525	10,443

$$Z_{1t} = \sum_{i=0}^{3} X_{t-i} = (X_t + X_{t-1} + X_{t-2} + X_{t-3})$$

$$Z_{2t} = \sum_{i=1}^{3} iX_{t-i} = (X_{t-1} + 2X_{t-2} + 3X_{t-3})$$

$$Z_{3t} = \sum_{i=1}^{3} i^2 X_{t-i} = (X_{t-1} + 4X_{t-2} + 9X_{t-3})$$

(b) Regressing Y on the Zs, we get

$$\hat{Y}_t = 171.80 + 0.44\,Z_{1t} + 0.27\,Z_{2t} - 0.15\,Z_{3t} \qquad R^2 = 0.99$$
$$\phantom{\hat{Y}_t = 171.80 + } (2.20) \qquad (0.56) \qquad (-0.99)$$

(c)
$$\hat{\alpha} = 171.80$$
$$\hat{b}_0 = \hat{c}_0 = 0.44$$
$$\hat{b}_1 = (\hat{c}_0 + \hat{c}_1 + \hat{c}_2) = (0.44 + 0.27 - 0.15) = 0.56$$
$$\hat{b}_2 = (\hat{c}_0 + 2\hat{c}_1 + 4\hat{c}_2) = (0.44 + 0.54 - 0.60) = 0.38$$
$$\hat{b}_3 = (\hat{c}_0 + 3\hat{c}_1 + 9\hat{c}_2) = (0.44 + 0.81 - 1.35) = -0.10$$

so that $\hat{Y}_t = 171.80 + 0.44\,X_t + 0.56\,X_{t-1} + 0.38\,X_{t-2} - 0.10\,X_{t-3}$
$$\phantom{so that \hat{Y}_t = 171.80 + } (2.20) \qquad (3.41) \qquad (2.31) \qquad (-0.47)$$

where the standard errors of the lagged values of X have been found by

$$\sqrt{\text{var}\,\hat{b}_i} = \sqrt{\text{var}\,(\hat{c}_0 + \hat{c}_1 i + \hat{c}_2 i^2)} \tag{8.17}$$

FORECASTING

8.16   (*a*) What is meant by *forecasting*? *Conditional forecast*? *Prediction*?   (*b*) What are the possible sources of errors in forecasting?   (*c*) What is the forecast-error variance? What is an unbiased estimate of the forecast-error variance? What do they depend on?   (*d*) How is the value of $\hat{Y}_F$ found? The 95% confidence interval of the forecast, Y_F?

(*a*)   *Forecasting* refers to the estimation of the value of a dependent variable, given the actual or projected value of the independent variable(s).   When the forecast is based on an estimated or projected (rather than on an actual) value of the independent variable, we have a *conditional forecast*.   *Prediction* is often used interchangeably with forecasting.   At other times, *prediction* refers to estimating an intrasample value of the dependent variable.   *Forecasting*, then, refers to estimating a future value of the dependent variable.

(*b*)   Forecasting errors arise because of   (1) the random nature of the error term,   (2) estimated unbiased parameters equal the true parameters only on the average,   (3) errors in projecting the independent variables, and   (4) incorrect model specification.

(*c*)   The forecast-error variance σ_F^2 is given by

$$\sigma_F^2 = \sigma_u^2 \left[1 + \frac{1}{n} + \frac{(X_F - \bar{X})^2}{\sum (X_i - \bar{X})^2} \right] \tag{8.14}$$

where n is the number of observations and σ_u^2 is the variance of u.   An unbiased estimate of the forecast-error variance s_F^2 is given by

$$s_F^2 = s^2 \left[1 + \frac{1}{n} + \frac{(X_F - \bar{X})^2}{\sum (X_i - \bar{X})^2} \right] \tag{8.15}$$

where s^2 is an unbiased estimate of σ_u^2 given by

$$s^2 = \frac{\sum (Y_i - \hat{Y}_i)^2}{n - 2} = \frac{\sum e_i^2}{n - 2} \tag{6.12}$$

The larger is n, the smaller is σ_F^2 (or s_F^2), σ_u^2 (or s^2), and the difference between X_F and $\bar{X}$.

(*d*)   The value of $\hat{Y}_F$ is found by substituting the actual or projected value of X_F into the estimated regression equation:

$$\hat{Y}_F = \hat{b}_0 + \hat{b}_1 X_F$$

The 95% confidence interval of the forecast Y_F is given by

$$\hat{Y}_F \pm t_{0.025} s_F$$

where t refers to the t distribution with $n - 2$ degrees of freedom.

8.17   Find the 95% confidence interval of the forecast for Y in Prob. 6.30 for   (*a*) $X = 15\%$ and (*b*) $X = 11.5\%$.

(*a*)   In Prob. 6.30, we found that $\hat{Y}_i = 59.13 - 2.60 X_i$, $n = 15$, $\bar{X} = 11.00$, $\sum x_i^2 = 442$, and $s^2 = 2872.8535/13 \cong 220.99$.   For $X = 11\%$, we obtain

$$s_F^2 = 220.99 \left(1 + \frac{1}{15} + \frac{(15 - 11)^2}{442} \right) \cong 243.72 \qquad s_F \cong 15.61$$

$$\hat{Y}_F = 59.13 - 2.60(15) = 20.13$$

Then the 95% confidence interval for Y_F is

$$20.13 \pm (2.18)(15.61) \qquad \text{or between } -13.90 \text{ and } 54.16$$

where $2.18 = t_{0.025}$ with df $= 13$.

(*b*)   For $X = 11.5\%$

$$s_F^2 = 220.99 \left(1 + \frac{1}{15} + \frac{(11.5 - 11)^2}{442} \right) \cong 235.85 \quad \text{and} \quad s_F \cong 15.35$$

$$\hat{Y}_F = 59.13 - 2.60(11.5) = 29.23$$

Then the 95% confidence interval for Y_F is

$$29.23 \pm (2.18)(15.35) \quad \text{or between} \quad -4.23 \text{ and } 62.69$$

Note that the range of the 95% confidence interval for Y_F is less here than in part a because the difference between the projected value of X and $\bar{X}$ is smaller here.

8.18   Draw a graph showing a hypothetical positively sloped estimated OLS regression line, the 95% confidence interval for Y_F for a given X_F, and the 95% confidence interval bands for Y_F.

See Fig. 8-2. Note that the 95% confidence bands are closest at $X_F = \bar{X}$.

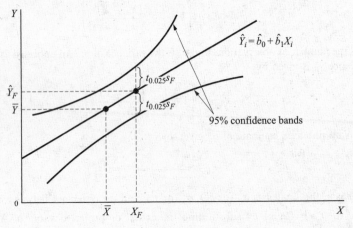

Fig. 8-2

8.19   Find the 95% confidence interval of Y_F for $X_{1F} = 35$ and $X_{2F} = 25$ in 1981, given that $\hat{Y}_i = 31.98 + 0.65X_{1i} + 1.11X_{2i}$, $\bar{X}_1 = 18$, $\bar{X}_2 = 12$ (from Example 7.1), $s^2 = \sum e_i^2/(n-k) = 13.67/7 \cong 1.95$, $s_{b_1}^2 \cong 0.06$, $s_{b_2}^2 \cong 0.07$ (from Example 7.2), $s_{b_0}^2 \cong 2.66$, $\text{cov}(\hat{b}_1, \hat{b}_2) = s_{\hat{b}_1, \hat{b}_2} \cong 0.07$ (from the computer), and if

$$s_F^2 = s^2 + s_{b_0}^2 + s_{b_1}^2 (X_{1F} - \bar{X}_1)^2 + s_{b_2}^2 (X_{2F} - \bar{X}_2)^2 + s_{\hat{b}_1, \hat{b}_2}(X_{1F} - \bar{X}_1)(X_{2F} - \bar{X}_2) \qquad (8.18)$$

$$= 1.95 + 2.66 + 0.06(35 - 18)^2 + 0.07(25 - 12)^2 + (-0.07)(35 - 18)(25 - 12)$$

$$\cong 18.31 \quad \text{and} \quad s_F \cong 4.28$$

$$\hat{Y}_F = 31.98 + 0.65(35) + 1.11(25) = 82.48$$

The 95% confidence interval for Y_F in 1981 is then $82.48 \pm (2.37)(4.28)$ or between 73.34 and 92.62.

BINARY CHOICE MODELS

8.20   (a) Derive the log-likelihood function for the probit model.   (b) Give two alternative representations of the log-likelihood function.   (b) How would the log-likelihood function differ for the logit model?

(a) Since this is a probit model, we know that u_i is normally distributed in the model of the underlying propensity of Y:

$$Y_i^* = b_0 + b_1 X_i + u_i$$

where $Y_i = 1$ if $Y_i^* \geq 0$ and $Y_i = 0$ if $Y_i^* < 0$. If we see an observed value of $Y = 1$, we know that $Y_i^* \geq 0$, or alternatively, $u_i > -b_0 - b_1 X_i$. The probability of u_i being in this range is $1 - \Phi(-b_0 - b_1 X_i)$, where $\Phi(\ )$ is the cumulative probability for the normal distribution. Since the normal distribution is symmetrical, we can also write this as

$$P(Y = 1) = \Phi(b_0 + b_1 X_i)$$

Similarly, the probability of observing $Y = 0$ for a single observation is

$$P(Y = 0) = P(Y_i^* < 0) = P(u_i < -b_0 - b_1 X_i) = \Phi(-b_0 - b_1 X_i)$$

We know from Sec. 3.2, rule 4 that if the observations are independent, then the joint probability of observing more than one event simultaneously is equal to the product of their individual probabilities. For a given set of data, the joint probability of the observed combination of $Y = 1$ and $Y = 0$ is the likelihood function L:

$$L = \Pi_1[\Phi(b_0 + b_1 X_i)] \cdot \Pi_0[\Phi(-b_0 - b_1 X_i)]$$

(Π_1 and Π_0 indicate multiplication of all probabilities for those data points where $Y_i = 1$ and 0, respectively). Taking logs yields the log-likelihood function

$$\ln L = \Sigma_1[\ln(\Phi(b_0 + b_1 X_i))] + \Sigma_0[\ln(\Phi(-b_0 - b_1 X_i))]$$

(b) Since it is awkward to write the summations for $Y = 1$ and $Y = 0$ separately, for notational convenience the log-likelihood function may be written

$$\ln L = \Sigma[(Y_i) \ln(\Phi(b_0 + b_1 X_i)) + (1 - Y_i) \ln(\Phi(-b_0 - b_1 X_i))]$$

or $\qquad \ln L = \Sigma[\ln(\Phi[(2Y_i - 1)(b_0 + b_1 X_i)])]$

(c) For the logit model, the only difference would be the substitution of the cumulative probability for the logistic distribution $[(\Lambda(z) = 1/(1 + e^{-z})]$ for $\Phi(\ )$.

8.21 Compare and contrast the logit and probit models.

Both the logit and probit models are based on the same underlying threshold model, but because a threshold model is based on the probability of observing the error term in a certain range, a distribution must be specified for estimation. The probit model specifies the *normal distribution*, which is a common distribution that appears often in nature. The logit model specifies the *logistic distribution*, which is similar to the normal distribution in appearance (it is close to a t distribution with df = 7). The benefit of the logistic distribution is in the ease of calculations since there are no tables required to find the cumulative probability. Both models will yield similar results. As a rule of thumb, $b_L = 1.6 b_P$, where L and P indicate coefficients for the logit and probit model, respectively.

8.22 By hand, graph the value of the log-likelihood function in Example 8.5 with $\ln L$ on the vertical axis (ordinate) and b_1 on the horizontal axis (abscissa) for $b_0 = -2$, and $b_1 =$ (a) -0.001 (b) 0 (c) 0.001 (d) 0.002 (e) 0.003.

All calculations are given in Table 8.17; the graph is presented in Fig. 8-3. As can be seen in Fig. 8-3, the maximum point of the log-likelihood function is at $b_1 = 0.001$, which was the estimated value from Example 8.5.

8.23 Estimate Example 8.5 using the logit model.

Logit estimation gives the following results:

$$\hat{Y}^* = -3.6050 + 0.0018 X \qquad s_{\hat{b}_0} = 1.6811, s_{\hat{b}_1} = 0.0009, \ln L = -6.7664$$

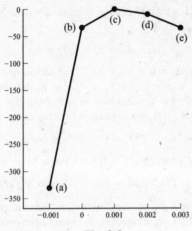

Fig. 8-3

To test significance, we have

$t_{\hat{b}_0} = b_0/s_{\hat{b}_0} = -3.6050/1.6811 = -2.14 < -1.96$; therefore significant at the 5% level

$t_{\hat{b}_1} = b_1/s_{\hat{b}_1} = 0.0018/0.0009 = 2 > 1.96$; therefore significant at the 5% level

The coefficients are proportionally higher in absolute value than in the probit model, but the marginal effects and significance should be similar.

INTERPRETATION OF BINARY CHOICE MODELS

8.24   (a) Explain the difference between the following pairs of terms in the context of binary choice models: (a) coefficient and marginal effect, (b) R^2 and likelihood ratio index, (c) predicted Y and observed Y.

   (a)   The coefficient in a binary choice model gives only the relationship between X and Y^*, the unobservable propensity of Y. Therefore, the coefficient has an ambiguous interpretation, and cannot be compared across different models, or between probit and logit. The marginal effect is the effect of X on the probability of observing a success for Y. Since Y is observable, the interpretation of the marginal effect is clearer, and the marginal effect should be robust across models.

   (b) R^2 is the ratio of explained sum of squares to total sum of squares in a regression, which cannot be defined in a model with an unobservable dependent variable. The likelihood ratio index uses the ratio of log-likelihood values to achieve a similar measure, but its interpretation is not as straightforward. It is bounded by 0 and 1, but achieves 1 only in the limit, and rarely takes on large values.

   (c)   Predicted Y values are successes of Y that are predicted by the binary choice model, usually by having a probability of $Y = 1$ greater than 50%. Observed Y values are the successes and failures of Y from the data set.

8.25   Find the marginal effects, LRI, and predicted values for Y for the logit model in Prob. 8.23. How do the results compare with Example 8.6?

   The marginal effect of GDP/cap on the probability of a country being open to trade is

$$\lambda(b_0 + b_1\bar{X}) \cdot b_1 = \frac{e^{-3.6050+0.0018(3469.5)}}{(1 + e^{-3.6050+0.0018(3469.5)})^2}(0.0018) = 0.0001$$

This can also be interpreted as the marginal effect of GDP/cap on the expected value of Y:

$$\text{LRI} = 1 - \frac{\ln L}{\ln L_0} = 1 - (-6.7664/-13.8629) = 0.51 \quad (\ln L_0 = -13.8629)$$

Predicted probabilities are listed in Table 8.18.

Table 8.17 Log-Likelihood Values for the Probit Model

Country	Y_i	X	$\ln(P(Y=Y_i))$ $b_1=-0.001$	$\ln(P(Y=Y_i))$ $b_1=0$	$\ln(P(Y=Y_i))$ $b_1=0.001$	$\ln(P(Y=Y_i))$ $b_1=0.002$	$\ln(P(Y=Y_i))$ $b_1=0.002$
Burundi	0	569	−0.0051	−0.0230	−0.0793	−0.2161	−0.4857
Chad	0	408	−0.0081	−0.0230	−0.0573	−0.1258	−0.2470
Congo	0	2240	0.0000	−0.0230	−0.9035	−5.0254	−13.6495
Egypt	0	1869	−0.0001	−0.0230	−0.5940	−3.1916	−8.7726
Hong Kong	1	16,471	< −69.0776	−3.7832	0.0000	0.0000	0.0000
India	0	1282	−0.0005	−0.0230	−0.2697	−1.2504	−3.4282
Indonesia	1	2102	−10.7955	−3.7832	−0.6150	−0.0139	0.0000
Ivory Coast	0	1104	−0.0010	−0.0230	−0.2047	−0.8732	−2.3564
Kenya	0	914	−0.0018	−0.0230	−0.1494	−0.5651	−1.4738
Malaysia	1	5746	−32.9756	−3.7832	0.0000	0.0000	0.0000
Morocco	1	2173	−11.1048	−3.7832	−0.5644	−0.0095	0.0000
Nigeria	0	978	−0.0015	−0.0230	−0.1665	−0.6587	−1.7421
Rwanda	0	762	−0.0029	−0.0230	−0.1141	−0.3813	−0.9482
Singapore	1	12,653	< −69.0776	−3.7832	0.0000	0.0000	0.0000
South Africa	1	3068	−15.4182	−3.7832	−0.1540	0.0000	0.0000
Tunisia	1	3075	−15.4550	−3.7832	−0.1522	0.0000	0.0000
Uganda	1	547	−5.2153	−3.7832	−2.6158	−1.7012	−1.0222
Uruguay	1	5185	−28.7146	−3.7832	−0.0007	0.0000	0.0000
Venezuela	1	7082	< −69.0776	−3.7832	0.0000	0.0000	0.0000
Zimbabwe	0	1162	−0.0008	−0.0230	−0.2244	−0.9863	−2.6789
			$\ln L = \sum \ln(P(Y=Y_i))$ $\ln L \approx -326.9333$	$\ln L = \sum \ln(P(Y=Y_i))$ $\ln L \approx -38.0620$	$\ln L = \sum \ln(P(Y=Y_i))$ $\ln L \approx -6.8651$	$\ln L = \sum \ln(P(Y=Y_i))$ $\ln L \approx -14.9985$	$\ln L = \sum \ln(P(Y=Y_i))$ $\ln L \approx -36.8047$

The model predicts 18 out of 20 countries correctly, or 90%. The marginal effect and predictions are virtually identical to the probit model, giving an indication of why the logit model was used almost exclusively before computers were readily available.

Table 8.18   Predicted Probabilities for Logit Model

Country $P(Y = 1)$	Burundi 0.07	Chad 0.05	Congo 0.61	Egypt 0.44	Hong Kong > 0.99
Country $P(Y = 1)$	India 0.21	Indonesia 0.54	Ivory Coast 0.17	Kenya 0.12	Malaysia > 0.99
Country $P(Y = 1)$	Morocco 0.58	Nigeria 0.14	Rwanda 0.10	Singapore > 0.99	South Africa 0.87
Country $P(Y = 1)$	Tunisia 0.87	Uganda 0.07	Uruguay > 0.99	Venezuela > 0.99	Zimbabwe 0.18

Predicted

		$Y = 0$	$Y = 1$
Actual	$Y = 0$	9	1
	$Y = 1$	1	9

Supplementary Problems

FUNCTIONAL FORM

8.26 Transform the following nonlinear functions into linear functions: (a) $Y = b_0 e^{b_1 X} e^u$,   (b) $Y = b_0 + b_1 \ln X + u$,   (c) $Y = b_0 - b/X + u$, and   (d) $Y = b_0 + b_1 X - b_2 X^2 + b_3 X^3 + u$.
Ans.   (a) $\ln Y = \ln b_0 + b_1 X + u$ (b) $Y = b_0 + b_1 R + u$, where $R = \ln X$ (c) $Y = b_0 - b_0 Z + u$, where $Z = 1/X$ (d) $Y = b_0 + b_1 X - b_2 W + b_3 T + u$, where $W = X^2$ and $T = X^3$

8.27 Fit a double-log function to the data in Table 6.12.
Ans. $$\ln Y = 2.64 + 0.72 \ln X \qquad R^2 = 83.26\%$$
    $$(14.69) \quad (6.31)$$

8.28 Fit a semilog function of the form $Y = b_0 + b_1 \ln X + u$ to the data in Table 6.12.
Ans. $$Y = 2.62 + 27.12 \ln X \qquad R^2 = 81.29\%$$
    $$(0.36) \quad (5.90)$$

8.29 (a) Fit a polynomial function of the form $Y = b_0 + b_1 X - b_2 X^2 + u$ to the data in Table 6.12.   (b) Which gives a better fit for the data in Table 6.12, the linear form of Probs. 6.34, 6.37, 6.38, and 6.40; the semilog form of Prob. 8.28; or the polynomial form of part a?

Ans.   (a) $$Y = -2.25 + 13.67X - 0.77X^2 \qquad R^2 = 80.75\%$$
    $$(1.99) \quad (-1.14) \qquad F_{2,7} = 14.68$$

(b) The fit with the semilog function is better than the fit with the linear and polynomial forms.

DUMMY VARIABLES

8.30   For the data in Table 8.2   (a) run regression Eq. (8.6).   (b) Is the slope coefficient significantly different in wartime than in peacetime?   (c) What is the slope coefficient in peacetime? In wartime?

Ans.   (a) $$\hat{Y} = -2.89 + 0.17X - 0.11XD \qquad R^2 = 0.95$$
$$(11.88) \quad (-7.56)$$

(b) Yes   (c) $b_1 = 0.17$ in peacetime and $b_1 = 0.06$ wartime

8.31   For the data in Table 8.2,   (a) run regression Eq. (8.7).   (b) Is the intercept significantly different in wartime than in peacetime?   (c) Is the slope coefficient significantly different in wartime than in peacetime?

Ans.   (a) $$\hat{Y} = -3.34 + 0.17X + 14.59D - 0.18XD \qquad R^2 = 0.95$$
$$(11.58) \quad (0.67) \qquad (-1.64)$$

(b) No   (c) No

8.32   Table 8.19 gives the aggregate reserves of U.S. depository institutions R from the first quarter of 1995 to the third quarter of 2000.   (a) Test for a linear trend in reserves and for seasonal effects.   (b) What is the value of the intercept for each season (use the 10% significance level)?

Ans.   (a) Assigning a trend value T that equals $1, 2, 3, \ldots, 23$ consecutively to each quarter and letting $D_1 = 1$ for the second quarter and 0 otherwise, $D_2 = 1$ for the third quarter and 0 otherwise, and $D_3 = 1$ for the fourth quarter and 0 otherwise, we get

Table 8.19   Aggregate Reserves of U.S. Depository Institutions (in Millions of Dollars)

Year	Quarter			
	I	II	III	IV
1995	57,571	57,031	57,162	57,896
1996	54,878	53,742	51,045	51,174
1997	47,551	46,606	46,060	47,919
1998	45,591	45,094	44,199	45,209
1999	43,229	42,331	41,314	41,655
2000	39,752	39,217	39,257	

Source: Federal Reserve Board of Governors.

$$\hat{R} = 58{,}370.70 - 956.30T + 153.10D_1 + 104.60D_2 + 1875.50D_3 \qquad R^2 = 0.98$$
$$(-17.56) \quad (0.18) \qquad (0.12) \qquad (2.12)$$

(b) Since only D_3 is statistically significant at the 10% level, $\hat{b}_0 = 58{,}370.70$ in quarters I, II, and III, while $\hat{b}_0 = 60{,}246.20$ in quarter IV.

8.33   Table 8.20 gives the per capita disposable income Y in thousands of dollars and the percentage of college graduates in the population 25 years of age or older X for the eastern United States in 1998.   (a) Run a regression of Y on X and on dummies to take regional effects into account.   (b) What is the value of the intercept for each region (use the 10% significance level)?

Ans.   (a) Taking South Atlantic as the base, $D_1 = 1$ for New England states and 0 otherwise and $D_2 = 1$ for Mid-Atlantic states and 0 otherwise, we get

Table 8.20 Disposable Income and Percent of College Graduates in the East in 1998

Disposable income, %	19.76	24.99	20.77	26.72	23.02	30.22	26.06	28.31	22.79
Percent with college degree, %	19.2	26.6	27.1	31.0	27.8	31.4	26.8	30.1	22.1
State	ME	NH	VT	MA	RI	CN	NY	NJ	PA
Region	New England						Mid-Atlantic		
Disposable income, %	24.96	24.9	23.0	17.12	20.49	18.52	21.27	22.06	
Percent with college degree, %	25.1	31.8	30.3	16.3	23.3	21.3	20.7	22.5	
State	DE	MD	VA	WV	NC	SC	GA	FL	
Region	South Atlantic								

Source: Statistical Abstract of the United States.

$$\hat{Y} = 8.16 + 0.56X + 0.88D_1 + 2.83D_2 \qquad R^2 = 0.86$$
$$\quad\quad\quad\quad (5.07) \quad (0.79) \quad\quad (2.10)$$

(b) $\hat{b}_0 = 8.16$ for New England and South Atlantic states, while $\hat{b}_0 = 10.99$ for Mid-Atlantic states.

DISTRIBUTED LAG MODELS

8.34 What are the problems in estimating (a) Equation (8.9)? (b) Equation (8.10)? (c) Equation (8.13)?
Ans. (a) One observation is lost for each lagged value of X and the Xs are likely to be related to each other
(b) The rigidly geometrically declining lag structure and the violation of two assumptions of OLS leading to biased and inconsistent estimators (c) The number of coefficients to be estimated is not reduced as much as in Eq. (8.10) and the period and the form of the lag may not be known

8.35 Table 8.21 gives the business expenditures for new plant equipment of public utilities Y and the gross national product X, both in billions of dollars, for the United States from 1960 to 1979. (a) Estimate the Koyck model [i.e., Eq. (8.10)]. (b) What are the values of $\hat{\lambda}$ and $\hat{a}$?

Ans. (a) $$\hat{Y}_t = -1.92 + 0.01X_t + 0.40Y_{t-1} \qquad R^2 = 0.99$$
$$\quad\quad\quad\quad\quad (4.55) \quad\quad (2.63)$$

(b) $\hat{\lambda} = 0.40$ and $\hat{a} = -3.20$

Table 8.21 Business Expenditures for New Plant Equipment of Public Utilities and the Gross National Product: United States, 1960–1979 (in Billions of Dollars)

Year	1960	1961	1962	1963	1964	1965	1966	1967	1968	1969
Y	5.2	5.0	4.9	5.0	5.5	6.3	7.4	8.7	10.2	11.6
X	506.0	523.3	563.8	594.7	635.7	688.1	753.0	796.3	868.5	935.5
Year	1970	1971	1972	1973	1974	1975	1976	1977	1978	1979
Y	13.1	15.3	17.0	18.7	20.6	20.1	22.3	25.8	29.5	33.2
X	982.4	1063.4	1171.1	1306.6	1412.9	1528.8	1702.2	1899.5	2127.6	2368.5

Source: Economic Report of the President, U.S. Government Printing Office, Washington, DC, 1980, pp. 203, 255.

8.36 Table 8.22 gives the total personal consumption expenditures Y and the total disposable personal income, X, both in billions of dollars, for the United States from 1960 to 1979. (a) Estimate the Almon lag model assuming a three-period lag taking the form of a second-degree polynomial. (b) Does this model fit the data well?

Table 8.22   Consumption and Disposable Income (in Billions of Dollars): United States, 1960–1979

Year	1960	1961	1962	1963	1964	1965	1966	1967	1968	1969
Y	324.9	335.9	355.2	374.6	400.4	430.2	464.8	490.4	535.9	579.7
X	349.4	362.9	383.9	402.8	437.0	472.2	510.4	544.5	588.1	630.4
Year	1970	1971	1972	1973	1974	1975	1976	1977	1978	1979
Y	618.8	668.2	733.0	809.9	889.6	979.1	1089.9	1210.0	1350.8	1509.8
X	685.9	742.8	801.3	901.7	984.6	1086.7	1184.5	1305.1	1458.4	1623.2

Source: *Economic Report of the President*, U.S. Government Printing Office, Washington, DC, 1980, p. 229.

Ans. (a) $\hat{Y} = -19.08 + 1.94X_t + 0.77X_{t-1} + 0.14X_{t-2} + 0.04X_{t-3}$ $R^2 = 0.09$

$\qquad\qquad\qquad\quad$ (0.98)   (2.62)   (0.36)   (0.13)

(b) Since only the coefficient of X_{t-1} (i.e., $\hat{b}_1$) is statistically significant at the 5% level and its value exceeds the value of $\hat{b}_0$, this model does not fit the data well. The Koyck model or another form of the Almon model might be more appropriate.

FORECASTING

8.37 For $X = 4$ in Prob. 6.44, find (a) s_F^2, (b) $\hat{Y}_F$, and (c) the 95% confidence interval for Y_F.
$\quad$ *Ans.* (a) $s_F^2 \cong 1.19$ (b) $\hat{Y}_F = 4.78$ (c) 4.78 ± 2.43

8.38 For Prob. 7.29 and $X_{1F} = 2$ and $X_{2F} = 1250$ for 2000 (a) find s_F^2 and (b) the 95% confidence interval for Y_F, given that $\hat{Y} = 82.27 - 5.11X_1 + 0.02X_2$, $\bar{X}_1 = 6$, $\bar{X}_2 = 1100$, $s^2 = \sum e^2/n - k = 226.32/12 \cong 18.86$, $s_{b_1}^2 \cong 1.41$; $s_{b_2}^2 \cong 0.01$, $s_{b_0}^2 \cong 238.19$, and $s_{\hat{b}_1\hat{b}_2} \cong 0.01$.
$\quad$ *Ans.* (a) $s_F^2 \cong 468.61$ (b) $97.05 \pm (2.18)(21.65)$, or between 49.85 and 144.25

BINARY CHOICE MODELS

8.39 Calculate the log-likelihood values for the logit model in Prob. 8.23 for $b_0 = -3.6$ and $b_1 = (a)$ 0, (b) 0.001, (c) 0.002.
$\quad$ *Ans.* (a) $\ln L = -36.59$ (b) $\ln L = -9.70$ (c) $\ln L = -6.91$

8.40 Calculate the log-likelihood values for the logit model in Prob. 8.23 for $b_1 = 0.0018$ and $b_0 = (a)$ -3.8 (b) -3.6 (c) -3.4.
$\quad$ *Ans.* (a) $\ln L = -6.80$ (b) $\ln L = -6.77$ (c) $\ln L = -6.82$

INTERPRETATION OF BINARY CHOICE MODELS

8.41 Should coefficients be the same between probit and logit models?
$\quad$ *Ans.* No, logit coefficients should be proportionally greater than probit coefficients.

8.42 Should marginal effects be the same between probit and logit models?
$\quad$ *Ans.* Yes, marginal effects should differ only slightly.

Problems in Regression Analysis

9.1 MULTICOLLINEARITY

Multicollinearity refers to the case in which two or more explanatory variables in the regression model are highly correlated, making it difficult or impossible to isolate their individual effects on the dependent variable. With multicollinearity, the estimated OLS coefficients may be statistically insignificant (and even have the wrong sign) even though R^2 may be "high." Multicollinearity can sometimes be overcome or reduced by collecting more data, by utilizing a priori information, by transforming the functional relationship (see Prob. 9.3), or by dropping one of the highly collinear variables.

EXAMPLE 1. Table 9.1 gives the growth rate of imports Y, gross domestic product X_1, and inflation X_2 for the United States from 1985 to 1999 (the reason for using growth rates is explained in Chap. 11). It is expected that the level of imports will be greater as GDP and domestic prices increase. Regressing Y on X_1 and X_2, we get

$$\hat{Y} = 0.0015 + 1.39X_1 + 0.09X_2 \qquad R^2 = 0.42$$
$$\qquad\quad (1.46) \qquad (1.85) \qquad r_{12} = 0.38$$

Table 9.1   Growth Rate of Imports, GDP and Inflation in the United States from 1985 to 1999

Year	1985	1986	1987	1988	1989	1990	1991	1992
Y	0.0540	0.0656	0.1475	0.0686	0.0455	0.0827	−0.0157	0.0753
X_1	0.0709	0.0505	0.0780	0.0750	0.0627	0.0464	0.0399	0.0640
X_2	−0.1593	−0.2683	0.4801	0.1348	−0.0218	0.1612	−0.2511	−0.2611
Year	1993	1994	1995	1996	1997	1998	1999	
Y	0.0841	0.1540	0.0578	0.0918	0.0949	0.0555	0.1593	
X_1	0.0503	0.0621	0.0432	0.0600	0.0623	0.0585	0.0652	
X_2	0.0527	−0.1500	0.0251	−0.1119	−0.0131	−0.3613	0.2579	

Source: St. Louis Federal Reserve (Bureau of Economic Analysis).

Neither $\hat{b}_1$ nor $\hat{b}_2$ is statistically significant at the 5% level. $\hat{b}_2$ is significant at the 10% level, but the R^2 indicates that 42% of the variation in Y is explained by the model even though none of the independent variables stand out individually. The correlation is positive correlation X_1 and X_2, as indicated by r_{12}. Reestimating the regression without either X_2 or X_1, we get

$$\hat{Y} = -0.04 + 2.06\ X_1 \qquad R^2 = 0.26$$
$$(2.13)$$

$$\hat{Y} = 0.09 + 0.11\ X_2 \qquad R^2 = 0.32$$
$$(2.48)$$

In simple regressions, the significance of both X_1 and X_2 increases, with X_1 almost significant at the 5% level and X_2 significant at more than the 5% level, indicating that the original regression exhibited multicollinearity. However, dropping either variable from the regression leads to biased OLS estimates, because economic theory suggests that both GDP and prices should be included in the import function.

9.2 HETEROSCEDASTICITY

If the OLS assumption that the variance of the error term is constant for all observations does not hold, we face the problem of *heteroscedasticity*. This leads to unbiased but inefficient (i.e., larger than minimum variance) estimates of the coefficients, as well as biased estimates of the standard errors (and, thus, incorrect statistical tests and confidence intervals).

One test for heteroscedasticity involves arranging the data from small to large values of the independent variable X and running two regressions, one for small values of X and one for large values, omitting, say, one-fifth of the middle observations. Then, we test that the ratio of the error sum of squares (ESS) of the second regression to the first regression is significantly different from zero, using the F table with $(n - d - 2k)/2$ degrees of freedom, where n is the total number of observations, d is the number of omitted observations, and k is the number of estimated parameters.

If the error variance is proportional to X^2 (often the case), heteroscedasticity can be overcome by dividing every term of the model by X and then reestimating the regression using the transformed variables.

EXAMPLE 2. Table 9.2 gives average wages Y and the number of workers employed X by 30 firms in an industry. Regressing Y on X for the entire sample, we get

$$\hat{Y} = \ \ 7.5 \ \ + 0.009\,X \qquad R^2 = 0.90$$
$$(40.27) \ \ (16.10)$$

The results of regressing Y on X for the first 12 and for the last 12 observations are, respectively

$$\hat{Y} = \ \ 8.1 + 0.006X \qquad R^2 = 0.66$$
$$(39.4) \ \ (4.36) \qquad ESS_1 = 0.507$$

$$\hat{Y} = \ \ 6.1 + 0.013X \qquad R^2 = 0.60$$
$$(4.16) \ \ (3.89) \qquad ESS_2 = 3.095$$

Table 9.2   Average Wages and Number of Workers Employed

Average Wages						Workers Employed
8.40	8.40	8.60	8.70	8.90	9.00	100
8.90	9.10	9.30	9.30	9.40	9.60	200
9.50	9.80	9.90	10.30	10.30	10.50	300
10.30	10.60	10.90	11.30	11.50	11.70	400
11.60	11.80	12.10	12.50	12.70	13.10	500

Since $\text{ESS}_2/\text{ESS}_1 = 3.095/0.507 = 6.10$ exceeds $F_{10,10} = 2.97$ at the 5% level of significance (see App. 7), the hypothesis of heteroscedasticity is accepted. Reestimating the transformed model to correct for heteroscedasticity, we get

$$\frac{\hat{Y}}{X} = \underset{(14.43)}{0.008} + \underset{(76.58)}{7.8} \left(\frac{1}{X}\right) \qquad R^2 = 0.99$$

Note that the slope coefficient is now given by the intercept (i.e., 0.008), and this is smaller than before the adjustment (i.e., 0.009).

9.3   AUTOCORRELATION

When the error term in one time period is positively correlated with the error term in the previous time period, we face the problem of (positive first-order) *autocorrelation*. This is common in time-series analysis and leads to downward-biased standard errors (and, thus, to incorrect statistical tests and confidence intervals).

The presence of first-order autocorrelation is tested by utilizing the table of the Durbin-Watson statistic (App. 8) at the 5 or 1% levels of significance for n observations and k' explanatory variables. If the calculated value of d from Eq. (9.1) is smaller than the tabular value of d_L (lower limit), the hypothesis of positive first-order autocorrelation is accepted:

$$d = \frac{\sum_{t=2}^{n}(e_t - e_{t-1})^2}{\sum_{t=1}^{n} e_t^2} \qquad (9.1)$$

The hypothesis is rejected if $d > d_U$ (upper limit), and the test is inconclusive if $d_L < d < d_U$. (For negative autocorrelation, see Prob. 9.8.)

One way to correct for autocorrelation is to first estimate ρ (Greek letter rho) from Eq. (9.2)

$$Y_t = b_0(1 - \rho) + \rho Y_{t-1} + b_1 X_t - b_1 \rho X_{t-1} + v_t \qquad (9.2)$$

and then reestimate the regression on the transformed variables:

$$(Y_t - \hat{\rho} Y_{t-1}) = b_0(1 - \hat{\rho}) + b_1(X_t - \hat{\rho} X_{t-1}) + (u_t - \hat{\rho} u_{t-1}) \qquad (9.3)$$

To avoid losing the first observation in the differencing process, $Y_1\sqrt{1 - \hat{\rho}^2}$ and $X_1\sqrt{1 - \hat{\rho}^2}$ are used for the first transformed observations of Y and X, respectively. When $\hat{\rho} \cong 1$, autocorrelation can be corrected by rerunning the regression in difference form and omitting the intercept term (see Prob. 9.12).

EXAMPLE 3. Table 9.3 gives the level of inventories Y and sales S, both in billions of dollars, in U.S. manufacturing from 1979 to 1998. Regressing Y on X, we get

$$\hat{Y}_t = \underset{(16.68)}{126.06} + 1.03 X_t \qquad \begin{array}{l} R^2 = 0.94 \\ d = 0.58 \end{array}$$

Table 9.3   Inventory and Sales (Both in Billions of Dollars) in U.S. Manufacturing 1979–1998

Year	1979	1980	1981	1982	1983	1984	1985	1986	1987	1988
Y	242	265	283	312	312	340	335	323	338	369
X	144	154	168	163	172	191	194	195	206	225
Year	1989	1990	1991	1992	1993	1994	1995	1996	1997	1998
Y	391	405	391	383	384	405	431	437	456	467
X	237	243	240	250	261	279	300	310	327	338

Source: Economic Report of the President.

Since $d = 0.58 < d_L = 1.20$ at the 5% level of significance with $n = 20$ and $k' = 1$ (from App. 8), there is evidence of autocorrelation.   An estimate of ρ is given by the coefficient of Y_{t-1} in the following regression:

$$\hat{Y}_t = 66.88 + 0.58\ Y_{t-1} + 0.88\ X_t - 0.50\ X_{t-1} \qquad R^2 = 0.97$$
$$\qquad\qquad\quad (3.43)\qquad\quad (2.36)\qquad (-1.04)$$

Utilizing $\hat{\rho} = 0.58$ to transform the original variables (it is a coincidence here that $\hat{\rho} = d$), as in Eq. (9.3), and using $242\sqrt{1 - 0.58^2} = 197.14$ and $144\sqrt{1 - 0.58^2} = 117.30$ for the first transformed observations of Y and X, respectively, we rerun the regression on the transformed variables (denoted by the asterisk) and get

$$\hat{Y}_t^* = 65.68 + 0.94 X_t^* \qquad R^2 = 0.83$$
$$\qquad\qquad (9.34)\qquad\quad d = 1.78$$

Since now $d = 1.78 > d_U = 1.41$ (from App. 8), there is no evidence of autocorrelation.   Note that the t value of X_t^* is less than for X_t (but is still highly significant) and R^2 is also lower.

9.4   ERRORS IN VARIABLES

Errors in variables refer to the case in which the variables in the regression model include measurement errors.   Measurement errors in the dependent variable are incorporated into the disturbance term and do not create any special problem.   However, errors in the explanatory variables lead to biased and inconsistent parameter estimates.

One method of obtaining consistent OLS parameter estimates is to replace the explanatory variable subject to measurement errors with another variable (called an *instrumental variable*) that is highly correlated with the original explanatory variable but is independent of the error term.   This is often difficult to do and somewhat arbitrary.   The simplest instrumental variable is usually the lagged explanatory variable in question (see Example 4).   Another method used when only X is subject to measurement errors involves regressing X on Y (inverse least squares; see Prob. 9.15).

EXAMPLE 4.   Table 9.4 gives inventories Y, actual sales X, and hypothetical values of X that include measurement error X', all in billions of dollars, in U.S. retail trade from 1979 to 1998. X and Y are assumed to be error-free.   Regressing Y_t on X_t, we get

$$\hat{Y}_t = 2.92 + 1.53\ X_t \qquad R^2 = 0.99$$
$$\qquad (0.72)\quad (56.67)$$

Regressing Y_t on X_t' (if X_t is not available), we get

$$\hat{Y}_t = 6.78 + 1.46\ X_t' \qquad R^2 = 0.99$$
$$\qquad (1.70)\quad (56.23)$$

Table 9.4   Inventories and Sales (in Billions of Dollars) in U.S. Retail Trade, 1979–1998

Year	1979	1980	1981	1982	1983	1984	1985	1986	1987	1988
Y	111	121	133	135	148	168	182	187	208	219
X	75	80	87	89	98	107	115	121	128	138
X'	76	82	89	91	100	109	118	124	132	142
Year	1989	1990	1991	1992	1993	1994	1995	1996	1997	1998
Y	237	240	243	252	269	294	310	321	330	341
X	147	154	155	163	174	188	197	209	218	229
X'	152	159	160	169	180	195	204	217	226	238

Source: Economic Report of the President.

Note that $\hat{b}_1' < \hat{b}_1$; furthermore, $\hat{b}_1$ falls outside the 95% confidence interval of b_1' (1.40 to 1.51). Using X_{t-1}' as an instrumental variable for X_t' (if X_t' is suspected to be correlated with u_t), we get

$$\hat{Y} = 13.88 + 1.50\ X_{t-1}' \qquad R^2 = 0.99$$
$$\quad (2.48) \quad (40.19)$$

The coefficient on X_{t-1}' is closer to the true one ($\hat{b}_1$ falls in the 95% confidence interval of 1.42 to 1.57), and is consistent.

Solved Problems

MULTICOLLINEARITY

9.1 (a) What is meant by *perfect multicollinearity*? What is its effect? (b) What is meant by *high*, but not perfect, *multicollinearity*? What problems may result? (c) How can multicollinearity be detected? (d) What can be done to overcome or reduce the problems resulting from multicollinearity?

(a) Two or more independent variables are *perfectly collinear* if one or more of the variables can be expressed as a linear combination of the other variable(s). For example, there is perfect multicollinearity between X_1 and X_2 if $X_1 = 2X_2$ or $X_1 = 5 - (1/3)X_2$. If two or more explanatory variables are perfectly linearly correlated, it will be impossible to calculate OLS estimates of the parameters because the system of normal equations will contain two or more equations that are not independent.

(b) *High*, but not perfect, *multicollinearity* refers to the case in which two or more independent variables in the regression model are highly correlated. This may make it difficult or impossible to isolate the effect that each of the highly collinear explanatory variables has on the dependent variable. However, the OLS estimated coefficients are still unbiased (if the model is properly specified). Furthermore, if the principal aim is prediction, multicollinearity is not a problem if the same multicollinearity pattern persists during the forecasted period.

(c) The classic case of multicollinearity occurs when none of the explanatory variables in the OLS regression is statistically significant (and some may even have the wrong sign), even though R^2 may be high (say, between 0.7 and 1.0). In the less clearcut cases, detecting multicollinearity may be more difficult. High, simple, or partial correlation coefficients among explanatory variables are sometimes used as a measure of multicollinearity. However, serious multicollinearity can be present even if simple or partial correlation coefficients are relatively low (i.e., less than 0.5).

(d) Serious multicollinearity may sometimes be corrected by (1) extending the size of the sample data, (2) utilizing a priori information (e.g., we may know from a previous study that $b_2 = 0.25b_1$), (3) transforming the functional relationship, or (4) dropping one of the highly collinear variables (however, this may lead to specification bias or error if theory tells us that the dropped variable should be included in the model).

9.2 Table 9.5 gives the output in tons Q, the labor input in worker-hours L, and the capital input in machine-hours K, of 15 firms in an industry. (a) Fit a Cobb-Douglas production function of the form $Q = b_0 L^{b_1} K^{b_2} e^u$ to the data and find $\bar{R}^2$ and the simple correlation coefficient between $\ln L$ and $\ln K$. (b) Regress $\ln Q$ on $\ln L$ only. (c) Regress $\ln Q$ on $\ln K$ only. (d) What can be concluded from the results with regard to multicollinearity?

Table 9.5   Output, Labor, and Capital Inputs of 15 Firms in an Industry

Firm	1	2	3	4	5	6	7	8	9	10	11	12	13	14	15
Q	2350	2470	2110	2560	2650	2240	2430	2530	2550	2450	2290	2160	2400	2490	2590
L	2334	2425	2230	2463	2565	2278	2380	2437	2446	2403	2301	2253	2367	2430	2470
K	1570	1850	1150	1940	2450	1340	1700	1860	1880	1790	1480	1240	1660	1850	2000

(a) Transforming the data into natural log form as shown in Table 9.6 and then regressing $\ln Q$ on $\ln L$ and $\ln K$, we get

$$R^2 = 0.969$$
$$\ln Q = 0.50 + 0.76 \ln L + 0.19 \ln K \qquad \bar{R}^2 = 0.964$$
$$ (1.07) \qquad (1.36)$$

$$r_{\ln L \ln K} = 0.992$$

(b)
$$\ln Q = -5.50 \quad + \quad 1.71 \ln L \qquad R^2 = 0.964$$
$$ (-7.74) \quad (18.69)$$

(c)
$$\ln Q = 5.30 + 0.34 \ln K \qquad R^2 = 0.966$$
$$ (4.78) \quad (19.19)$$

(d) Since neither $\hat{b}_1$ nor $\hat{b}_2$ in part a is statistically significant at the 5% level (i.e., they have unduly large standard errors) while $R^2 = 0.97$, there is clear indication of serious multicollinearity.   Specifically, large firms tend to use both more labor and more capital than do small firms.   This is confirmed by the very high value of 0.99 for the simple correlation coefficient between $\ln L$ and $\ln K$.   In parts b and c, simple regressions were reestimated with either $\ln L$ or $\ln K$ as the only explanatory variable.   In these simple regressions, both $\ln L$ and $\ln K$ are statistically significant at much more than the 1% level with R^2 exceeding 0.96.   However, dropping either $\ln K$ or $\ln L$ from the multiple regression leads to a biased

Table 9.6   Output, Labor, and Capital Inputs in Original and Log Form

Firm	Q	L	K	$\ln Q$	$\ln L$	$\ln K$
1	2350	2334	1570	7.76217	7.75534	7.35883
2	2470	2425	1850	7.81197	7.79359	7.52294
3	2110	2230	1150	7.65444	7.70976	7.04752
4	2560	2463	1940	7.84776	7.80914	7.57044
5	2650	2565	2450	7.88231	7.84971	7.80384
6	2240	2278	1340	7.71423	7.73105	7.20042
7	2430	2380	1700	7.79565	7.77486	7.43838
8	2530	2437	1860	7.83597	7.79852	7.52833
9	2550	2446	1880	7.84385	7.80221	7.53903
10	2450	2403	1790	7.80384	7.78447	7.48997
11	2290	2301	1480	7.73631	7.74110	7.29980
12	2160	2253	1240	7.67786	7.72002	7.12287
13	2400	2367	1660	7.78322	7.76938	7.41457
14	2490	2430	1850	7.82004	7.79565	7.52294
15	2590	2470	2000	7.85941	7.81197	7.60090

OLS slope estimate for the retained variable because economic theory postulates that both labor and capital should be included in the production function.

9.3   How can the multicollinearity difficulty faced in Prob. 9.2 be overcome if it is known that constant returns to scale (i.e., $b_1 + b_2 = 1$) prevail in this industry?

With constant returns to scale, the Cobb-Douglas production function can be rewritten as

$$Q = b_0 L^{b_1} K^{1-b_1} e^u$$

Expressing this production function in double-log form and rearranging it, we get

$$\ln Q = \ln b_0 + b_1 \ln L + (1 - b_1) \ln K + u$$
$$\ln Q - \ln K = \ln b_0 + b_1 (\ln L - \ln K) + u$$

Setting $\ln Q^* = \ln Q - \ln K$ and $\ln L^* = \ln L - \ln K$ and then regressing $\ln Q^*$ on $\ln L^*$, we get

$$\ln Q^* = \underset{(9.26)}{0.07} + \underset{(39.81)}{0.83} \ln L^* \qquad R^2 = 0.992$$

Then $\hat{b}_2 = 1 - \hat{b}_1 = 1 - 0.83 = 0.17$.

HETEROSCEDASTICITY

9.4   (a) What is meant by *heteroscedasticity*?   (b) Draw a figure showing homoscedastic disturbances and the various forms of heteroscedastic disturbances.   (c) Why is heteroscedasticity a problem?

(a)   *Heteroscedasticity* refers to the case in which the variance of the error term is not constant for all values of the independent variable; that is, $E(X_i u_i) \neq 0$, so $E(u_i)^2 \neq \sigma_u^2$. This violates the third assumption of the OLS regression model (see Prob. 6.4). It occurs primarily in cross-sectional data. For example, the error variance associated with the expenditures of low-income families is usually smaller than for high-income families because most of the expenditures of low-income families are on necessities, with little room for discretion.

(b)   Figure 9-1a shows homoscedastic (i.e., constant variance) disturbances, while Fig. 9-1b, c, and d shows heteroscedastic disturbances. In Fig. 9-1b, σ_u^2 increases with X_i. In Fig. 9-1c, σ_u^2 decreases with X_i. In Fig. 9-1d, σ_u^2 first decreases and then increases as X_i increases. In economics, the heteroscedasticity shown in Fig. 9-1b is the most common, so the discussion that follows refers to that.

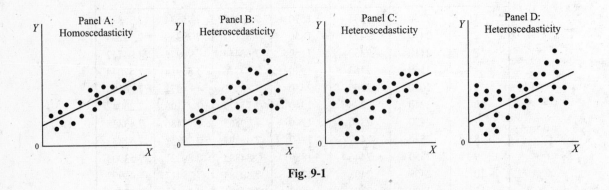

Fig. 9-1

(c)   With heteroscedasticity, the OLS parameter estimates are still unbiased and consistent, but they are inefficient (i.e., they have larger than minimum variances). Furthermore, the estimated variances of the parameters are biased, leading to incorrect statistical tests for the parameters and biased confidence intervals.

9.5   (a) How is the presence of heteroscedasticity tested?   (b) How can heteroscedasticity be corrected?

(a)   The presence of heteroscedasticity can be tested by arranging the data from small to large values of the independent variable X_i and then running two separate regressions, one for small values of X_i and one for large values of X_i, omitting some (say, one-fifth) of the middle observations. Then the ratio of the error sum of squares of the second regression to the error sum of squares of the first regression (i.e., ESS_2/ESS_1) is tested to see if it is significantly different from zero. The F distribution is used for this test with $(n - d - 2k)/2$ degrees of freedom, where n is the total number of observations, d is the number of omitted observations, and k is the number of estimated parameters. This is the *Goldfeld-Quandt test for heteroscedasticity* and is most appropriate for large samples (i.e., for $n \geq 30$). If no middle observations are omitted, the test is still correct, but it will have a reduced power to detect heteroscedasticity.

(b)   If it is assumed (as often is the case) that $\operatorname{var} u_i = CX_i^2$, where C is a nonzero constant, we can correct for heteroscedasticity by dividing (i.e., weighting) every term of the regression by X_i and then reestimating the regression using the transformed variables. In the two-variable case, we have

$$\frac{Y_i}{X_i} = \frac{b_0}{X_i} + b_1 + \frac{u_i}{X_i} \tag{9.4}$$

The transformed error term is now homoscedastic:

$$\operatorname{var} u_i = \operatorname{var} \frac{u_i}{X_i} = \frac{1}{X_i^2} \operatorname{var} u_i = C \frac{X_i^2}{X_i^2} = C$$

Note that the original intercept has become a variable in Eq. (9.4), while the original slope parameter, b_1, is now the new intercept. However, care must be used to correctly interpret the results of the transformed or weighted regression. Since in Eq. (9.4) the errors are homoscedastic, the OLS estimates are not only unbiased and consistent, but also efficient. In the case of a multiple regression, each term of the regression is divided (i.e., weighted) by the independent variable (say, X_{2i}) that is thought to be associated with the error term, so we have

$$\frac{Y_i}{X_{2i}} = \frac{b_0}{X_{2i}} + b_1 \frac{X_{1i}}{X_{2i}} + b_2 + \frac{u_i}{X_{2i}} \tag{9.5}$$

In Eq. (9.5), the original intercept, b_0, has become a variable, while b_2 has become the new intercept term. We can visually determine whether it is X_{2i} or X_{1i} that is related to the u_i by plotting X_{2i} and X_{1i} against the regression residuals, e_i.

9.6   Table 9.7 gives the consumption expenditures C and disposable income Y_d for 30 families. (a) Regress C on Y_d for the entire sample and test for heteroscedasticity. (b) Correct for heteroscedasticity if it is found in part a.

Table 9.7   Consumption and Income Data for 30 Families (in U.S. Dollars)

Consumption			Income
10,600	10,800	11,100	12,000
11,400	11,700	12,100	13,000
12,300	12,600	13,200	14,000
13,000	13,300	13,600	15,000
13,800	14,000	14,200	16,000
14,400	14,900	15,300	17,000
15,000	15,700	16,400	18,000
15,900	16,500	16,900	19,000
16,900	17,500	18,100	20,000
17,200	17,800	18,500	21,000

(a) Regressing C on Y_d for the entire sample of 30 observations, we get

$$\hat{C} = 1480.0 + 0.788\ Y_d \qquad R^2 = 0.97$$
$$\quad\ (3.29) \quad (29.37)$$

To test for heteroscedasticity, we regress C on Y_d for the first 12 and for last 12 observations, leaving the middle 6 observations out, and we get

$$\hat{C} = 846.7 + 0.837 Y_d \qquad\qquad R^2 = 0.91$$
$$\quad\ (0.74) \quad (9.91) \qquad\quad ESS_1 = 1,069,000$$

$$\hat{C} = 2,306.7 + 0.747 Y_d \qquad\quad R^2 = 0.71$$
$$\quad\ (0.79) \qquad (5.00) \qquad\quad ESS_2 = 3,344,000$$

Since $ESS_2/ESS_1 = 3,344,000/1,069,000 = 3.13$ exceeds $F = 2.97$ with $(30 - 6 - 4)/2 = 10$ degrees of freedom in the numerator and denominator at the 5% level of significance (see App. 7), we accept the hypothesis of heteroscedasticity.

(b) Assuming that the error variance is proportional to Y_d^2, and then reestimating the regression using the transformed variables of Table 9.8 to correct for heteroscedasticity, we get (in the last column of Table 9.8; 0.833333E-04 = 0.0000833333) the following:

$$\frac{\hat{C}}{Y_d} = \underset{(31.51)}{0.792} + \underset{(3.59)}{1421.3}\frac{1}{Y_d} \qquad R^2 = 0.32$$

Note that the marginal propensity to consume is now given by the intercept (i.e., 0.792) and is larger than before the adjustment (i.e., 0.788).   The statistical significance of both estimated parameters is now even higher than before.   The R^2 of the weighted regression (i.e., 0.32) is much lower but not directly comparable with the R^2 of 0.97 before the transformation because the dependent variables are different (Y/X as opposed to Y).

9.7 Table 9.9 gives the level of inventories I and sales S, both in millions of dollars, and borrowing rates for 35 firms in an industry.   It is expected that I will be directly related to S but inversely related to R.   (a) Regress I on S and R for the entire sample and test for heteroscedasticity. (b) Correct for heteroscedasticity if it is found in part a, assuming that the error variance is proportional to S^2.

(a) Regressing I on S and R for the entire sample of 35 firms, we get

$$\hat{I} = -6.17 + \underset{(12.39)}{0.20\ S} - \underset{(-2.67)}{0.25\ R} \qquad R^2 = 0.98$$

To test for heteroscedasticity, we regress I on S and R for the first 14 and for the last 14 observations, leaving the middle 7 observations out, and we get

$$\hat{I} = -2.23 + \underset{(1.90)}{0.16S} - \underset{(-0.81)}{0.22R} \qquad\quad R^2 = 0.94$$
$$\qquad\qquad\qquad\qquad\qquad\qquad ESS_1 = 0.908$$

$$= 16.10 + \underset{(3.36)}{0.11S} - \underset{(-3.35)}{1.40R} \qquad\quad R^2 = 0.96$$
$$\qquad\qquad\qquad\qquad\qquad\qquad ESS_2 = 5.114$$

Since $ESS_2/ESS_1 = 5.114/0.908 = 5.63$ exceeds $F_{11,11} = 2.82$ at the 5% level of significance (see App. 7), we accept the hypothesis of heteroscedasticity.

(b) Assuming that the error variance is proportional to S^2 and reestimating the regression using the transformed variable to correct for heteroscedasticity, we get

$$\frac{\hat{I}}{S} = \underset{(12.34)}{0.21} - 8.45(1/S) - \underset{(-2.98)}{0.18\ (R/S)} \qquad R^2 = 0.93$$

Table 9.8   Consumption C and Disposable Income (Y_d) in Original and Transformed Form

Family	C, \$	Y_d, \$	C/Y_d, %	$1/Y_d$, %
1	10,600	12,000	0.883333	0.833333E-04
2	10,800	12,000	0.900000	0.833333E-04
3	11,100	12,000	0.925000	0.833333E-04
4	11,400	13,000	0.876923	0.769231E-04
5	11,700	13,000	0.900000	0.769231E-04
6	12,100	13,000	0.930769	0.769231E-04
7	12,300	14,000	0.878571	0.714286E-04
8	12,600	14,000	0.900000	0.714286E-04
9	13,200	14,000	0.942857	0.714286E-04
10	13,000	15,000	0.866667	0.666667E-04
11	13,300	15,000	0.886667	0.666667E-04
12	13,600	15,000	0.906667	0.666667E-04
13	13,800	16,000	0.862500	0.625000E-04
14	14,000	16,000	0.875000	0.625000E-04
15	14,200	16,000	0.887500	0.625000E-04
16	14,400	17,000	0.847059	0.588235E-04
17	14,900	17,000	0.876471	0.588235E-04
18	15,300	17,000	0.900000	0.588235E-04
19	15,000	18,000	0.833333	0.555556E-04
20	15,700	18,000	0.872222	0.555556E-04
21	16,400	18,000	0.911111	0.555556E-04
22	15,900	19,000	0.836842	0.526316E-04
23	16,500	19,000	0.868421	0.526316E-04
24	16,900	19,000	0.889474	0.526316E-04
25	16,900	20,000	0.845000	0.500000E-04
26	17,500	20,000	0.875000	0.500000E-04
27	18,100	20,000	0.905000	0.500000E-04
28	17,200	21,000	0.819048	0.476190E-04
29	17,800	21,000	0.847619	0.476190E-04
30	18,500	21,000	0.880952	0.476190E-04

$b_0 = 0.21$ is now the slope coefficient associated with the variable S (instead of 0.16 before the transformation), while $b_2 = -0.18$ is the slope coefficient associated with the variable R (instead of -0.25 before the transformation). Both these slope coefficients remain highly significant before and after the transformation, as does R^2. The new constant is -8.45 instead of -6.17.

AUTOCORRELATION

9.8   (a) What is meant by *autocorrelation*?   (b) Draw a figure showing positive and negative first-order autocorrelation.   (c) Why is autocorrelation a problem?

(a)   *Autocorrelation* or *serial correlation* refers to the case in which the error term in one time period is correlated with the error term in any other time period. If the error term in one time period is correlated with the error term in the *previous* time period, there is *first-order* autocorrelation. Most of the applications in econometrics involve first rather than second- or higher-order autocorrelation. Even though *negative* autocorrelation is possible, most economic time series exhibit *positive*

Table 9.9   Inventories, Sales, and Borrowing Rates for 35 Firms

Firm	1	2	3	4	5	6	7	8	9	10	11	12	13	14	15	16	17	18
I	10	10	10	11	11	11	12	12	12	12	12	13	13	13	14	14	14	15
S	100	101	103	105	106	106	108	109	111	111	112	113	114	114	116	117	118	120
R	17	17	17	16	16	16	15	15	14	14	14	14	13	13	12	12	12	11
Firm	19	20	21	22	23	24	25	26	27	28	29	30	31	32	33	34	35	
I	15	15	15	16	16	16	17	17	17	17	18	18	19	19	19	20	20	
S	122	123	125	128	128	131	133	134	135	136	139	143	147	151	157	163	171	
R	11	11	11	10	10	10	10	9	9	9	8	8	8	8	8	7	7	

autocorrelation.   Positive, first-order serial or autocorrelation means that $E_{u_t u_{t-1}} > 0$, thus violating the fourth OLS assumption (see Prob. 6.4).   This is common in time-series analysis.

(b) Figure 9-2a shows positive and Fig. 9-2b shows negative first-order autocorrelation.   Whenever several consecutive residuals have the same sign as in Fig. 9-2a, there is positive first-order autocorrelation. However, whenever consecutive residuals change sign frequently, as in Fig. 9-2b, there is negative first-order autocorrelation.

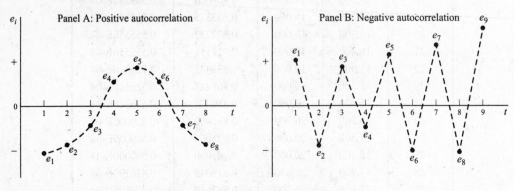

Fig. 9-2

(c) With autocorrelation, the OLS parameter estimates are still unbiased and consistent, but the standard errors of the estimated regression parameters are biased, leading to incorrect statistical tests and biased confidence intervals.   With positive first-order autocorrelation, the standard errors of the estimated regression parameters are biased downward, thus exaggerating the precision and statistical significance of the estimated regression parameters.

9.9   (a) How is the presence of positive or negative first-order autocorrelation tested?   (b) How can autocorrelation be corrected?

(a) The presence of autocorrelation can be tested by calculating the Durbin-Watson statistic d given by Eq. (9.1).   This is routinely given by most computer programs such as SAS:

$$d = \frac{\sum_{t=2}^{n}(e_t - e_{t-1})^2}{\sum_{t=1}^{n} e_t^2} \tag{9.1}$$

The calculated value of d ranges between 0 and 4, with no autocorrelation when d is in the neighborhood of 2. The values of d indicating the presence or absence of positive or negative first-order autocorrelation, and for which the test is inconclusive, are summarized in Fig. 9-3. When the lagged dependent appears as an explanatory variable in the regression, d is biased toward 2 and its power to detect autocorrelation is hampered.

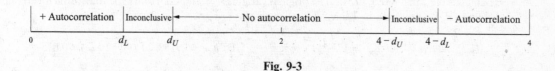

Fig. 9-3

(b) One method to correct positive first-order autocorrelation (the usual type) involves first regressing Y on its value lagged one period, the explanatory variable of the model, and the explanatory variable lagged one period:

$$Y_t = b_0(1 - \rho) + \rho Y_{t-1} + b_1 X_t - b_1 \rho X_{t-1} + v_t \qquad (9.2)$$

(The preceding equation is derived by multiplying each term of the original OLS model lagged one period by ρ, subtracting the resulting expression from the original OLS model, transposing the term ρY_{t-1} from the left to the right side of the equation, and defining $v_t = u_t - \rho u_{t-1}$.) The second step involves using the value of ρ found in Eq. (9.2) to transform all the variables of the original OLS model, as indicated in Eq. (9.3), and then estimating Eq. (9.3):

$$Y_t - \hat{\rho} Y_{t-1} = b_0(1 - \hat{\rho}) + b_1(X_t - \hat{\rho} X_{t-1}) + \varepsilon_t \qquad (9.3)$$

The error term, ε_t, in Eq. (9.3) is now free of autocorrelation. This procedure, known as the *Durbin two-stage method*, is an example of generalized least squares. To avoid losing the first observation in the differencing process, $Y_1\sqrt{1 - \hat{\rho}^2}$ and $X_1\sqrt{1 - \hat{\rho}^2}$ are used for the first transformed observation of Y and X, respectively. If the autocorrelation is due to the omission of an important variable, wrong functional form, or improper model specification, these problems should be removed first, before applying the preceding correction procedure for autocorrelation.

9.10 Table 9.10 gives the level of U.S. imports M and GDP (both seasonally adjusted in billions of dollars) from 1980 to 1999. (a) Regress M on GDP and test for autocorrelation at the 5% level of significance. (b) Correct for autocorrelation if it is found in part a.

(a)
$$\hat{M}_t = -201.80 + 0.14\ GDP_t \qquad R^2 = 0.98$$
$$\phantom{\hat{M}_t =} (-6.48) \quad (29.44) \qquad d = 0.54$$

Since $d = 0.54 < d_L = 1.20$ at the 5% level of significance with $n = 20$ and $k' = 1$ (from App. 8), there is evidence of positive first-order autocorrelation.

Table 9.10 Seasonally Adjusted U.S. Imports and GDP (Both in Billions of Dollars) from 1980 to 1999

Year	1980	1981	1982	1983	1984	1985	1986	1987	1988	1989
M	299.2	319.4	294.9	358.0	416.4	438.9	467.7	536.7	573.5	599.6
GDP	2918.8	3203.1	3315.6	3688.8	4033.5	4319.3	4537.5	4891.6	5258.3	5588.0
Year	1990	1991	1992	1993	1994	1995	1996	1997	1998	1999
M	649.2	639.0	687.1	744.9	859.6	909.3	992.8	1087.0	1147.3	1330.1
GDP	5847.3	6080.7	6469.8	6795.5	7217.7	7529.3	7981.4	8478.6	8974.9	9559.7

Source: St. Louis Federal Reserve (Bureau of Economic Analysis).

(b) To correct for autocorrelation, first the following regression is run:

$$\hat{M}_t = -103.21 + 0.82\, M_{t-1} + 0.36\, \text{GDP}_t - 0.33\, \text{GDP}_{t-1} \qquad R^2 = 0.98$$
$$\qquad\qquad (4.72) \qquad\quad (4.68) \qquad\quad (-4.23)$$

Then, using $\hat{\rho} = 0.82$ (the coefficient on M_{t-1} in the preceding regression), we transform the original variables as indicated in Eq. (9.3). The original variables (M and GDP) and the transformed variables (M^* and GDP*) are given in Table 9.11.

$$M^*_{1980} = 299.2\sqrt{1 - 0.82^2} = 171.251 \qquad \text{and} \qquad \text{GDP}^*_{1980} = 2918.8\sqrt{1 - 0.82^2} = 1670.615$$

Table 9.11   U.S. Imports and GDP in Original and Transformed Form

Year	M	GDP	M^*	GDP*
1980	299.2	2918.8	171.250	1670.610
1981	319.4	3203.1	74.056	809.684
1982	294.9	3315.6	32.992	689.058
1983	358.0	3688.8	116.182	970.008
1984	416.4	4033.5	122.840	1008.684
1985	438.9	4319.3	97.452	1011.830
1986	467.7	4537.5	107.802	995.674
1987	536.7	4891.6	153.186	1170.850
1988	573.5	5258.3	133.406	1247.188
1989	599.6	5588.0	129.330	1276.194
1990	649.2	5847.3	157.528	1265.140
1991	639.0	6080.7	106.656	1285.914
1992	687.1	6469.8	163.120	1483.626
1993	744.9	6795.5	181.478	1490.264
1994	859.6	7217.7	248.782	1645.390
1995	909.3	7529.3	204.428	1610.786
1996	992.8	7981.4	247.174	1807.374
1997	1087.0	8478.6	272.904	1933.852
1998	1147.3	8974.9	255.960	2022.448
1999	1330.1	9559.7	389.314	2200.282

Regressing M^* on GDP*, we get

$$\hat{M}^*_t = 579.53 + 4.75\, \text{GDP}^*_t \qquad R^2 = 0.88$$
$$\qquad (7.79) \quad (11.91) \qquad\qquad d = 1.69$$

Since now $d = 1.69 > d_U = 1.41$ at the 5% level of significance with $n = 20$ and $k' = 1$ (from App. 8), there is no evidence of autocorrelation. Note that though GDP$^*_t$ remains highly significant, its t value is lower than the t value of GDP$_t$. In addition, $R^2 = 0.88$ now, as opposed to $R^2 = 0.98$ before the correction for autocorrelation.

9.11 Table 9.12 gives gross private domestic investment (GPDI) and GDP, both in seasonally adjusted billions of 1996 dollars, and the GDP deflator price index P for the United States from 1980 to 1999. (a) Regress GPDI on GDP and P and test for autocorrelation at the 5% level of significance. (b) Correct for autocorrelation if it is found in part a.

Table 9.12   U.S. GPDI, GDP (Both in Seasonally Adjusted Billions of 1996 Dollars), and GDP Deflator Price Index, 1982–1999

Year	1982	1983	1984	1985	1986	1987	1988	1989	1990
GPDI	571.1	762.2	876.9	887.8	838.2	929.3	916.7	922.9	849.6
GDP	4915.6	5286.8	5583.1	5806.0	5969.5	6234.4	6465.2	6633.5	6664.2
P	67.44	69.75	72.24	74.40	76.05	78.46	81.36	84.24	87.76
Year	1991	1992	1993	1994	1995	1996	1997	1998	1999
GPDI	864.2	941.6	1015.6	1150.5	1152.4	1283.7	1438.5	1609.9	1751.6
GDP	6720.9	6990.6	7168.7	7461.1	7621.9	7931.3	8272.9	8654.5	9084.1
P	90.47	92.56	94.79	96.74	98.79	100.63	102.49	103.69	105.31

Source: St. Louis Federal Reserve (Bureau of Economic Analysis).

(*a*)
$$\widehat{\text{GPDI}}_t = -199.71 + 0.56\ \text{GDP}_t - 29.70\ P_t \qquad R^2 = 0.97$$
$$(10.61) \qquad (-6.07) \qquad d = 0.56$$

Since $d = 0.56 < d_L = 1.05$ at the 5% level of significance with $n = 18$ and $k' = 2$ (from App. 8), there is evidence of autocorrelation.

(*b*)   To correct for autocorrelation, first, the following regression is run:

$$\widehat{\text{GPDI}}_t = -291.79 + 0.74\ \text{GPDI}_{t-1} + 0.76\ \text{GDP}_t - 0.73\ \text{GDP}_{t-1} + 1.91\ P_t + 1.40\ P_{t-1}$$
$$(2.99) \qquad\qquad (7.12) \qquad\qquad (-4.28) \qquad\qquad (0.06) \qquad (0.06)$$
$$R^2 = 0.99$$

Then, using $\hat{\rho} = 0.74$ (the coefficient on GPDI_{t-1} in the preceding regression), we transform the original variables as indicated in Eq. (*9.3*). The original and the transformed variables (the latter indicated by an asterisk) are given in Table 9.13.

$$\text{GPDI}^*_{1982} = 571.1\sqrt{1 - 0.74^2} = 384.126$$
$$\text{GDP}^*_{1982} = 4915.6\sqrt{1 - 0.74^2} = 3306.266$$
$$P^*_{1982} = 67.44\sqrt{1 - 0.74^2} = 45.361$$

Regressing GPDI^*_t on GDP^*_t and P^*_t, we get

$$\widehat{\text{GPDI}}^*_t = 31.05 + 0.52\ \text{GDP}^*_t - 30.02 P^*_t \qquad R^2 = 0.88$$
$$(9.81) \qquad (-6.54) \qquad d = 1.77$$

Since $d = 1.77 > d_U = 1.53$ at the 5% level of significance with $n = 18$ and $k' = 2$ (from App. 8), there is no evidence of autocorrelation. Both variables remain highly significant, and R^2 falls.

9.12   Table 9.14 gives personal consumption expenditures C and disposable personal income Y, both in billions of dollars, for the United States from 1982 to 1999. (*a*) Regress C_t on Y_t and test for autocorrelation. (*b*) Correct for autocorrelation if it is found in part *a*.

(*a*)
$$\hat{C}_t = -293.46 + 0.97 Y_t \qquad R^2 = 0.99$$
$$(-6.58) \quad (99.65) \qquad d = 0.58$$

Since $d = 0.58$, there is evidence of autocorrelation at both the 5 and 1% levels of significance.

Table 9.13 GPDI, GDP, and P in Original and Transformed Form

Year	GPDI	GDP	P	GPDI*	GDP*	P^*
1980	662.2	4936.6	59.16	384.126	3306.266	45.3610
1981	708.8	4997.1	64.10	218.772	1344.016	20.3216
1982	571.1	4915.6	67.44	46.588	1217.746	20.0060
1983	762.2	5286.8	69.75	339.586	1649.256	19.8444
1984	876.9	5583.1	72.24	312.872	1670.868	20.6250
1985	887.8	5806.0	74.40	238.894	1674.506	20.9424
1986	838.2	5969.5	76.05	181.228	1673.060	20.9940
1987	929.3	6234.4	78.46	309.032	1816.970	22.1830
1988	916.7	6465.2	81.36	229.018	1851.744	23.2996
1989	922.9	6633.5	84.24	244.542	1849.252	24.0336
1990	849.6	6664.2	87.76	166.654	1755.410	25.4224
1991	864.2	6720.9	90.47	235.496	1789.392	25.5276
1992	941.6	6990.6	92.56	302.092	2017.134	25.6122
1993	1015.6	7168.7	94.79	318.816	1995.656	26.2956
1994	1150.5	7461.1	96.74	398.956	2156.262	26.5954
1995	1152.4	7621.9	98.79	301.030	2100.686	27.2024
1996	1283.7	7931.3	100.63	430.924	2291.094	27.5254
1997	1438.5	8272.9	102.49	488.562	2403.738	28.0238
1998	1609.9	8654.5	103.69	545.410	2532.554	27.8474
1999	1751.6	9084.1	105.31	560.274	2679.770	28.5794

Table 9.14 U.S. Consumption Expenditures and Disposable Income (in Billions of Dollars), 1982-1999

Year	1982	1983	1984	1985	1986	1987	1988	1989	1990
C	2079.3	2286.4	2498.4	2712.6	2895.2	3105.3	3356.6	3596.7	3831.5
Y	2406.8	2586.0	2887.6	3086.5	3262.5	3459.5	3752.4	4016.3	4293.6

Year	1991	1992	1993	1994	1995	1996	1997	1998	1999
C	3971.2	4209.7	4454.7	4716.4	4969.0	5237.5	5524.4	5848.6	6254.9
Y	4474.8	4754.6	4935.3	5165.4	5422.6	5677.7	5982.8	6286.2	6639.2

Source: *Economic Report of the President.*

(b) To correct for autocorrelation, first the following regression is run:

$$\hat{C}_t = 93.90 + 1.23\,C_{t-1} + 0.40\,Y_t - 0.60\,Y_{t-1} \qquad R^2 = 0.99$$
$$\qquad\qquad (5.18) \qquad (1.79) \qquad (-3.08)$$

Since $\hat{\rho} \cong 1$ (the coefficient on C_{t-1} in the preceding regression), we rerun the regression on the first differences of the original variables (i.e., ΔC_t and ΔY_t), omitting the intercept, and get

$$\Delta\hat{C}_t = 0.97\Delta Y_t \qquad R^2 = 0.98$$
$$\quad (25.88) \qquad d = 1.75$$

The new value of d indicates no evidence of autocorrelation at either the 1 or at the 5% level of significance. (*Note*: R^2 is not well defined in regression with no intercept and therefore is not comparable with the previous regressions. For a more in-depth study of procedure when $\rho = 1$, see Sec. 11.3.)

ERRORS IN VARIABLES

9.13 (a) What is meant by *errors in variables*? (b) What problems do errors in variables create? (c) Is there any test to detect the presence of errors in variables? (d) How can the problems created by the existence of errors in variables be corrected?

(a) *Errors in variables* refer to the case in which the variables in the regression model include measurement errors. These are probably very common in view of the way most data are collected and elaborated.

(b) Measurement errors in the dependent variable are incorporated into the disturbance term leaving unbiased and consistent (although inefficient or larger than minimum variance) OLS parameter estimates. However, with measurement errors in the explanatory variables, the fifth of the OLS assumption of independence of the explanatory variables and error term is violated (see Prob. 6.4), leading to biased and inconsistent OLS parameter estimates. In a simple regression, $\hat{b}_1$ is biased downward, while $\hat{b}_0$ is biased upward.

(c) There is no formal test to detect the presence of errors in variables. Only economic theory and knowledge of how the data were gathered can sometimes give some indication of the seriousness of the problem.

(d) One method of obtaining consistent (but still biased and inefficient) OLS parameter estimates is to replace the explanatory variable subject to measurement errors with another variable that is highly correlated with the explanatory variable in question but which is independent of the error term. In the real world, it might be difficult to find such an instrumental variable, and one could never be sure that it would be independent of the error term. The most popular instrumental variable is the lagged value of the explanatory variable in question. Measurement errors in the explanatory variable only also can be corrected by inverse least squares. This involves regressing X on Y. Then, $\hat{b}_0 = -\hat{b}_0'/\hat{b}_1'$ and $\hat{b}_1 = 1/\hat{b}_1'$, where $\hat{b}_0$ and $\hat{b}_1$ are consistent estimates of the intercept and slope parameter of the regression of Y_t on X_t.

9.14 Table 9.15 gives inventories Y, actual sales X, and hypothetical values of X that include measurement errors, X', all in billions of dollars, in U.S. manufacturing from 1983 to 1998. Y and X are assumed to be free of measurement errors. (a) Regress Y_t on X_t. (b) Regress Y_t on X_t' (on the assumption that X is not available). What type of bias results in the estimates in using X' instead of X? (c) Use instrumental variables to obtain consistent parameter estimates, on the assumption that X_t is correlated with u_t. How do these parameter estimates compare with those obtained in part b?

(a)
$$\hat{Y}_t = 169.69 + 0.90\ X_t \qquad R^2 = 0.95$$
$$(11.66)\ \ (16.46)$$

Table 9.15   Inventory and Sales (Both in Billions of Dollars) in U.S. Manufacturing, 1983–1998

Year	1983	1984	1985	1986	1987	1988	1989	1990
Y	312	340	335	323	338	369	391	405
X	172	191	194	195	206	225	237	243
X'	176	195	199	200	212	232	245	252
Year	1991	1992	1993	1994	1995	1996	1997	1998
Y	391	383	384	405	431	437	456	467
X	240	250	261	279	300	310	327	338
X'	251	263	276	296	320	333	352	366

Source: *Economic Report of the President.*

(b) Regressing Y_t on X_t' (if X_t is not available), we get

$$\hat{Y}_t = 182.50 + 0.78\, X_t' \qquad R^2 = 0.94$$
$$(13.38) \quad (15.23)$$

Note that $\hat{b}_1' < \hat{b}_1$; furthermore, b_1 falls outside the 95% confidence interval of b_1' (0.67 to 0.89).

(c) Using X_{t-1}' as an instrumental variable for X_t' (if X_t' is believed to be correlated with u_t), we get

$$\hat{Y}_t = 187.90 + 0.80\, X_{t-1}' \qquad R^2 = 0.92$$
$$(11.44) \quad (12.57)$$

The coefficient on X_{t-1}' is closer to the true one ($\hat{b}_1$ falls in the 95% confidence interval of 0.66 to 0.94), and is consistent. Of course, in the real world it is rarely known what error of measurement might be present (otherwise, the errors could be corrected before running the regression). It is also difficult or impossible to establish whether X_t' is correlated with u_t.

9.15 Using the data in Table 9.15, (a) regress X_t' on Y_t in order to overcome errors in measuring X_t. (b) How do these results compare with those in Prob. 9.14(c)?

(a) Since only X_t (i.e., the explanatory variable) is subject to measurement errors, inverse least squares is another method for obtaining consistent parameter estimates. Regressing X_t' on Y_t, we get

$$\hat{X}_t' = -206.10 + 1.21\, Y_t \qquad\qquad R^2 = 0.94$$
$$(-6.68) \quad (15.23)$$

$$\hat{b}_0 = -\frac{\hat{b}_0'}{\hat{b}_1'} = -\frac{(-206.10)}{1.21} = 170.33 \quad\text{and}\quad \hat{b}_1 = \frac{1}{\hat{b}_1'} = \frac{1}{1.21} = 0.83$$

where $\hat{b}_0$ and $\hat{b}_1$ are consistent (but still biased) estimates of the intercept and slope parameters of the regression of Y_t on X_t.

(b) Using inverse least squares gives better results in this case compared to the instrumental-variable method [see Prob. 9.14(c)]. With instrumental variables, both the estimated intercept and slope parameter are farther from the true values. However, the results may very well differ in other cases. In any event, in the real world we seldom know what types of errors are present, what type of adjustment is appropriate, and how close the adjusted parameters are to the true parameter values.

Supplementary Problems

MULTICOLLINEARITY

9.16 Why can the following consumption function not be estimated?

$$C_t = b_0 + b_1 Y_{dt} + b_2 Y_{dt-1} + b_3\, \Delta Y_{dt} + u_t$$

where $\Delta Y_{dt} = Y_{dt} - Y_{dt-1}$.

Ans. Because there is a perfect multicollinearity between ΔY_{dt} on one hand and Y_{dt} and Y_{dt-1} on the other. As a result, there are only three independent normal equations and four coefficients to estimate, and so no unique solution is possible.

9.17 Table 9.16 gives hypothetical data on consumption expenditures C, disposable income Y_d, and wealth W, all in thousands of dollars, for a sample of 15 families. (a) Regress C on Y_d and W and find $\bar{R}^2$ and $r_{Y_d W}$. (b) Regress C on Y_d only. (c) Regress C on W only. (d) What can you conclude from the preceding with regard to multicollinearity?

Table 9.16   Consumption Expenditures, Disposable Income, and Wealth for 15 Families

Family	1	2	3	4	5	6	7	8	9	10	11	12	13	14	15
C	32	11	15	17	16	13	18	20	14	17	41	17	33	20	18
Y_d	36	12	16	18	17	14	20	23	15	18	50	19	37	22	19
W	144	47	63	70	67	52	79	90	58	70	204	76	149	86	76

Ans. (*a*)
$$\hat{C} = 1.54 + 1.41\,Y_d - 0.15\,W$$
$$(1.94) \quad (-0.83)$$
$$R^2 = 0.994$$
$$\bar{R}^2 = 0.993$$
$$r_{Y_d W} = 0.995$$

(*b*)
$$\hat{C} = 2.13 + 0.80\,Y_d \qquad R^2 = 0.994$$
$$(4.98) \quad (46.25)$$

(*c*)
$$\hat{C} = 2.92 + 0.19\,W \qquad R^2 = 0.992$$
$$(6.37) \quad (41.46)$$

(*d*) Serious multicollinearity is present.

9.18   (*a*) How can a priori information that $b_2 = 0.25 b_1$ be utilized to overcome the multicollinearity problem in Prob. 9.17? (*b*) Reestimate the regression of Prob. 9.17, incorporating the a priori information (as indicated in part *a*) to overcome the multicollinearity problem. (*c*) What is the value of $\hat{b}_1$? Of $\hat{b}_2$?
Ans. (*a*) By estimating $C = b_0 + b_1 Z$, where $Z = Yd + 0.25W$.

(*b*)
$$\hat{C} = 2.53 + 0.39\,Z \qquad R^2 = 0.993$$
$$(5.75) \quad (44.10)$$

(*c*)
$$\hat{b}_1 = 0.39 \quad \text{and} \quad \hat{b}_2 = 0.10$$

HETEROSCEDASTICITY

9.19   Table 9.17 gives gross fixed capital formation Y_i and sales X_i, both in thousands of dollars, for 35 firms in an industry. Regress Y_i on X_i (*a*) for all the data, (*b*) for the first 14 observations only and record the error sum of squares (ESS$_1$), (*c*) for the last 14 observations only and record the error sum of squares (ESS$_2$). (*d*) Test for the presence of heteroscedasticity.

Table 9.17   Gross Fixed Capital Formation and Sales for 35 Firms

Gross Fixed Capital Formation							Sales
30.2	30.5	30.5	30.7	30.9	31.2	31.2	100
31.5	31.5	31.9	32.3	32.8	33.4	33.4	150
35.1	35.7	36.3	36.9	37.4	37.4	37.8	200
38.4	39.1	40.2	40.8	42.1	42.9	43.2	250
44.3	44.9	45.2	45.9	46.5	47.7	48.5	300

Ans. (*a*)
$$\hat{Y}_i = 21{,}637 + 0.079\,X_i \qquad R^2 = 0.94$$
$$(28.50) \quad (22.00)$$

(*b*)
$$\hat{Y}_i = 27{,}429 + 0.033 X_i \qquad R^2 = 0.66$$
$$(31.51) \quad (4.85) \qquad \text{ESS}_1 = 4.897$$

(c)
$$\hat{Y}_i = 15,029 + 0.104X_i \qquad R^2 = 0.73$$
$$(2.99) \quad (5.71) \qquad ESS_2 = 34.694$$

(d) Since $ESS_2/ESS_1 = 7.08$ exceeds $F_{11,11} = 2.82$ at the 5% level of significance, heteroscedasticity is present.

9.20 Assuming that the error variance is proportional to X_i^2 in Prob. 9.19, (a) correct for heteroscedasticity. (b) What is the value of the new intercept and the new slope parameter associated with the variable X_i? How do they compare with the corresponding values before the transformation?

Ans. (a)
$$\frac{\hat{Y}_i}{X_i} = 0.074 + 23,187\left(\frac{1}{X_i}\right) \qquad R^2 = 0.98$$
$$(20.41) \quad (42.16)$$

(b) The value of the new intercept is 23,187 (instead of 21,637), and the new slope parameter associated with the variable X_i is now 0.074 (instead of 0.079).

9.21 Table 9.18 gives the level of gross fixed capital formation Y, sales X_1, both in thousands of dollars, and a productivity index X_2, for 35 firms in an industry. It is expected that Y will be directly related to both X_1 and X_2. Regress Y on X_1 and X_2 for (a) the entire sample, (b) the 14 observations with the smallest values of X_2 and record ESS_1, and (c) the 14 observations with the largest values of X_2 and record ESS_2. (d) Test for the presence of heteroscedasticity.

Table 9.18 Gross Fixed Capital Formation, Sales, and Productivity in 35 Firms

Firm	1	2	3	4	5	6	7	8	9	10	11	12
Y	30.9	31.5	43.2	36.9	44.3	30.5	32.3	42.9	31.2	39.1	35.7	40.8
X_1	135	150	300	225	310	105	170	285	145	250	205	275
X_2	10.3	10.8	16.4	12.9	16.7	10.0	10.9	15.9	10.6	14.6	12.1	15.5
Firm	13	14	15	16	17	18	19	20	21	22	23	24
Y	31.2	42.1	32.8	36.3	37.4	30.5	33.4	37.4	44.9	33.4	45.2	30.2
X_1	140	280	180	215	235	110	190	230	315	195	320	100
X_2	10.5	15.6	10.9	12.5	13.8	10.0	11.1	13.1	17.1	11.3	17.3	9.9
Firm	25	26	27	28	29	30	31	32	33	34	35	
Y	45.9	46.8	35.1	40.2	47.9	30.7	38.1	49.3	31.9	37.8	31.5	
X_1	330	345	200	260	350	120	250	355	165	245	150	
X_2	17.5	17.9	11.5	14.9	18.3	10.1	14.1	18.5	10.8	13.9	10.7	

Ans. (a)
$$\hat{Y} = 12,089 + 0.017\,X_1 + 1.608\,X_2 \qquad R^2 = 0.99$$
$$(2.53) \quad (8.93)$$

(b)
$$\hat{Y} = 33,332 + 0.044\,X_1 - 0.784\,X_2 \qquad R^2 = 0.95$$
$$(3.91) \quad (-0.99) \qquad ESS_1 = 0.658$$

(c)
$$\hat{Y} = 5874 + 0.010\,X_1 + 2.115\,X_2 \qquad R_2 = 0.99$$
$$(0.30) \quad (2.46) \qquad ESS_2 = 2.126$$

(d) Since $ESS_2/ESS_1 = 3.23$ exceeds $F_{11,11} = 2.82$ at the 5% level of significance, heteroscedasticity is present.

9.22   (a) Assuming that the error variance is proportional to X_2^2 in Prob. 9.21,   (a) correct for heteroscedasticity. (a) What is the value of the new intercept and the slope coefficients associated with X_1 and X_2? How do they compare with the corresponding values before the transformation?

Ans.   (a) $$\frac{\hat{Y}}{X_2} = \underset{(10.53)}{1.622} + \underset{(2.85)}{0.016}\left(\frac{X_1}{X_2}\right) + 12{,}200\left(\frac{1}{X_2}\right)   \qquad R^2 = 0.94$$

(b) The new intercept term is 12,200 (instead of 12,089), while the new slope parameter associated with the variable X_1 is 0.016 (instead of 0.017) and the slope parameter associated with variable X_2 is 1.622 (instead of 1.608).

AUTOCORRELATION

9.23   Table 9.19 gives fixed private investment Y, GDP X_1, both seasonally adjusted in billions of dollars, and the commercial paper interest rate X_2 for the United States from 1982 to 1999. (a) Regress Y on X_1. Is there evidence of autocorrelation at the 5 and 1% levels of significance? (b) Regress Y_t on Y_{t-1}, X_{1t}, and X_{1t-1}. What is the value of ρ? (c) Regress Y_t^* on X_{1t}^* to correct for autocorrelation, where Y_t^* and X_{1t}^* are the transformed variables. Is there any evidence of autocorrelation at the 1% level of significance? At the 5% level of significance?

Table 9.19   Private Fixed Investment, GDP (Both Seasonally Adjusted in Billions of Dollars), and Commercial Paper Interest Rate in the United States, 1982–1999

Year	1982	1983	1984	1985	1986	1987	1988	1989	1990
Y	523.3	615.6	695.7	729.2	749.8	768.5	822.9	850.1	824.2
X_1	3315.6	3688.8	4033.5	4319.3	4537.5	4891.6	5258.3	5588.0	5847.3
X_2	11.84	8.87	10.07	7.94	6.61	6.74	7.58	9.11	8.15
Year	1991	1992	1993	1994	1995	1996	1997	1998	1999
Y	801.1	889.6	978.8	1071.6	1135.4	1250.9	1369.3	1524.1	1651.0
X_1	6080.7	6469.8	6795.5	7217.7	7529.3	7981.4	8478.6	8974.9	9559.7
X_2	5.89	3.71	3.17	4.43	5.93	5.43	5.54	5.43	5.12

Source: St. Louis Federal Reserve (Bureau of Economic Analysis) (for Y and X_1 values); Federal Reserve Board of Governors (for X_2 values).

Ans.   (a) $$\hat{Y}_t = \underset{(-0.65)}{-43.95} + \underset{(15.41)}{0.16 X_{1t}}   \qquad \begin{array}{l} R^2 = 0.94 \\ d = 0.23 \end{array}$$

Since $d = 0.23$, there is evidence of autocorrelation at both the 5 and 1% levels of significance.

(b) To correct for autocorrelation, first the following regression is run:

$$\hat{Y}_t = -101.69 + \underset{(7.94)}{0.88 Y_{t-1}} + \underset{(4.62)}{0.51 X_{1t}} - \underset{(-4.55)}{0.49 X_{1t-1}}   \qquad \begin{array}{l} R^2 = 0.99 \\ \hat{\rho} \cong 0.88 \end{array}$$

(c) $$\hat{Y}_t^* = \underset{(-2.22)}{-70.67} + \underset{(8.16)}{0.23 X_{1t}^*}   \qquad \begin{array}{l} R^2 = 0.81 \\ d = 1.17 \end{array}$$

There is no evidence of autocorrelation at the 1% level of significance, but the test is inconclusive at the 5% level of significance.

9.24   Using the data in Table 9.19,   (a) Regress Y_t on X_{1t} and X_{2t}. Is there any evidence of autocorrelation at the 5 and 1% levels of significance?   (b) If evidence of autocorrelation is found in part a, find the value of ρ to be used to transform the variables in order to adjust for autocorrelation. (c) If evidence of autocorrelation is found in part a, regress Y_t^* on X_{1t}^* and X_{2t}^* to correct for autocorrelation. Is there any evidence of remaining autocorrelation at the 1% level of significance? At the 5% level of significance?

Ans.   (a) $\hat{Y}_t = -356.28 + 0.19X_{1t} + 25.62X_{2t}$ $R^2 = 0.95$
                                              (13.48)   (2.26) $d = 0.49$

Since $d = 0.49$, there is evidence of autocorrelation at both the 5 and 1% levels of significance.

(b) $\hat{\rho} \cong 0.94$

(c) $\hat{Y}_t^* = -95.01 + 0.30X_{1t}^* - 0.84X_{2t}^*$ $R^2 = 0.78$
                                         (6.80)   (-0.14) $d = 1.02$

Although d is closer to 2, there is still evidence of autocorrelation at the 5% level, and the test is inconclusive at the 1% level of significance.

9.25   Using the data in Table 9.19, (a) regress ΔY_t on ΔX_{1t} and ΔX_{2t}. (b) Is there evidence of autocorrelation at the 1 and 5% levels of significance? (c) Why is this transformation valid?

Ans.   (a) $\Delta \hat{Y}_t = -99.04 + 0.45 \Delta X_{1t} - 0.91 \Delta X_{2t}$ $R^2 = 0.73$
                                         (6.02)   (-0.18) $d = 1.51$

(b) There is now no evidence of autocorrelation at either the 5% or the 1% level of significance. (c) A regression of ΔY_t on ΔX_{1t} would be less valid since $\hat{\rho}$ is not as close to 1.

ERRORS IN VARIABLES

9.26   Table 9.20 gives inventories Y, actual shipments X, and hypothetical values of X that include measurement errors X', all in billions of dollars, in U.S. durable-goods industries from 1983 to 1998. Y and X are assumed to be free of measurement errors. (a) Regress Y_t on X_t. (b) Regress Y_t on X_t' (on the assumption that X is not available). What type of bias results in the estimates in using X' instead of X? (c) Use instrumental variables to obtain consistent parameter estimates, on the assumption that X_t is correlated with u_t. How do these parameter estimates compare with those of part b?

Table 9.20   Inventories and Shipments (in Billions of Dollars) in the U.S. Durable-Goods Industries, 1983–1998

Year	1983	1984	1985	1986	1987	1988	1989	1990
Y	85.48	97.94	101.28	103.24	108.13	118.46	123.16	123.78
X	199.85	221.33	218.19	211.00	220.80	242.47	257.51	263.21
X'	205.21	207.10	217.16	228.78	228.06	225.49	218.77	213.90
Year	1991	1992	1993	1994	1995	1996	1997	1998
Y	121.00	128.49	135.89	149.13	160.59	167.01	179.89	189.67
X	250.02	238.11	239.33	253.62	268.35	273.82	286.37	295.34
X'	222.48	250.22	263.72	274.88	282.01	288.99	301.30	311.24

Source: St. Louis Federal Reserve (Department of Commerce, Census Bureau).

Ans.   (a) $\hat{Y}_t = -124.23 + 1.04 \, X_t$ $R^2 = 0.86$
                                              (-4.56)   (9.42)

(b) $$\hat{Y}_t = -70.98 + 0.82\, X_t' \qquad R^2 = 0.91$$
$$(-4.06)\ (11.66)$$

With errors of measurement in the value of shipments, $\hat{b}_1' < \hat{b}_1$.

(c) Using X_{t-1}' as an instrument for X_t', we get

$$\hat{Y}_t = -77.82 + 0.88\, X_{t-1}' \qquad R^2 = 0.93$$
$$(-4.93)\ (13.51)$$

The new parameter estimates are closer to the true ones than those obtained in part b.

9.27 Using the data in Table 9.20, (a) regress X_t' on Y_t in order to overcome errors in measuring X_t. When is this method appropriate? (b) How do these results compare with those in Prob. 9.26(c)?

Ans.   (a) $$\hat{X}_t' = 101.49 + 1.11\, Y_t \qquad R^2 = 0.91$$
$$(7.98)\ (11.66)$$

Consistent parameter estimates of the regression of Y_t on X_t are $\hat{b}_0 = -91.43$ and $\hat{b}_1 = 0.90$. Inverse least squares is appropriate when only the explanatory variable includes measurement errors.

(b) Using inverse least squares gives better results in this case compared to the instrumental-variable method [see Prob. 9.26(c)].

CHAPTER 10

Simultaneous-Equations Methods

10.1 SIMULTANEOUS-EQUATIONS MODELS

When the dependent variable in one equation is also an explanatory variable in some other equation, we have a *simultaneous-equations system* or *model*. The dependent variables in a system of simultaneous equations are called *endogenous variables*. The variables determined by factors outside the model are called *exogenous variables*. There is one *behavioral or structural equation* for each endogenous variable in the system (see Example 1). Using OLS to estimate the structural equations results in biased and inconsistent parameter estimates. This is referred to as *simultaneous-equations bias*. To obtain consistent parameter estimates, the *reduced-form equations* of the model must first be obtained. These express each endogenous variable in the system only as a function of the exogenous variable of the model (see Example 2).

EXAMPLE 1. The following two equations represent a simple macroeconomic model:

$$M_t = a_0 + a_1 Y_t + u_{1t}$$
$$Y_t = b_0 + b_1 M_t + b_2 I_t + u_{2t}$$

where M_t is money supply in time period t, Y is income, and I is investment. Since M depends on Y in the first equation and Y depends on M (and I) in the second equation, M and Y are jointly determined, so we have a simultaneous-equations model. M and Y are the endogenous variables, while I is exogenous or determined outside the model. A change in u_{1t} affects M_t in the first equation. This, in turn, affects Y_t in the second equation. As a result, Y_t and u_{1t} are correlated, leading to biased and inconsistent OLS estimates of the M (and Y) equation.

EXAMPLE 2. The first reduced-form equation can be derived by substituting the second equation into the first and rearranging:

$$M_t = a_0 + a_1(b_0 + b_1 M_t + b_2 I_t + u_{2t}) + u_{1t}$$
$$= \frac{a_0 + a_1 b_0}{1 - a_1 b_1} + \frac{a_1 b_2}{1 - a_1 b_1} I_t + \frac{u_{1t} + a_1 u_{2t}}{1 - a_1 b_1}$$

or
$$M_t = \pi_0 + \pi_1 I_t + v_{1t}$$

The second reduced-form equation can be derived by substituting the first equation into the second and rearranging:

$$Y_t = b_0 + b_1(a_0 + a_1 Y_t + u_{1t}) + b_2 I_t + u_{2t}$$

$$= \frac{a_0 b_1 + b_0}{1 - a_1 b_1} + \frac{b_2}{1 - a_1 b_1} I_t + \frac{b_1 u_{1t} + u_{2t}}{1 - a_1 b_1}$$

or

$$Y_t = \pi_2 + \pi_3 I_t + v_{2t}$$

10.2 IDENTIFICATION

Identification refers to the possibility of calculating the structural parameters of a simultaneous-equations model from the reduced-form parameters. An equation of a system is *exactly identified* if the number of excluded exogenous variables from the equation is equal to the number of endogenous variables in the equation minus 1. However, an equation of a system is *overidentified* (or *underidentified*) if the number of excluded exogenous variables from the equation exceeds (or is smaller than) the number of endogenous variables included in the equation minus 1 (see Example 3). Although this is only a necessary rather than a sufficient condition for identification, it usually gives the correct answer (see Prob. 10.5). Unique structural coefficients can be calculated from the reduced-form coefficients only for an exactly identified equation (see Example 4).

EXAMPLE 3. The money supply (M) equation of Example 1 is exactly identified because it excludes one exogenous variable (I) and includes two endogenous variables (M and Y). However, the income, Y, equation is underidentified because it excludes no exogenous variable. If this second equation had included the additional exogenous variable G (government expenditures), the first, or M, equation would have been overidentified because the number of excluded exogenous variables would then have exceeded the number of endogenous variables minus 1.

EXAMPLE 4. A unique value of the structural parameters of the exactly identified M equation of Example 1 can be calculated from the reduced-form parameters of Example 2 as follows:

$$a_1 = \frac{\pi_1}{\pi_3} = \frac{\dfrac{a_1 b_2}{1 - a_1 b_1}}{\dfrac{b_2}{1 - a_1 b_1}} \quad \text{and} \quad a_0 = \pi_0 - a_1 \pi_2 = \frac{a_0(1 - a_1 b_1)}{1 - a_1 b_1}$$

10.3 ESTIMATION: INDIRECT LEAST SQUARES

Indirect least squares (ILS) is a method of calculating structural-parameter values for exactly identified equations. ILS involves using OLS to estimate the reduced-form equations of the system and then using the estimated coefficients to calculate the structural parameters. However, it is not easy to calculate the standard errors of the structural parameters, nor can ILS be used in cases of overidentification.

EXAMPLE 5. Table 10.1 gives the money supply (M = currency plus demand deposits), GDP Y, gross private domestic investment I, and government purchases of goods and services G, all seasonably adjusted in billions of dollars, for the United States from 1982 to 1999 (G will be used in Example 6).

The estimated reduced-form equations of Example 2 are

$$\hat{M}_t = 312.0608 + 0.5693 I_t \quad R^2 = 0.67$$
$$\quad\quad (2.98) \quad\quad (5.65)$$

$$\hat{Y}_t = 852.3203 + 5.3522 I_t \quad R^2 = 0.93$$
$$\quad\quad (2.17) \quad\quad (14.18)$$

$$\hat{a}_1 = \frac{\hat{\pi}_1}{\hat{\pi}_3} = \frac{0.5693}{5.3522} = 0.1064$$

and

**Table 10.1   Money Supply, GDP, Investments, and Government Expenditures
(Seasonably Adjusted in Billions of Dollars) in the United States, 1982–1999**

Year	1982	1983	1984	1985	1986	1987	1988	1989	1990
M	474.30	520.79	551.20	619.28	724.20	749.61	786.25	792.49	824.41
Y	3315.60	3688.80	4033.50	4319.30	4537.50	4891.60	5258.30	5588.00	5847.30
I	483.50	639.50	743.60	762.30	737.10	831.60	842.00	866.70	812.80
G	710.10	742.70	829.00	905.10	963.20	1019.30	1060.70	1123.90	1213.10
Year	1991	1992	1993	1994	1995	1996	1997	1998	1999
M	896.34	1024.31	1129.69	1150.08	1126.80	1081.06	1073.94	1097.37	1122.96
Y	6080.70	6469.80	6795.50	7217.70	7529.30	7981.40	8478.60	8974.90	9559.70
I	832.10	909.80	995.80	1146.10	1155.60	1284.30	1434.50	1590.80	1723.70
G	1239.50	1281.80	1307.10	1344.00	1374.50	1438.90	1508.20	1567.20	1688.80

Source: St. Louis Federal Reserve (Bureau of Economic Analysis).

$$\hat{a}_0 = \hat{\pi}_0 - a_1\hat{\pi}_3 = 312.0608 - 0.1064(852.3203) = 221.3739$$

Thus the M equation of Example 1 estimated by ILS is

$$\hat{M}_t = 221.3739 + 0.1064Y_t$$

The same equation estimated by OLS (inappropriately) is

$$\hat{M}_t = 162.7044 + 0.1159Y_t \qquad R^2 = 0.85$$
$$\quad (2.13) \qquad (9.70)$$

10.4   ESTIMATION: TWO-STAGE LEAST SQUARES

Two-stage least-squares (2SLS) is a method of estimating consistent structural parameters for over-identified equations (for exactly identified equations, 2SLS gives the same results as ILS, but it also gives the standard errors of the estimated structural parameters).   2SLS involves regressing each endogenous variable on all the exogenous variables of the system and then using the predicted values of the endogenous variables to estimate the structural equations of the model.

EXAMPLE 6.   If the second, or Y, equation of Example 1 now includes G (government expenditures) as an additonal explanatory variable, then the first, or M, equation is overidentified (see Example 3) and can be estimated by 2SLS.   The first stage is

$$\hat{Y}_t = -1007.5346 + 1.7471I_t + 4.5794G_t \qquad R^2 = 0.99$$
$$\quad (-5.71) \qquad (6.10) \qquad (13.57)$$

The second stage is

$$\hat{M}_t = 166.5660 + 0.1153\hat{Y}_t \qquad\qquad R^2 = 0.84$$
$$\quad (2.07) \qquad (9.19)$$

$\hat{a}_1 = 0.1153$ is a consistent estimate of a_1.

Solved Problems

SIMULTANEOUS-EQUATIONS MODELS

10.1 What is meant by (*a*) *Simultaneous-equations system or model*? (*b*) *Endogenous variables*? (*c*) *Exogenous variables*? (*d*) *Structural equations*? (*e*) *Simultaneous-equations bias*? (*f*) *Reduced-form equations*?

(*a*) A *simultaneous-equations system or model* refers to the case in which the dependent variable in one or more equations is also an explanatory variable in some other equation of the system. Specifically, not only are the *Y*s determined by the *X*s, but some of the *X*s are, in turn, determined by the *Y*s, so that the *Y*s and the *X*s are jointly or simultaneously determined.

(*b*) The *endogenous variables* are the dependent variables in the system of simultaneous equations. These are the variables that are determined by the system, even though they also appear as explanatory variables in some other equation of the system.

(*c*) *Exogenous variables* are those variables which are determined outside of the model. These also include the lagged endogenous variables, since their values are already known in any given period. The exogenous variables and the lagged endogenous variables are sometimes called *predetermined variables*.

(*d*) *Structural* or *behavioral equations* describe the structure of an economy or the behavior of some economic agents such as consumers or producers. There is one structural equation for each endogenous variable of the system. The coefficients of the structural equations are called *structural parameters* and express the *direct effect* of each explanatory variable on the dependent variable.

(*e*) *Simultaneous-equations bias* refers to the overestimation or underestimation of the structural parameters obtained from the application of OLS to the structural equations of a simultaneous-equations model. This bias results because those endogenous variables of the system which are also explanatory variables are correlated with the error terms, thus violating the fifth assumption of OLS (see Prob. 6.4).

(*f*) *Reduced-form equations* are obtained by solving the system of structural equations so as to express each endogenous variable of the system as a function only of the exogenous or predetermined variables of the system. Since the exogenous variables of the system are uncorrelated with the error terms, OLS gives consistent reduced-form parameter estimates. These measure the total *direct and indirect effects* of a change in the exogenous variables on the endogenous variables and may be used to obtain consistent structural parameters.

10.2 The following two structural equations represent a simple demand-supply model:

$$\text{Demand:} \quad Q_t = a_0 + a_1 P_t + a_2 Y_t + u_{1t} \quad a_1 < 0 \quad \text{and} \quad a_2 > 0$$
$$\text{Supply:} \quad Q_t = b_0 + b_1 P_t + u_{2t} \quad b_1 > 0$$

where *Q* is quantity, *P* is price, and *Y* is consumers' income. It is assumed that the market is cleared in every year so that Q_t represents both quantity bought and sold in year *t*. (*a*) Why is this a simultaneous-equations model? (*b*) Which are the endogenous and exogenous variables of the system? (*c*) Why would the estimation of the demand and supply function by OLS give biased and inconsistent parameter estimates?

(*a*) The given demand-supply model represents a simple simultaneous-equations market system because *Q* and *P* are mutually or jointly determined. If price were below equilibrium, the quantity demanded would exceed the quantity supplied, and vice versa. At equilibrium, the (negatively sloped) demand curve crosses the (positively sloped) supply curve, jointly or simultaneously determining (the equilibrium) *Q* and *P*.

(*b*) The endogenous variables of the model are *Q* and *P*. These are the variables determined within the model. *Y* is the only exogenous variable of the model (i.e., determined outside the model).

(*c*) Since the endogenous variable *P* is also an explanatory variable in both the demand and supply equations, *P* is correlated with u_{1t} in the demand equation and with u_{2t} in the supply equation. This violates the fifth assumption of OLS, which requires that the explanatory variable be uncorrelated with the error term. As a result, estimating the demand and supply functions by OLS results in parameter estimates that are not only biased but also inconsistent (i.e., that do not converge on the true parameters even as the sample size is increased).

10.3 (a) Find the reduced-form equations corresponding to the structural equations of Prob. 10.2. (b) Why are these reduced-form equations important? What do the reduced-form coefficients measure in this market model?

(a) To find the reduced-form equations, the structural equations of Prob. 10.2 are solved for Q and P (the endogenous variables) as a function of only Y (the exogenous variable). Converting the supply equation into a function of P and substituting into the demand equation, we get

$$P_t = \frac{1}{b_1}(Q_t - b_0 - u_{2t})$$

$$Q_t = a_0 + \frac{a_1}{b_1}(Q_t - b_0 - u_{2t}) + a_2 Y_t + u_{1t}$$

$$Q_t\left(\frac{b_1 - a_1}{b_1}\right) = \left(\frac{a_0 b_1 - a_1 b_0}{b_1}\right) + a_2 Y_t + \left(\frac{b_1 u_{1t} - a_1 u_{2t}}{b_1}\right)$$

$$Q_t = \left(\frac{a_0 b_1 - a_1 b_0}{b_1 - a_1}\right) + \left(\frac{b_1 a_2}{b_1 - a_1}\right) Y_t + \left(\frac{b_1 u_{1t} - a_1 u_{2t}}{b_1 - a_1}\right)$$

$$Q_t = \pi_0 + \pi_1 Y_t + v_{1t}$$

where $\pi_0 = \dfrac{a_0 b_1 - a_1 b_0}{b_1 - a_1}$ $\pi_1 = \dfrac{b_1 a_2}{b_1 - a_1}$ $v_{1t} = \dfrac{b_1 u_{1t} - a_1 u_{2t}}{b_1 - a_1}$

Substituting the demand equation into the supply equation as a function of P, we get

$$P_t = \frac{1}{b_1}(a_0 + a_1 P_t + a_2 Y_t + u_{1t} - b_0 - u_{2t})$$

$$P_t\left(\frac{b_1 - a_1}{b_1}\right) = \frac{1}{b_1}(a_0 + a_2 Y_t + u_{1t} - b_0 - u_{2t})$$

$$P_t = \left(\frac{a_0 - b_0}{b_1 - a_1}\right) + \left(\frac{a_2}{b_1 - a_1}\right) Y_t + \left(\frac{u_{1t} - u_{2t}}{b_1 - a_1}\right)$$

$$P_t = \pi_2 + \pi_3 Y_t + v_{2t}$$

where $\pi_2 = \dfrac{a_0 - b_0}{b_1 - a_1}$ $\pi_3 = \dfrac{a_2}{b_1 - a_1}$ $v_{2t} = \dfrac{u_{1t} - u_{2t}}{b_1 - a_1}$

(b) Reduced-form equations

$$Q_t = \pi_0 + \pi_1 Y_t + v_{1t}$$
$$P_t = \pi_2 + \pi_3 Y_t + v_{2t}$$

are important because Y_t is uncorrelated with v_{1t} and v_{2t}, so that consistent estimates of reduced-form parameters π_0, π_1, π_2, and π_3 can be obtained by applying OLS to the reduced-form equations. π_1 and π_3 give, respectively, the total of the direct and indirect effects of a change in Y on Q and P. A change in Y causes a shift in the demand curve, which affects both the equilibrium P and Q.

10.4 Given the following three-equations system, (a) explain why this is not a simultaneous-equations model. (b) Could OLS be used to estimate each equation of this system? Why?

$$Y_{1t} = a_0 + a_1 X_t + u_{1t}$$
$$Y_{2t} = b_0 + b_1 Y_{1t} + b_2 X_t + u_{2t}$$
$$Y_{3t} = c_0 + c_1 Y_{2t} + c_2 X_t + u_{3t}$$

(a) The preceding system is not simultaneous because although Y_2 is a function of Y_1, Y_1 is not a function of Y_2. Similarly, although Y_3 is a function of Y_2, Y_2 is not a function of Y_3. Thus the line of causation runs only in one rather than in both directions. Once Y_1 has been estimated in the first equation, Y_1 can be used (together with X) to estimate Y_2 in the second equation. Similarly, once Y_2 has been estimated in the second equation, Y_2 can be used (together with X) to estimate Y_3 in the third equation. Models of this nature are *recursive* rather than simultaneous.

(b) In the first equation, exogenous variable X is uncorrelated with error term u_1, so that OLS gives unbiased parameter estimates for the first equation. In the second equation, X and Y are uncorrelated with u_2 (i.e., Y_1 is correlated with u_1 but not with u_2), so that OLS gives unbiased parameter estimates for the second equation. The same is true for the third equation. Thus recursive models can be estimated by the sequential application of OLS.

IDENTIFICATION

10.5 (a) What is meant by *identification*? (b) When is an equation of a system exactly identified? (c) Overidentified? (d) Underidentified? (e) Are these rules sufficient for identification?

(a) *Identification* refers to the possibility or impossibility of obtaining the structural parameters of a simultaneous-equations system from the reduced-form parameters. An equation of a system can be exactly identified, overidentified, or underidentified. The system as a whole is exactly identified if all its equations are exactly identified.

(b) An equation of a system is *just or exactly identified* if the number of excluded exogenous variables from the equation is equal to the number of endogenous variables in the equation minus 1. For an exactly identified equation, a unique value of the structural parameters can be *calculated* from the reduced-form parameters.

(c) An equation of a system is *overidentified* if the number of excluded exogenous variables from the equation exceeds the number of endogenous variables in the equation minus 1. For an overidentified equation, more than one numerical value can be calculated from some of the structural parameters of the equation from the reduced-form parameters.

(d) An equation of a system is *underidentified or unidentified* if the number of excluded variables from the equation is smaller than the number of endogenous variables excluded from the equation minus 1. In this case, no structural parameters can be calculated from the reduced-form parameters.

(e) The preceding rules for identification (called the *order condition*) are necessary but not sufficient. However, since these rules do give the correct result in most cases, they are the only ones actually used here. A sufficient condition for identification is given by the *rank condition*, which states that in a system of G equations, any particular equation is identified if and only if it is possible to obtain one nonzero determinant of order $G - 1$ from the coefficients of the variables excluded from that particular equation but included in the other equations of the model. When this rank condition is satisfied, the order condition is automatically satisfied. However, the reverse is not true.

10.6 Given the following demand-supply model (a) determine if the demand and/or supply is exactly identified, overidentified, or underidentified.

$$\text{Demand:} \quad Q_t = a_0 + a_1 P_t + u_{1t} \qquad a_1 < 0$$
$$\text{Supply:} \quad Q_t = b_0 + b_1 P_t + u_{2t} \qquad b_1 > 0$$

(b) What would a regression of Q_t on P_t indicate?

(a) Since this demand-supply model does not include any exogenous variable, both the demand and supply functions are underidentified. In this case, there are no reduced-form equations, and no structural parameters can be calculated. Each price-quantity observation represents the equilibrium quantity bought and sold at the given price and corresponds to the interception of an (unknown) demand and supply curve.

(b) Regressing Q_t on P_t gives neither a demand curve nor a supply curve, but rather a hybrid of demand and supply, which should be referred to simply as a regression line.

10.7 With reference to the demand-supply model in Prob. 10.2   (*a*) determine if the demand and/or supply function is exactly identified, overidentified, or underidentified.   (*b*) Give a graphical interpretation of your answer to part *a*.   (*c*) Derive the formula for the structural coefficients from the reduced-form coefficients.

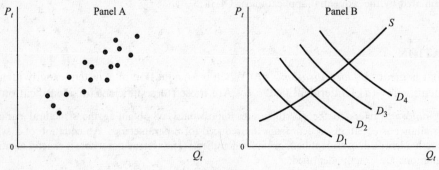

Fig. 10-1

(*a*) The demand function is underidentified because it does not exclude any exogenous variable.   However, since there is one excluded exogenous variable from the supply equation (that is, Y) and two included endogenous variables (i.e., Q and P), the supply function is exactly identified.

(*b*) Changes in Y cause shifts in the demand curve, thus tracing the supply curve.   Figure 10-1*a* shows a hypothetical scatter of points resulting from changes in Y and the error terms, while Fig. 10-1*b* shows the resulting supply curve that could be generated.

(*c*) Unique values of the structural coefficients of the supply equation (the exactly identified equation) can be calculated from the reduced-form coefficients in Prob. 10.3 as follows:

$$b_1 = \frac{\pi_1}{\pi_3} = \frac{\dfrac{b_1 a_2}{b_1 - a_1}}{\dfrac{a_2}{b_1 - a_1}}$$

$$b_1 = \pi_0 - b_1 \pi_2 = \frac{a_0 b_1 - a_1 b_0}{b_1 - a_1} - \frac{b_1 a_0 + b_0 b_1}{b_1 - a_1} = \frac{b_0(b_1 - a_1)}{b_1 - a_1}$$

The formula for the structural coefficients of the demand function cannot be derived from the reduced-form coefficients because the demand function in this model is underidentified.

10.8 With reference to the demand-supply model given below, (*a*) determine if the demand and/or supply functions are exactly identified, overidentified, or underidentified.   (*b*) Find the reduced-form equations.   (*c*) Derive the formula for the structural parameters.

$$\text{Demand:} \quad Q_t = a_0 + a_1 P_t + a_2 Y_t + u_{1t} \qquad a_1 < 0, \qquad a_2 > 0$$
$$\text{Supply:} \quad Q_t = b_0 + b_1 P_t + b_2 T + u_{2t} \qquad b_1 > 0, \qquad b_2 \lessgtr 0$$

where T = trend.

(*a*) The supply equation is exactly identified (as in Prob. 10.7) because it excludes one exogenous variable (Y) and includes two endogenous variables (P and Q).   The demand equation is now also exactly identified because it excludes one exogenous variable (T) and includes two endogenous variables (P and Q).

(*b*) The reduced-form equations can be obtained as in Prob. 10.3(*a*):

$$Q_t = \left(\frac{a_0 b_1 - a_1 b_0}{b_1 - a_1}\right) + \left(\frac{a_2 b_1}{b_1 - a_1}\right) Y_t + \left(\frac{-a_1 b_2}{b_1 - a_1}\right) T + \left(\frac{b_1 u_{1t} - a_1 u_{2t}}{b_1 - a_1}\right)$$

$$P_t = \left(\frac{a_0 - b_0}{b_1 - a_1}\right) + \left(\frac{a_2}{b_1 - a_1}\right) Y_t + \left(\frac{-b_2}{b_1 - a_1}\right) T + \left(\frac{u_{1t} - u_{2t}}{b_1 - a_1}\right)$$

or,

$$Q_t = \pi_0 + \pi_1 Y_t + \pi_2 T + v_{1t}$$
$$P_t = \pi_3 + \pi_4 Y_t + \pi_5 T + v_{2t}$$

where

$$\pi_0 = \frac{a_0 b_1 - a_1 b_0}{b_1 - a_1} \qquad \pi_1 = \frac{a_2 b_1}{b_1 - a_1} \qquad \pi_2 = \frac{-a_1 b_2}{b_1 - a_1} \qquad v_{1t} = \frac{b_1 u_{1t} - a_1 u_{2t}}{b_1 - a_1}$$

$$\pi_3 = \frac{a_0 - b_0}{b_1 - a_1} \qquad \pi_4 = \frac{a_2}{b_1 - a_1} \qquad \pi_5 = \frac{-b_2}{b_1 - a_1} \qquad v_{2t} = \frac{u_{1t} - u_{2t}}{b_1 - a_1}$$

(c) $\qquad a_1 = \dfrac{\pi_2}{\pi_5} \qquad$ and $\qquad b_1 = \dfrac{\pi_1}{\pi_4}$

$$a_2 = \pi_4(b_1 - a_1) = \pi_4\left(\frac{\pi_1}{\pi_4} - \frac{\pi_2}{\pi_5}\right) \qquad \text{and} \qquad b_2 = -\pi_5(b_1 - a_1) = \pi_5\left(\frac{\pi_2}{\pi_5} - \frac{\pi_1}{\pi_4}\right)$$

$$a_0 = \pi_3(b_1 - a_1) + b_0 = \pi_3\left(\frac{\pi_0}{\pi_3} - \frac{\pi_2}{\pi_5}\right) \qquad \text{and} \qquad b_0 = -\pi_3(b_1 - a_1) + a_0 = \pi_3\left(\frac{\pi_0}{\pi_3} - \frac{\pi_1}{\pi_4}\right)$$

10.9 With reference to the demand-supply model given below, (a) determine if the demand and/or supply equation is exactly identified, overidentified, or underidentified. (b) Calculate the structural slope parameters.

$$\text{Demand:} \quad Q_t = a_0 + a_1 P_t + a_2 Y_t + a_3 W_t + u_{1t}$$
$$\text{Supply:} \quad\;\; Q_t = b_0 + b_1 P_t + u_{2t}$$

where W_t is wealth and the expectation is that $a_3 > 0$.

(a) The demand equation is underidentified because it does not exclude any exogenous variable. However, since there are two excluded exogenous variables from the supply equation (i.e., Y and W) and two included endogenous variables (i.e., Q and P), the supply function is overidentified.

(b) In order to calculate the structural slope parameters, the reduced-form equations must be found. They are obtained as in Prob. 10.7(c) and are

$$Q_t = \pi_0 + \pi_1 Y_t + \pi_2 W_t + v_{1t}$$
$$P_t = \pi_3 + \pi_4 Y_t + \pi_5 W_t + v_{2t}$$

where

$$\pi_0 = \frac{a_0 b_1 - a_1 b_0}{b_1 - a_1} \qquad \pi_1 = \frac{a_2 b_1}{b_1 - a_1} \qquad \pi_2 = \frac{a_3 b_1}{b_1 - a_1}$$

$$\pi_3 = \frac{a_0 - b_0}{b_1 - a_1} \qquad \pi_4 = \frac{a_2}{b_1 - a_1} \qquad \pi_5 = \frac{a_3}{b_1 - a_1}$$

The value of b_1 can be calculated from

$$\frac{\pi_1}{\pi_4} = b_1 \qquad \text{or} \qquad \frac{\pi_2}{\pi_5} = b_1$$

These two estimates of b_1 will generally be different, reflecting the fact that the supply equation is now overidentified. As in Prob. 10.7(c), the structural coefficients of the demand function cannot be calculated from the reduced-form coefficients because the demand function in this model is underidentified.

ESTIMATION: INDIRECT LEAST SQUARES

10.10 (a) When can indirect least squares be used? (b) What does it involve? (c) What are some of the shortcomings of using indirect least squares?

(a) *Indirect least squares* (ILS) is a method of calculating consistent structural parameter values for the exactly identified equations in a system of simultaneous equations.

(b) ILS involves using OLS to estimate the reduced-form equations of the system and then using the estimated reduced-form parameters to calculate unique and consistent structural parameter estimates, as indicated in Probs. 10.7(c), 10.8(c), and 10.9(b).

(c) One disadvantage of using ILS is that it does not give the standard error of the calculated structural parameters, and it is rather complicated (and beyond the scope of this book) to calculate them. Another disadvantage of ILS is that it cannot be used to calculate unique and consistent structural-parameter estimates from the reduced-form coefficients for the overidentified equations of a simultaneous-equations model.

10.11 Table 10.2 gives the index of crop output Q (indexed to 1992), crop prices P (indexed to 1991–1992), and disposable income per capita Y (in 1996 dollars), in the United States from 1975 to 1996. Assume that the market is cleared in every year so that Q_t represents both the quantity bought and sold in year t. (a) Estimate by OLS the reduced-form equations given in Prob. 10.3(a). (b) Calculate the supply structural parameters from the reduced-form coefficients. (c) How do these compare with the structural parameters obtained by regressing Q_t on P_t directly?

Table 10.2   Index of Crop Output, Prices, and Disposable Income per Capita in 1996 Dollars: United States, 1975–1996

Year	1975	1976	1977	1978	1979	1980	1981	1982	1983	1984	1985
Q	68	68	74	76	83	75	87	87	68	85	89
P	88	87	83	89	98	107	111	98	108	111	98
Y	14,236	14,653	15,010	15,627	15,942	15,944	16,154	16,250	16,564	17,687	18,120
Year	1986	1987	1988	1989	1990	1991	1992	1993	1994	1995	1996
Q	84	86	75	86	92	92	100	90	106	96	103
P	87	86	104	109	103	101	101	102	105	112	127
Y	18,536	18,790	19,448	19,746	19,967	19,892	20,359	20,354	20,675	21,032	21,385

Source: *Economic Report of the President, 2000.*

(a) The estimated reduced-form equations [from Prob. 10.3(a)] are

$$\hat{Q}_t = 14.2802 + 0.0039\,Y_t \qquad R^2 = 0.67$$
$$\qquad\quad (1.26) \qquad (6.31)$$

$$\hat{P}_t = 54.1671 + 0.0026\,Y_t \qquad R^2 = 0.30$$
$$\qquad\quad (3.36) \qquad (2.91)$$

(b) $$\hat{b}_1 = \frac{\hat{\pi}_1}{\hat{\pi}_3} = \frac{0.0039}{0.0026} = 1.5000 \quad [\text{see Prob. 10.7(c)}]$$

$$\hat{b}_0 = \hat{\pi}_0 - b_1\hat{\pi}_3 = 14.2802 - 1.5000(54.1671) = -66.9705$$

where $\hat{b}_0$ and $\hat{b}_1$ are consistent estimators of b_0 and b_1, respectively, and the structural supply equation (estimated by ILS) is

$$\hat{Q}_t = -66.9705 + 1.5000\,P_t$$

(c) Regressing Q_t on P_t directly, we get

$$\hat{Q}_t = 33.1984 + 0.5145\,P_t \qquad R^2 = 0.26$$
$$\qquad\quad (1.67) \quad (2.63)$$

The values of $\hat{b}_0$ and $\hat{b}_1$ obtained by regressing Q_t on P_t are biased and inconsistent estimates of the supply parameters.

10.12 With reference to the demand-supply model of Prob. 10.8 and using the data in Table 10.2 and trend values $T = 1, 2, 3, \ldots, 30$, (a) calculate consistent structural parameters for the demand equation. (b) How do these compare with the structural parameters obtained by estimating the demand equation directly by OLS?

(a) Since the demand equation is exactly identified [see Prob. 10.8(a)], we can use ILS to obtain consistent demand structural-parameter values. The estimated reduced-form equations [from Prob. 10.8(b)] are

$$\hat{Q}_t = 102.6080 - 0.0024Y_t + 2.2520T \qquad R^2 = 0.70$$
$$\qquad\quad (1.73) \qquad (-0.57) \qquad (1.51)$$

$$\hat{P}_t = 211.3674 - 0.0087Y_t + 4.0079T \qquad R^2 = 0.41$$
$$\qquad\quad (2.58) \qquad (-1.49) \qquad (1.95)$$

where
$$\hat{\pi}_0 = 102.6080, \hat{\pi}_1 = -0.0024, \hat{\pi}_2 = 2.2520$$
$$\hat{\pi}_3 = 211.3674, \hat{\pi}_4 = -0.0087, \hat{\pi}_5 = 4.0079$$

Using the formulas given in Prob. 10.8(c), we get

$$\hat{a}_1 = \frac{\hat{\pi}_2}{\hat{\pi}_5} = \frac{2.2520}{4.0079} = 0.5619$$

$$\hat{a}_2 = \hat{\pi}_4\left(\frac{\hat{\pi}_1}{\hat{\pi}_4} - \frac{\hat{\pi}_2}{\hat{\pi}_5}\right) = (-0.0087)\left(\frac{(-0.0024)}{(-0.0087)} - \frac{2.2520}{4.0079}\right) = 0.0025$$

$$\hat{a}_0 = \hat{\pi}_3\left(\frac{\hat{\pi}_0}{\hat{\pi}_3} - \frac{\hat{\pi}_2}{\hat{\pi}_5}\right) = 211.3674\left(\frac{102.6080}{211.3674} - \frac{2.2520}{4.0079}\right) = -16.1573$$

Thus the demand equation estimated by ILS (and showing consistent parameter estimates) is

$$\hat{Q}_t = -16.1573 + 0.5619P_t + 0.0025T$$

(b) The OLS estimation of the demand function is

$$\hat{Q}_t = 9.4529 + 0.0891P_t + 0.0037T \qquad R^2 = 0.67$$
$$\qquad\quad (0.66) \qquad (0.56) \qquad (4.89)$$

The values of $\hat{a}_0$, $\hat{a}_1$, and $\hat{a}_2$ estimated by OLS are biased and inconsistent. Indeed, $\hat{a}_1$ is less than 20% of the ILS estimate, and $\hat{a}_0$ even has the wrong sign (but is not statistically significant).

ESTIMATION: TWO-STAGE LEAST SQUARES

10.13 When can 2SLS be used? (b) What does it involve? (c) What are the advantages of 2SLS with respect to ILS?

(a) *Two-stage least squares* (2SLS) is a method of estimating consistent structural-parameter values for the exactly identified or overidentified equations of a simultaneous-equations system. For exactly identified equations, 2SLS gives the same result as ILS.

(b) 2SLS estimation involves the application of OLS in two stages. In the first stage, each endogenous variable is regressed on all the predetermined variables of the system. These are now the reduced-form equations. In the second stage, the predicted rather than the actual values of the endogenous variables are used to estimate the structural equations of the model. The predicted values of the endogenous variables are obtained by substituting the observed values of the exogenous variables into the reduced-form equations. The predicted values of the endogenous variables are uncorrelated with the error terms, leading to consistent 2SLS structural-parameter estimates.

(c) One advantage of 2SLS over ILS is that 2SLS can be used to obtain consistent structural-parameter estimates for the overidentified as well as for the exactly identified equations in a system of simultaneous equations. Another important advantage is that 2SLS (but not ILS) gives the standard error of the estimated structural parameters directly. Since most identified models are in fact overidentified, 2SLS is very useful.   Indeed, 2SLS is the simplest and one of the best and most common of all simultaneous-equations estimators.

10.14 For the demand-supply model in Prob. 10.8 and using the data in Table 10.2 to estimate the demand equation, (a) show the first-stage result of 2SLS estimation. (b) Show the second-stage result of 2SLS estimation.   (c) How do these results compare with the ILS estimation of the demand equation found in Prob. 10.12(a)?

(a)   The first-stage result of the 2SLS estimation of the demand equation is

$$\hat{P}_t = 211.3674 - 0.0087Y_t + 4.0079T \quad R^2 = 0.41$$
$$\phantom{\hat{P}_t =}(2.58) \quad\;\; (-1.49) \quad\;\; (1.95)$$

(b)   The second-stage result of 2SLS estimation of the demand equation is

$$\hat{Q}_t = -16.16 + 0.56\hat{P}_t - 0.0025Y_t \quad R^2 = 0.70$$
$$\phantom{\hat{Q}_t =}(-0.70) \quad (2.18) \quad\;\; (1.51)$$

(c)   Since the demand equation in Prob. 10.8 is exactly identified, 2SLS estimation gives identical results to ILS estimation [see Prob. 10.12(a)].   However, with 2SLS estimation (as opposed to ILS), we also get the standard errors of the estimated structural parameters directly.

10.15 Table 10.3 includes the additional variable wealth W, measured here by total liquid assets, in billions of dollars, to the data in Table 10.2 for the United States for the years 1975 to 1996.   For the demand-supply model in Prob. 10.9, estimate the supply equation by (a) 2SLS and (b) OLS.

Table 10.3   Index of Crop Output, Prices, Disposable Income per Capita, and Total Liquid Assets in Billions of Dollars in the United States, 1975–1996

Year	1975	1976	1977	1978	1979	1980	1981	1982	1983	1984	1985
Q	68	68	74	76	83	75	87	87	68	85	89
P	88	87	83	89	98	107	111	98	108	111	98
Y	14,236	14,653	15,010	15,627	15,942	15,944	16,154	16,250	16,564	17,687	18,120
W	1366.5	1516.7	1705.4	1911.3	2121.2	2330.0	2601.8	2846.0	3150.7	3518.7	3827.1
Year	1986	1987	1988	1989	1990	1991	1992	1993	1994	1995	1996
Q	84	86	75	86	92	92	100	90	106	96	103
P	87	86	104	109	103	101	101	102	105	112	127
Y	18,536	18,790	19,448	19,746	19,967	19,892	20,359	20,354	20,675	21,032	21,385
W	4122.4	4340.0	4663.7	4893.2	4977.5	5008.0	5081.4	5173.3	5315.8	5702.3	6083.6

Source: Economic Report of the President, 2000.

(a)   Since the supply equation in Prob. 10.9 is overidentified, 2SLS is an appropriate estimating technique to obtain consistent structural parameters.   The first stage is

$$\hat{P}_t = 197.51 - 0.01Y_t + 0.02W_t \quad R^2 = 0.36$$
$$\phantom{\hat{P}_t =}(1.84) \quad (-1.05) \quad (1.35)$$

The second stage is

$$\hat{Q}_t = -32.11 + 1.16\hat{P}_t \quad R^2 = 0.47$$
$$\phantom{\hat{Q}_t =}(-1.15) \quad (4.21)$$

(b)   The (inappropriate) OLS estimation of the supply equation is

$$\hat{Q}_t = -33.20 + 0.51\hat{P}_t \quad R^2 = 0.26$$
$$\phantom{\hat{Q}_t =}(-1.67) \quad (2.63)$$

Supplementary Problems

SIMULTANEOUS-EQUATIONS MODELS

10.16 The following two equations represent a simple wage-price model:

$$W_t = a_0 + a_1 P_t + a_2 Q_t + u_{1t}$$
$$P_t = b_0 + b_1 W_t + u_{2t}$$

where W_t is the wage in time period t, P represents prices, and Q is productivity. (a) Why is this a simultaneous-equations model? (b) Which are the endogenous and exogenous variables? (c) Why would the estimation of W and P equations by OLS give biased and inconsistent parameter estimates? *Ans.* (a) This two-equations model is simultaneous in nature because $W = f(P)$ and $P = f(W)$; thus W and P are jointly determined. (b) The endogenous variables are W and P. The exogenous variable is Q. (c) The estimation of the W function by OLS gives biased and inconsistent parameter estimates because P is correlated with u_1. Similarly, estimating the second, or P, equation by OLS also gives biased and inconsistent parameter estimates because W and u_2 are correlated.

10.17 (a) Find the reduced-form equations for the model in Prob. 10.16. (b) Why are they important? (c) What do the reduced-form coefficients measure in this macro model?

Ans.   (a) $\quad W_t = \dfrac{a_0 + a_1 b_0}{1 - a_1 b_1} + \dfrac{a_2}{1 - a_1 b_1} Q_t + \dfrac{u_{1t} + a_1 u_{2t}}{1 - a_1 b_1}$ or $W_t = \pi_0 + \pi_1 Q_t + v_{1t}$

$\qquad\qquad\qquad P_t = \dfrac{b_0 + a_0 b_1}{1 - a_1 b_1} + \dfrac{a_2 b_1}{1 - a_1 b_1} Q_t + \dfrac{b_1 u_{1t} + u_{2t}}{1 - a_1 b_1}$ or $P_t = \pi_2 + \pi_3 Q_t + v_{2t}$

(b) The reduced-form equations are important because they express each endogenous variable of the model as a function of the exogenous variable(s) only, so that OLS gives consistent parameter estimates. (c) The reduced-form parameters give the total direct and indirect effects of a change in any exogenous variable of the model on each endogenous variable of the model.

10.18 (a) What type of model is the following? (b) How can the equations of this model be estimated?

$$Y_{1t} = a_0 + a_1 X_{1t} + u_{1t}$$
$$Y_{2t} = b_0 + b_1 Y_{1t} + b_2 X_{2t} + u_{2t}$$
$$Y_{3t} = c_0 + c_1 Y_{1t} + c_2 Y_{2t} + c_3 X_{3t} + u_{3t}$$

Ans. (a) The model is recursive. (b) The equations of the model can be estimated by applying OLS sequentially, starting with the first equation.

IDENTIFICATION

10.19 If the simple macroeconomic model in Prob. 10.16 did not include the variable Q_t, (a) would the first equation be exactly identified, overidentified, or underidentified? (b) What about the second equation? *Ans.* (a) The first equation would be underidentified. (b) The second equation also would be underidentified.

10.20 For the macro model in Prob. 10.16, determine (a) if the first equation is exactly identified, overidentified, or underidentified. (b) What about the second equation? (c) What are the values of the structural parameters?
Ans. (a) The first equation is underidentified. (b) The second equation is exactly identified. (c) $b_1 = \pi_3/\pi_1$; $b_0 = \pi_2 - b_1 \pi_0$; a_1 and a_2 cannot be calculated from the reduced-form coefficients because the W equation is underidentified.

10.21 If the second equation of the macro model in Prob. 10.16 included the additional variable Y (GNP), (a) determine if the W and/or P equations are exactly identified, overidentified, or underidentified. (b) Find the reduced-form equations. (c) Derive the formula for the structural parameters.

Ans. (*a*) Both the first, or W, equation and the second, or P, equation are now exactly identified.

(*b*)
$$W_t = \frac{a_0 + a_1 b_0}{1 - a_1 b_1} + \frac{a_2}{1 - a_1 b_1} Q_t + \frac{a_1 b_2}{1 - a_1 b_1} Y_t + \frac{u_{1t} + a_1 u_{2t}}{1 - a_1 b_1}$$

$$P_t = \frac{a_0 b_1 + b_0}{1 - a_1 b_1} + \frac{a_2 b_1}{1 - a_1 b_1} Q_t + \frac{b_2}{1 - a_1 b_1} Y_t + \frac{b_1 u_{1t} + u_{2t}}{1 - a_1 b_1}$$

or
$$W_t = \pi_0 + \pi_1 Q_t + \pi_2 Y_t + v_{1t}$$

$$P_t = \pi_3 + \pi_4 Q_t + \pi_5 Y_t + v_{2t}$$

(*c*)
$$a_1 = \frac{\pi_2}{\pi_5} \quad \text{and} \quad b_1 = \frac{\pi_4}{\pi_1}$$

$$a_2 = \pi_2 \left(\frac{\pi_1}{\pi_2} - \frac{\pi_4}{\pi_5} \right) \quad \text{and} \quad b_2 = \pi_2 \left(\frac{\pi_5}{\pi_2} - \frac{\pi_4}{\pi_1} \right)$$

$$a_0 = \pi_3 \left(\frac{\pi_0}{\pi_3} - \frac{\pi_2}{\pi_5} \right) \quad \text{and} \quad b_0 = \pi_0 \left(\frac{\pi_3}{\pi_0} - \frac{\pi_4}{\pi_1} \right)$$

10.22 If the first equation in Prob. 10.16 included the additional variable P_{t-1} (price lagged 1 year), (*a*) would the equations be exactly identified, overidentified, or underidentified? (*b*) What is the value of the structural slope parameters?
Ans. (*a*) The first, or W, equation is underidentified, while the second, or P, equation is overidentified. (*b*) $b_1 = \pi_4/\pi_1$ or π_5/π_2, reflecting the fact that the P equation is now overidentified; a_1, a_2, and a_3 cannot be calculated because the W equation is underidentified.

ESTIMATION: INDIRECT LEAST SQUARES

10.23 Table 10.4 gives an index of hourly earnings W, consumer prices P, output per hour in nonfarm businesses, Q, and GDP in billions of dollars Y in the United States from 1980 to 1999. (*a*) Estimate the reduced-form equations of Prob. 10.17(*a*). (*b*) Calculate the structural coefficients of the P equation from the reduced-form coefficients. (*c*) How do these compare with the structural parameters obtained by regressing P on W directly?

Table 10.4 Earnings, Price Index, Productivity, and GDP: United States, 1980–1999

Year	1980	1981	1982	1983	1984	1985	1986	1987	1988	1989
W	56.6	61.5	65.7	68.0	71.1	74.8	78.5	81.4	84.8	87.2
P	86.4	94.1	97.7	101.4	105.5	109.5	110.8	115.7	120.8	126.4
Q	82.4	82.5	83.3	87.3	88.4	90.2	92.2	93.1	94.1	94.6
Y	2918.8	3203.1	3315.6	3688.8	4033.5	4319.3	4537.5	4891.6	5258.3	5588.0
Year	1990	1991	1992	1993	1994	1995	1996	1997	1998	1999
W	92.3	96.7	101.4	102.7	105.0	107.7	111.1	114.7	120.8	126.5
P	134.3	138.3	142.4	146.4	150.2	154.1	159.1	161.8	164.4	168.8
Q	94.4	94.4	101.5	101.3	102.4	103.6	105.9	108.1	111.2	115.8
Y	5847.3	6080.7	6469.8	6795.5	7217.7	7529.3	7981.4	8478.6	8974.9	9559.7

Source: St. Louis Federal Reserve (Bureau of Labor Statistics (W, P, Q values), Bureau of Economic Analysis (Y values)).

Ans. (*a*)
$$\hat{W}_t = -114.8528 + 2.1270 Q_t \qquad R^2 = 0.98$$
$$\quad\quad (-17.61) \quad\quad (31.62)$$

$$\hat{P}_t = -126.0632 + 2.6471 Q_t \qquad R^2 = 0.96$$
$$\quad\quad (-9.89) \quad\quad (20.14)$$

(*b*) $\hat{b}_1 = 1.2445$; $\hat{b}_0 = 16.8711$ (*c*) By OLS $\hat{b}_1 = 1.2550$ and $\hat{b}_0 = 15.9256$

10.24 For the model in Prob. 10.21,   (*a*) estimate the reduced-form equations, and   (*b*) calculate the structural coefficients of the W equation from the reduced-form coefficients.   (*c*) How do these compare with the structural coefficients of the W equation obtained by OLS?

Ans.   (*a*)
$$\hat{W}_t = -3.2144 + 0.4954Q_t + 0.0079Y_t \qquad R^2 = 0.99$$
$$\phantom{\hat{W}_t = }(-0.10) \qquad (1.11) \qquad (3.69)$$

$$\hat{P}_t = 114.3837 - 0.8671Q_t + 0.0169Y_t \qquad R^2 = 0.98$$
$$\phantom{\hat{P}_t = }(2.10) \qquad (-1.10) \qquad (4.47)$$

(*b*) $\hat{a}_0 = -56.6837$, $\hat{a}_1 = 0.4675$, and $\hat{a}_2 = 0.9007$ (*c*) By OLS, $\hat{a}_0 = -54.2209$, $\hat{a}_1 = 0.4810$, and $\hat{a}_2 = 0.8539$

10.25 For the model in Prob. 10.21, write the structural equation for the P equation estimated by   (*a*) ILS and   (*b*) OLS.

Ans.   (*a*)
$$\hat{P}_t = -108.7575 - 1.7503W_t + 0.0307Y_t$$

(*b*)
$$\hat{P}_t = 12.9178 + 1.3544W_t - 0.0010Y_t \qquad R^2 = 0.99$$
$$\phantom{\hat{P}_t = }(1.47) \qquad (1.87) \qquad (-0.36)$$

TWO-STAGE LEAST SQUARES

10.26 For the model in Prob. 10.21 and using the data in Table 10.4 to estimate the W equation, (*a*) show the first-stage results of 2SLS estimation, and   (*b*) show the second-stage results of 2SLS estimation.   (*c*) How do these results compare with ILS estimation of the W equation found in Prob. 10.24?

Ans.   (*a*)
$$\hat{P}_t = 114.3837 - 0.8671Q_t + 0.0169Y_t \qquad R^2 = 0.98$$
$$\phantom{\hat{P}_t = }(2.10) \qquad (-1.10) \qquad (4.47)$$

(*b*)
$$\hat{W}_t = -56.32 + 0.46\hat{P}_t + 0.90Q_t \qquad R^2 = 0.99$$
$$\phantom{\hat{W}_t = }(-3.39) \qquad (3.69) \qquad (2.66)$$

(*c*)   They are identical (there is a slight difference due to rounding); we also get the standard errors.   The structural parameters estimated by 2SLS and ILS are consistent.

10.27 For the model in Prob. 10.22 and the data in Table 10.4, estimate the P equation by   (*a*) 2SLS and   (*b*) OLS.

Ans.   (*a*)
$$\hat{P}_t = 16.07 + 1.25\hat{W}_t \qquad R^2 = 0.99$$
$$\phantom{\hat{P}_t = }(5.26) \qquad (38.63)$$

(*b*)
$$\hat{P}_t = 15.93 + 1.25W_t \qquad R^2 = 0.99$$
$$\phantom{\hat{P}_t = }(6.17) \qquad (45.03)$$

Time-Series Methods

11.1 ARMA

In Sec. 9.3, we discussed the problem of first-order autocorrelation in time series. Often, variables are exploited solely for their time series properties to achieve forecasts. These forecasts are not based on a theoretical model, but use past movements to predict future movements. High-frequency data (monthly, daily, etc.) can follow complex time-series processes that will change the appropriate method of estimation.

There are two main types of correlation:

1. Autoregressive of order $p[\mathrm{AR}(p)]$

$$y_t = \gamma_1 y_{t-1} + \gamma_2 y_{t-2} + \cdots + \gamma_p y_{t-p} + \varepsilon_t$$

2. Moving average of order $q[\mathrm{MA}(q)]$

$$y_t = \varepsilon_t - \theta_1 \varepsilon_{t-1} - \theta_2 \varepsilon_{t-2} - \cdots - \theta_q \varepsilon_{t-q}$$

Combining the two yields the $\mathrm{ARMA}(p, q)$ representation

$$y_t = \gamma_1 y_{t-1} + \gamma_2 y_{t-2} + \cdots + \gamma_p y_{t-p} + \varepsilon_t - \theta_1 \varepsilon_{t-1} - \theta_2 \varepsilon_{t-2} - \cdots - \theta_q \varepsilon_{t-q}$$

Estimation of $\mathrm{AR}(p)$ is simply a lag-dependent variable and can be estimated with OLS for large samples. Inclusion of the moving-average process yields nonlinear equations that can be estimated by computer as shown in Chap. 12.

EXAMPLE 1. Using the observations of ε_1 in Table 11.1, we generate $\mathrm{AR}(1)[\gamma_1 = 0.8]$, $\mathrm{MA}(1)[\theta_1 = -0.8]$, and $\mathrm{ARMA}(1,1)[\gamma_1 = 0.8; \theta_1 = -0.8]$ and graph the results in Fig. 11-1 (with $\varepsilon_0 = 0$ to start the processes).

As can be seen in Table 11.1 and in Fig. 11-1, the original series, ε_t, fluctuates around its mean (0). The $\mathrm{AR}(1)$ process also moves around 0 but retains part of the past values and does not revert back to 0 as quickly. The $\mathrm{MA}(1)$ process retains some memory of past values, but only for 1 period, and thus moves away from past values more quickly. The $\mathrm{ARMA}(1,1)$ process has some qualities of both $\mathrm{AR}(1)$ and $\mathrm{MA}(1)$.

11.2 IDENTIFYING ARMA

An AR process can be distinguished from an MA process by its persistence. Since autoregression is an iterative process, values of the random error fade away slowly as each year feeds to the next. The MA process is correlation of only the random component, so after q periods the random error is no longer in the system.

Table 11.1 Time-Series Observations

t	1	2	3	4	5	6	7	8	9	10
ε	−0.69	1.04	1.3	−0.15	−0.26	0.07	1.12	−0.3	−1.72	0.01
$y_t[\text{AR}(1)]$	−0.69	0.488	1.6904	1.2023	0.7019	0.6315	1.6252	1.0002	−0.9199	−0.7259
$y_t[\text{MA}(1)]$	−0.69	0.488	2.132	0.89	−0.38	−0.138	1.176	0.596	−1.96	−1.366
$y_t[\text{ARMA}(1,1)]$	−0.69	−0.064	2.0808	2.5546	1.6637	1.193	2.1304	2.3003	−0.1198	−1.4618

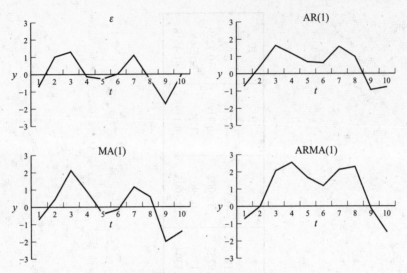

Fig. 11-1   Time-Series Processes

The persistence of error terms can be examined through the *autocorrelation function*

$$\mathrm{ACF}_s = \frac{\mathrm{cov}(y_t, y_{t-s})}{\sigma_y^2}$$

and the *partial autocorrelation function* (PACF_s), which is the coefficient on y_{t-s} in the regression

$$y_t = \gamma_1 y_{t-1} + \gamma_2 y_{t-2} + \cdots + \gamma_s y_{t-s} + \varepsilon_t$$

Once the degree of correlation is narrowed down, multiple possibilities can be estimated. One way to choose the best specification is to take the one which minimizes Aikake's information criteria (AIC)

$$\mathrm{AIC} = \ln\left(\frac{\mathrm{ESS}}{T}\right) + \frac{2j}{T}$$

where j is the number of parameters estimated and ESS is the sum of squared errors ($\sum e^2$).

To test the presence of correlations, the Box-Pierce statistic $Q = T \sum \mathrm{ACF}_s^2$ tests the null hypothesis that there are no correlations. Q follows the chi-square distribution with degrees of freedom equal to the highest lag calculated (usually the minimum of 40 and $T/2$).

EXAMPLE 2. A company is trying to aid their prediction of sales patterns by looking at the time-series properties of the past 4 years of weekly sales (208 weeks). Table 11.2 shows the ACF and PACF, and Fig. 11-2 reflects the correlations plotted against the number of lags, known as a *correlelogram*.

Analyzing the ACF, we see a large positive correlation at 4 lags, and subsequently smaller correlations at intervals of every 4 lags (8, 12, 16 lags). The persistence is consistent with an AR process. Looking at the PACF confirms this. There is a large partial correlation at 4 lags, but after accounting for this, 8, 12, and 16 lags no longer show correlation. Therefore our finding is of an AR(4) process. To see if it is significant, we calculate the Q statistic (here we use only 16 lags for simplicity)

$$Q = T \sum \mathrm{ACF}_s^2 = (208)(0.7994) = 166.28$$

The critical value for the chi-square distribution with 16 df at the 5% level of significance is 26.3. Since $Q = 166.28 > 26.3$, we reject the null hypothesis and there is no correlation; therefore the AR(4) process is statistically significant.

Table 11.2   ACF and PACF of Sales

s	ACF	PACF
1	−0.12678	−0.12678
2	0.10376	0.08912
3	−0.08842	−0.06674
4	0.67066	0.66278
5	−0.12356	0.01116
6	0.08703	−0.01332
7	−0.05300	0.01831
8	0.42026	−0.05191
9	−0.04892	0.10053
10	0.06823	0.01297
11	−0.05903	−0.05758
12	0.24187	−0.04140
13	0.04356	0.08821
14	0.10290	0.12567
15	−0.04299	0.03352
16	0.17137	0.05023

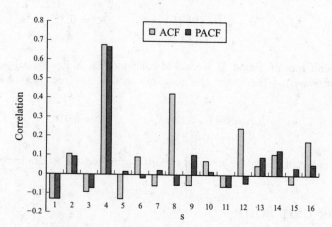

Fig. 11-2   ACF and PACF Correlelogram

11.3   NONSTATIONARY SERIES

For OLS estimation in general to be valid, the error term must be time-invariant, that is, stationary. A nonstationary series follows the form

$$Y_t = Y_{t-1} + \varepsilon_t$$

which is autoregressive with $\gamma = 1$, also called *unit root*, or integrated of order $1[I(1)]$.

Since the entire value from the previous period is carried forward to the current period, values of the random error never fade away.   The continuous buildup of the errors creates the problem that a nonstationary series will tend toward an infinite variance.   Furthermore, if the Y and X variables in a regression are both nonstationary, the model will have a spuriously significant result and high R^2 even if the two variables are unrelated.

Taking first differences will eliminate the autoregressive component, and the unit root:

$$Y_t - Y_{t-1} = \Delta Y_t = \varepsilon_t$$

EXAMPLE 3. The two series in Table 11.3, Y and X, are independently generated variables containing a unit root. There should be no statistical relationship between Y and X.

Table 11.3   Unit-Root Variables and First Differences

t	1	2	3	4	5	6	7	8	9	10
Y	−2.3356	−1.3109	0.7429	0.6579	2.0952	2.2506	0.6410	−0.7852	−1.3934	−0.3937
X	−0.3670	−0.5800	−0.6762	−1.9027	−4.0932	−4.5873	−4.7776	−5.9336	−5.5949	−7.6423
ΔY	—	1.0246	2.0538	−0.0850	1.4373	0.1553	−1.6095	−1.4263	−0.6082	0.9997
ΔX	—	−0.2130	−0.0961	−1.2265	−2.1904	−0.4940	−0.1903	−1.1560	0.3387	−2.0474

t	11	12	13	14	15	16	17	18	19	20
Y	−0.7470	0.0555	1.3462	2.6339	2.7433	2.7969	3.3475	4.4176	4.8743	6.6956
X	−6.3011	−7.9872	−7.5572	−9.2341	−9.2107	−9.0498	−7.4928	−7.8962	−8.2248	−7.6246
ΔY	−0.3532	0.8025	1.2907	1.2876	0.1094	0.0536	0.5506	1.0700	0.4567	1.8213
ΔX	1.3411	−1.6860	0.4300	−1.6768	0.0233	0.1609	1.5570	−0.4034	−0.3285	0.6001

Regressing Y and X yields

$$\hat{Y}_t = -1.16 - 0.45X_t \qquad R^2 = 0.32$$
$$(-2.84)$$

If we ignored the unit root of Y and X, we would conclude that X has a statistically significant effect on Y (at the 5% significance level). Taking the unit root into account and regressing ΔY in ΔX, we get reliable results:

$$\Delta \hat{Y}_t = 0.43 - 0.12\Delta X_t \qquad R^2 = 0.02$$
$$(-0.53)$$

Correcting for the unit root lowers the spurious t statistic of b_1 and the R^2 dramatically.

11.4   TESTING FOR UNIT ROOT

Stationary and nonstationary series can follow different patterns, many of which look similar when graphed. This makes testing for a unit root a tricky proposition.

Stationary		Nonstationary	
White noise:	$Y_t = \mu + \varepsilon_t$	Random walk:	$Y_t = Y_{t-1} + \varepsilon_t$
Autoregressive:	$Y_t = \mu + \gamma Y_{t-1} + \varepsilon_t (|\gamma| < 1)$	Random walk with drift:	$Y_t = \mu + Y_{t-1} + \varepsilon_t$
Trend stationary:	$Y_t = \mu + \beta t + \varepsilon_t (t = 1, 2, \ldots)$		

To distinguish a unit root, we can run the regression

$$\Delta Y_t = b_0 + \sum b_j \Delta Y_{t-j} + \beta t + \gamma Y_{t-1} + \mu_t$$

The regression includes enough lags of ΔY_t so that u_t contains no autocorrelation. The model may be run without t if a time trend is not necessary. If there is a unit root, differencing Y should result in a white-noise series (no correlation with Y_{t-1}). The *augmented Dickey-Fuller* (ADF) test of the null hypothesis of no unit root tests $H_0 : \beta = \gamma = 0$ if there is a trend (F test), and $H_0 : \gamma = 0$ if there is no trend (t test). If the null is accepted, we assume that there is a unit root and difference the data before running a regression. If the null is rejected, the data are stationary and can be used without differencing. Since a unit root biases the estimation of γ downward, special tables in App. 11 are used to find the critical value for the ADF test.

EXAMPLE 4. We test Y from Example 3 for a unit root at the 5% level of significance with and without a time trend.

Without trend:

$$\Delta \hat{Y}_t = 0.50 - 0.02 Y_{t-1} \qquad R^2 < 0.01$$
$$(-0.20)$$

Since $t_{b_1} = -0.20 > -3.33$ (from App. 11), we fail to reject the null hypothesis that there is a unit root. The correct procedure is then to take first differences of Y before using it in a regression.

With a trend:

$$\Delta \hat{Y}_t = -0.04 + 0.07t - 0.17 Y_{t-1} \qquad R^2 = 0.08$$
$$(1.13) \quad (-0.96) \qquad F = 0.66$$

Since $F = 0.66 < 7.24$, we again find a unit root.

11.5 COINTEGRATION AND ERROR CORRECTION

For a series which has a unit root, the best forecast of the next period's value is the current period's value. In some cases, even though two series have a unit root and follow a random walk individually, they move together in the long run. If $Y_t = Y_{t-1} + \varepsilon_{Yt}$ and $X_t = X_{t-1} + \varepsilon_{Xt}$, we see that Y and X have a unit root. If there is no unit root in the error term from the regression $Y_t = b_0 + b_1 X_t + u_t$, then Y and X are *cointegrated*.

If Y and X are cointegrated, then it is not enough to simply difference the variables to run a regression. One must also take into account the long-run relationship between the variables. When Y is above the level indicated by X, we would expect Y to fall, and vice versa. Therefore the deviations from the long-run relationship should be included as an explanatory variable in an *error-correction model*. First, the long-run relationship is estimated.

$$e_t = Y_t - \hat{b}_0 - \hat{b}_1 X_t$$

are the deviations from the long-run relationship.

Next, these differences are included as an additional variable

$$\Delta Y_t = c_0 + c_1 \Delta X_t + c_2 e_{t-1} + u_t$$

Since all variables in the error-correction model are stationary, OLS may be used.

EXAMPLE 5. A potential investor wishes to model consumption in Korea. Table 11.4 reports log of consumption Y and log of GDP X in Korea from 1953 to 1991 (both measured in 1985 international prices).

To ensure the validity of the results, we first test each series for a unit root:

$$\Delta \hat{Y}_t = 0.03 + 0.01 Y_{t-1} \qquad R^2 = 0.04$$
$$(1.24)$$

$$\Delta \hat{X}_t = 0.02 + 0.01 X_{t-1} \qquad R^2 = 0.08$$
$$(1.78)$$

Both accept the null of a unit root. To test that first differences are stationary:

Table 11.4   Log of Consumption and GDP in 1985 International Prices in Korea, 1953–1991

Year	Y	X	Year	Y	X	Year	Y	X	Year	Y	X
1953	2.5291	2.8062	1963	2.9921	3.2771	1973	3.9005	4.2500	1983	4.5313	4.9981
1954	2.5861	2.8423	1964	3.0988	3.3384	1974	3.9615	4.3487	1984	4.6068	5.0865
1955	2.6923	2.9255	1965	3.1563	3.4035	1975	4.0161	4.4062	1985	4.6650	5.1479
1956	2.7412	2.9618	1966	3.2181	3.5233	1976	4.1105	4.5185	1986	4.7397	5.2488
1957	2.8026	3.0434	1967	3.2817	3.5896	1977	4.1584	4.6173	1987	4.8115	5.3528
1958	2.8311	3.0732	1968	3.3758	3.7099	1978	4.2465	4.7359	1988	4.9042	5.4610
1959	2.8772	3.0966	1969	3.4438	3.8178	1979	4.3523	4.8255	1989	5.0142	5.5538
1960	2.8899	3.1081	1970	3.6410	3.9821	1980	4.3252	4.7699	1990	5.1114	5.6562
1961	2.9091	3.1499	1971	3.7531	4.0802	1981	4.3691	4.8233	1991	5.1934	5.7485
1962	2.9689	3.1870	1972	3.7829	4.1170	1982	4.4447	4.8941			

Source: Penn-World Tables 5.6.

$$\Delta\Delta\hat{Y}_t = 0.06 - 0.90\,\Delta Y_{t-1} \qquad R^2 = 0.45$$
$$(-5.36)$$

$$\Delta\Delta\hat{X}_t = 0.06 - 0.74\,\Delta X_{t-1} \qquad R^2 = 0.38$$
$$(-4.58)$$

Both ΔY and ΔX reject the null of a unit root. This establishes that Y and X both have unit roots; we now test for a long-run relationship (i.e., cointegration). Estimating residuals of the long-run relationship, we obtain

$$e_t = Y_t - 0.13 - 0.88X_t$$

Unit-root test of e_t yields

$$\Delta\hat{e}_t = 0.002 - 0.55e_{t-1} \qquad R^2 = 0.34$$
$$(-4.27)$$

Since we can reject the null of a unit root for e_t at the 5% level of significance, we conclude that Y and X are cointegrated. Therefore the correct model of consumption and GDP is an error-correction model:

$$\Delta\hat{Y}_t = 0.01 + 0.73\,\Delta X_t - 0.55e_{t-1} \qquad R^2 = 0.76$$
$$(9.49) \qquad (-4.42)$$

The results reveal that for a 1% increase in income there is a 0.73% increase in consumption (note that this is a double-log model). The negative coefficient on e_{t-1} indicates that if consumption is above its long-run relationship with GDP, it will decrease to return to equilibrium.

11.6   CAUSALITY

The usual OLS model only identifies the correlation between variables; it does not help in determining the direction of the relationship. While causality is an elusive concept that can never be proved with certainty, time-series econometrics can help sort out these timing issues. If changes in X precede changes in Y, we can rule out Y causing X. Using this logic, we can estimate the regression:

$$Y_t = b_0 + \sum b_j Y_{t-j} + \sum c_j X_{t-j} + u_t$$

If past values of X help determine current values of Y, we say X *Granger causes* Y. The test of $H_0 : c_i = 0$ can be carried out with an F test. The number of lags may be chosen using the AIC, adjusted R^2, or one may include the highest feasible number of lags. To calculate the magnitude of causality $\sum c_j$ represents a short-run effect of X_i. Since there is a feedback effect from lags of Y in the long run, the long-run effect is $\sum c_j / (1 - \sum b_j)$.

EXAMPLE 6. Using the data in Table 11.4, we want to test to see if either consumption or GDP leads the other. Since the two series are cointegrated, the correct procedure would be Granger causality in an error correction model. We use one lag of the variables, thus a t test can be used to test for Granger causality.

$$\Delta \hat{Y}_t = 0.06 - 0.19 \, \Delta Y_{t-1} + 0.34 \, \Delta X_{t-1} - 0.58 e_{t-1} \qquad R^2 = 0.25$$
$$\qquad\qquad (-0.70) \qquad\quad (0.19) \qquad\quad (-2.05)$$

Since the coefficient on ΔX_{t-1} is not significant at the 5% level, we conclude that X does not Granger-cause Y. We then test for reverse causality:

$$\Delta \hat{X}_t = 0.06 + 0.53 \Delta \, X_{t-1} - 0.35 \, \Delta Y_{t-1} + 0.10 e_{t-1} \qquad R^2 = 0.25$$
$$\qquad\quad (1.80) \qquad\quad (-1.13) \qquad\quad (0.30)$$

Since the coefficient on ΔY_{t-1} is not significant at the 5% level of significance, we conclude that Y does not Granger-cause X. Therefore there is no leading variable in the relation between X and Y, and we can conclude that the effect is contemporaneous.

Solved Problems

ARMA

11.1 (a) Explain the difference between an autoregressive and a moving-average process. (b) Why are AR and MA processes referred to as *stationary processes*?

(a) *Autoregression* is a process in which a proportion of y_t is carried forward to the next period, then a proportion of y_{t+1} is carried to the next, and so forth. Since some of y_t is in y_{t+1} when it is carried forward, we say that autoregression is long-lasting. A high observation at time t will be carried forward indefinitely in smaller and smaller proportions. The moving-average process, on the other hand, carries forward ε_t, the random component of y_t, so previous observations are not perpetuated.

(b) Both the AR process (given that $|\gamma_1| < 1$), and the MA process eventually revert back to their original means after a positive or negative shock. In the AR process, the shock eventually dies out. In the MA process, the shock leaves after a number of periods greater than the number of lags in the MA process. Since both of these processes stay around their means, they are stationary.

11.2 Show algebraically that (a) an AR(1) process is equivalent to an MA(∞) process and (b) an MA(1) process is equivalent to an AR(∞) process.

(a) An AR(1) process is defined as

$$y_t = \gamma_1 y_{t-1} + \varepsilon_t$$

Extending this process to y_{t-1} gives

$$y_{t-1} = \gamma_1 y_{t-2} + \varepsilon_{t-1}$$

Substituting into the equation for y_t yields

$$y_t = \gamma_1 (\gamma_1 y_{t-2} + \varepsilon_{t-1}) + \varepsilon_t = \gamma_1^2 y_{t-2} + \gamma_1 \varepsilon_{t-1} + \varepsilon_t$$

Similarly, substituting for y_{t-1}, we obtain

$$y_t = \gamma_1 (\gamma_1 (\gamma_1 y_{t-3} + \varepsilon_{t-2}) + \varepsilon_{t-1}) + \varepsilon_t = \gamma_1^3 y_{t-3} + \gamma_1^2 \varepsilon_{t-2} + \gamma_1 \varepsilon_{t-1} + \varepsilon_t$$

Recursively substituting for each y_{t-s} yields

$$y_t = \gamma_1^t \varepsilon_0 + \gamma_1^{t-1} \varepsilon_1 + \cdots + \gamma_1^s \varepsilon_{t-s} + \cdots + \gamma_1 \varepsilon_{t-1} + \varepsilon_t$$

As can be seen, an AR(1) process contains some part of each previous error term. Since γ_1 is a fraction, errors farther away are reflected in smaller proportions. Also, the preceding equation is equivalent to an MA(∞) process (as $t \to \infty$) with $-\theta_s = \gamma_1^s$ from $s = 1$ to t.

(b) Starting with the MA(1) process and performing a similar manipulation as in part a, we obtain

$$y_t = \varepsilon_t - \theta_1 \varepsilon_{t-1}$$
$$y_{t-1} = \varepsilon_{t-1} - \theta_1 \varepsilon_{t-2}$$

Solving for ε_{t-1}

$$\varepsilon_{t-1} = y_{t-1} + \theta_1 \varepsilon_{t-2}$$

Substituting into the equation for y_t

$$y_t = \varepsilon_t - \theta_1(y_{t-1} + \theta_1 \varepsilon_{t-2}) = \varepsilon_t - \theta_1 y_{t-1} - \theta_1^2 \varepsilon_{t-2}$$

Substituting for ε_{t-2}

$$y_t = \varepsilon_t - \theta_1(y_{t-1} + \theta_1(y_{t-2} + \theta_1 \varepsilon_{t-3})) = \varepsilon_t - \theta_1 y_{t-1} - \theta_1^2 y_{t-2} - \theta_1^3 \varepsilon_{t-3}$$

Recursively substituting for ε_{t-s}

$$y_t = -\theta_1^t y_0 - \theta_1^{t-1} y_1 - \cdots - \theta_1^s y_{t-s} - \cdots - \theta_1 y_{t-1} + \varepsilon_t$$

This is equivalent to an AR(∞) process (as $t \to \infty$) with $\gamma_s = -\theta_1^s$ for $s = 1$ to t.

11.3 For the randomly generated error terms in Table 11.5, calculate (a) AR(1), $\gamma_1 = -0.5$; (b) AR(1), $\gamma_1 = -0.1$; (c) AR(1), $\gamma_1 = 0.1$; (d) AR(1), $\gamma_1 = 0.5$. (e) When would one see positive and negative correlations?

Table 11.5 Randomly Generated, Standard Normal Distributed Variable

t	1	2	3	4	5	6	7	8	9	10
ε	1.4884	0.2709	−0.2714	−2.3637	−1.7548	0.0142	−0.3184	0.6471	0.7578	0.7866
t	11	12	13	14	15	16	17	18	19	20
ε	0.0231	−0.2975	2.0248	0.3581	−0.2191	0.5701	−0.4038	−0.2615	0.2056	0.6881

The calculations for the first four parts are given in Table 11.6. To carry out the calculations for an AR process, we will use as an example the calculations for part a since parts b, c, and d use the same method with only a change of γ_1. The formula for an AR process is

$$y_t = \gamma_1 y_{t-1} + \varepsilon_t$$

(for part a, $\gamma_1 = -0.5$). Starting at $t = 1$, $y_t = 1.4884$. (We assume here that $\varepsilon_0 = 0$ to get a starting value. Another commonly used method to deal with a starting value is to delete the first period after generating the series since it had no lag associated with it.)

$$y_2 = -0.5(1.4884) + 0.2709 = -0.4733$$
$$y_3 = -0.5(-0.4733) - 0.2714 = -0.03475$$
$$y_4 = -0.5(-0.03475) - 2.3637 = -2.346325$$

etc.

(e) Note that the autoregressive series with negative correlation moves around zero to the opposite direction of the previous value. Natural phenomena that can exhibit negative correlations are overshooting,

Table 11.6 Autoregressive Series

t	ε	(a) AR(1) $\gamma_1 = -0.5$	(b) AR(1) $\gamma_1 = -0.1$	(c) AR(1) $\gamma_1 = 0.1$	(d) AR(1) $\gamma_1 = 0.5$
1	1.4884	1.4884	1.4884	1.4884	1.4884
2	0.2709	−0.4733	0.1220	0.4917	1.0151
3	−0.2714	−0.0347	−0.2836	−0.2294	0.2361
4	−2.3637	−2.3463	−2.3353	−2.3866	−2.2456
5	−1.7548	−0.5816	−1.5212	−1.9934	−2.8776
6	0.0142	0.3050	0.1663	−0.1851	−1.4246
7	−0.3184	−0.4709	−0.3350	−0.3369	−1.0307
8	0.6471	0.8825	0.6806	0.6134	0.1317
9	0.7578	0.3165	0.6897	0.8191	0.8236
10	0.7866	0.6283	0.7176	0.8685	1.1984
11	0.0231	−0.2910	−0.0486	0.1099	0.6223
12	−0.2975	−0.1519	−0.2926	−0.2865	0.0136
13	2.0248	2.1007	2.0540	1.9961	2.0316
14	0.3581	−0.6922	0.1526	0.5577	1.3739
15	−0.2191	0.1270	−0.2343	−0.1633	0.4678
16	0.5701	0.5065	0.5935	0.5537	0.8040
17	−0.4038	−0.6570	−0.4631	−0.3484	−0.0017
18	−0.2615	0.0670	−0.2151	−0.2963	−0.2623
19	0.2056	0.1720	0.2271	0.1759	0.0744
20	0.6881	0.6020	0.6653	0.7056	0.7253

smoothing, and scarce resources.   Series with positive correlation move in the same direction as the previous values.   Examples of positive correlations are herding, learning, and spillovers.

11.4 For the randomly generated error terms in Prob. 11.3, calculate (a) MA(1), $\theta_1 = 0.5$;   (b) MA(1), $\theta_1 = 0.1$;   (c) MA(1), $\theta_1 = -0.1$;   (d) MA(1), $\theta_1 = -0.5$.

Calculations for parts a through d are listed in Table 11.7.   To carry out the calculations for an MA process, we will use as an example the calculations for part a since parts b, c, and d use the same method with only a change of θ_1.   The formula for an MA process is

$$y_t = \varepsilon_t - \theta_1 \varepsilon_{t-1}$$

(for part a, $\theta_1 = 0.5$). Starting at $t = 1$, $y_1 = 1.4884$.   (We assume again that $\varepsilon_0 = 0$ to get a starting value.)

$$y_2 = 0.2709 - 0.5(1.4884) = -0.4733$$
$$y_3 = -0.2714 - 0.5(0.2709) = -0.40685$$
$$y_4 = -2.3637 - 0.5(-0.2714) = -2.2280$$

etc.

IDENTIFYING ARMA

11.5 Compare the ACF for Prob. 11.3(d), y_1; Prob. 11.4(d), y_2; and the random error of Prob. 11.3, y_3. Calculate the ACF up to four lags.

Table 11.8 gives the variables and the first lags. The estimated covariances of the lags are

Table 11.7 Moving Average Series

t	ε	(a) MA(1) $\theta_1 = 0.5$	(b) MA(1) $\theta_1 = 0.1$	(c) MA(1) $\theta_1 = -0.1$	(d) MA(1) $\theta_1 = -0.5$
1	1.4884	1.4884	1.4884	1.4884	1.4884
2	0.2709	−0.4733	0.12206	0.41974	1.0151
3	−0.2714	−0.40685	−0.29849	−0.24431	−0.13595
4	−2.3637	−2.228	−2.33656	−2.39084	−2.4994
5	−1.7548	−0.57295	−1.51843	−1.99117	−2.93665
6	0.0142	0.8916	0.18968	−0.16128	−0.8632
7	−0.3184	−0.3255	−0.31982	−0.31698	−0.3113
8	0.6471	0.8063	0.67894	0.61526	0.4879
9	0.7578	0.43425	0.69309	0.82251	1.08135
10	0.7866	0.4077	0.71082	0.86238	1.1655
11	0.0231	−0.3702	−0.05556	0.10176	0.4164
12	−0.2975	−0.30905	−0.29981	−0.29519	−0.28595
13	2.0248	2.17355	2.05455	1.99505	1.87605
14	0.3581	−0.6543	0.15562	0.56058	1.3705
15	−0.2191	−0.39815	−0.25491	−0.18329	−0.04005
16	0.5701	0.67965	0.59201	0.54819	0.46055
17	−0.4038	−0.68885	−0.46081	−0.34679	−0.11875
18	−0.2615	−0.0596	−0.22112	−0.30188	−0.4634
19	0.2056	0.33635	0.23175	0.17945	0.07485
20	0.6881	0.5853	0.66754	0.70866	0.7909

Table 11.8 Variables and First Lags

t	y_1	y_2	y_3	y_{1t-1}	y_{2t-1}	y_{3t-1}
1	1.4884	1.4884	1.4884			
2	1.0151	1.0151	0.2709	1.4884	1.4884	1.4884
3	0.2361	−0.13595	−0.2714	1.0151	1.0151	0.2709
4	−2.2456	−2.4994	−2.3637	0.2361	−0.13595	−0.2714
5	−2.8776	−2.93665	−1.7548	−2.2456	−2.4994	−2.3637
6	−1.4246	−0.8632	0.0142	−2.8776	−2.93665	−1.7548
7	−1.0307	−0.3113	−0.3184	−1.4246	−0.8632	0.0142
8	0.1317	0.4879	0.6471	−1.0307	−0.3113	−0.3184
9	0.8236	1.08135	0.7578	0.1317	0.4879	0.6471
10	1.1984	1.1655	0.7866	0.8236	1.08135	0.7578
11	0.6223	0.4164	0.0231	1.1984	1.1655	0.7866
12	0.0136	−0.28595	−0.2975	0.6223	0.4164	0.0231
13	2.0316	1.87605	2.0248	0.0136	−0.28595	−0.2975
14	1.3739	1.3705	0.3581	2.0316	1.87605	2.0248
15	0.4678	−0.04005	−0.2191	1.3739	1.3705	0.3581
16	0.804	0.46055	0.5701	0.4678	−0.04005	−0.2191
17	−0.0017	−0.11875	−0.4038	0.804	0.46055	0.5701
18	−0.2623	−0.4634	−0.2615	−0.0017	−0.11875	−0.4038
19	0.0744	0.07485	0.2056	−0.2623	−0.4634	−0.2615
20	0.7253	0.7909	0.6881	0.0744	0.07485	0.2056
σ_y^2	1.460805	1.406931	0.899516	—	—	—

$\text{cov}(y_{1t}, y_{1t-1}) = 0.980011$ $\text{cov}(y_{1t}, y_{1t-2}) = 0.336492$ $\text{cov}(y_{1t}, y_{1t-3}) = -0.128358$ $\text{cov}(y_{1t}, y_{1t-4}) = -0.471731$

$\text{cov}(y_{2t}, y_{2t-1}) = 0.828574$ $\text{cov}(y_{2t}, y_{2t-2}) = 0.056333$ $\text{cov}(y_{2t}, y_{2t-3}) = -0.308114$ $\text{cov}(y_{2t}, y_{2t-4}) = -0.517564$

$\text{cov}(y_{3t}, y_{3t-1}) = 0.307029$ $\text{cov}(y_{3t}, y_{3t-2}) = 0.034227$ $\text{cov}(y_{3t}, y_{3t-3}) = -0.134893$ $\text{cov}(y_{3t}, y_{3t-4}) = -0.299672$

$$\text{ACF}_s = \frac{\text{cov}(y_t, y_{t-s})}{\sigma_y^2}$$

For the first series:

$$\text{ACF}_1 = 0.980011/1.460805 = 0.6709$$
$$\text{ACF}_2 = 0.336492/1.460805 = 0.2303$$
$$\text{ACF}_3 = -0.128358/1.460805 = -0.0879$$
$$\text{ACF}_4 = -0.471731/1.460805 = -0.3229$$

The correlation is high for the first lag, declines for the second, but is still positive, and then is close to zero at the third lag, indicating an AR process.

For the second series:

$$\text{ACF}_1 = 0.828574/1.406931 = 0.5889$$
$$\text{ACF}_2 = 0.056333/1.406931 = 0.0400$$
$$\text{ACF}_3 = -0.308114/1.406931 = -0.2190$$
$$\text{ACF}_4 = -0.517564/1.406931 = -0.3679$$

Correlation is high for the first lag, and then close to zero for the second, indicating an MA process.

For the third series:

$$\text{ACF}_1 = 0.307029/0.899516 = 0.3413$$
$$\text{ACF}_2 = -0.034227/0.899516 = -0.0381$$
$$\text{ACF}_3 = -0.134893/0.899516 = -0.1500$$
$$\text{ACF}_4 = -0.299672/0.899516 = -0.3331$$

All correlations are relatively low, indicating white noise.

11.6 Calculate the Q statistic for the three series in Prob. 11.5 up to four lags.

For the first series:

$$Q = T\sum \text{ACF}_s^2 = 20[0.6709^2 + 0.2303^2 + (-0.0879)^2 + (-0.3229)^2] = 20(0.6151) = 12.30$$

For the second series:

$$Q = T\sum \text{ACF}_s^2 = 20[0.5889^2 + 0.0400^2 + (-0.2190)^2 + (-0.3679)^2] = 20(0.5317) = 10.63$$

For the third series:

$$Q = T\sum \text{ACF}_s^2 = 20[0.3413^2 + (-0.0381)^2 + (-0.1500)^2 + (-0.3331)^2] = 20(0.2514) = 5.03$$

The critical value for the chi-square distribution with four degrees of freedom is 9.49 with a 5% level of significance. For the first two series, $Q > 9.49$; therefore we reject the null hypothesis that there is no time-series correlation. For the third series, $Q = 5.03 < 9.49$; therefore we accept the null hypothesis that it is white noise.

11.7 For the AR(1) series in Prob. 11.3(d), use the AIC to test between (a) white noise (no correlation), (b) AR(1), (c) AR(2), and (d) AR(3).

Since the AR process simply involves a lag-dependent variable, we use OLS to estimate the four possible models. For the four models the estimation yields

(a) $y_t = 0.1582$ $R^2 = N/A$
                                      (0.57) $ESS = 29.22$

$$AIC = \ln\left(\frac{ESS}{T}\right) + \frac{2j}{T} = \ln\left(\frac{29.22}{20}\right) + \frac{2(1)}{20} = 0.4791$$

(b) $y_t = 0.0054 + 0.6448y_{t-1}$ $R^2 = 0.44$
                                (0.02) (3.65) $ESS = 15.35$

$$AIC = \ln\left(\frac{ESS}{T}\right) + \frac{2j}{T} = \ln\left(\frac{15.35}{19}\right) + \frac{2(2)}{19} = -0.0030$$

(c) $y_t = 0.0380 + 0.8995y_{t-1} - 0.3714y_{t-2}$ $R^2 = 0.50$
                            (0.17) (3.71) (-1.58) $ESS = 13.14$

$$AIC = \ln\left(\frac{ESS}{T}\right) + \frac{2j}{T} = \ln\left(\frac{13.14}{18}\right) + \frac{2(3)}{18} = 0.0190$$

(d) $y_t = 0.0557 + 0.8697y_{t-1} - 0.2872y_{t-2} - 0.0815y_{t-3}$ $R^2 = 0.51$
                      (0.23) (3.11) (-0.80) (-0.30) $ESS = 13.02$

$$AIC = \ln\left(\frac{ESS}{T}\right) + \frac{2j}{T} = \ln\left(\frac{13.02}{17}\right) + \frac{2(4)}{17} = 0.2039$$

Since the AIC is at its minimum for the model in part b, we choose AR(1) as the appropriate specification. Note that for each additional lag, there is one fewer observation. An alternative method for model selection is to make the sample consistent for each model (i.e., 17 observations for each) so that the same data are used for each specification.

NONSTATIONARY SERIES

11.8 (a) What are the problems of a nonstationary series? (b) What types of variables are likely to be nonstationary?

(a) A nonstationary series invalidates the standard statistical tests because it has a time-varying variance. Without a specified variance, test statistics cannot be standardized. Also, nonstationary series tend to show a statistically significant spurious correlation when regressed even if they are independent.

(b) Variables quoted in levels rather than growth rates tend to possess a unit root since their next-period value is a function of their current value plus growth. Since the full current value carries forward in the stock, it is nonstationary.

11.9 Algebraically show that the variance of a unit root series increases with time.

The function of a unit root series is

$$Y_t = Y_{t-1} + \varepsilon_t$$

Tracing this series from its initial value yields

$$Y_1 = \varepsilon_1 \qquad\qquad\qquad \sigma_{Y_1}^2 = \sigma_\varepsilon^2$$
$$Y_2 = Y_1 + \varepsilon_2 = \varepsilon_1 + \varepsilon_2 \qquad\qquad \sigma_{Y_2}^2 = \sigma_\varepsilon^2 + \sigma_\varepsilon^2$$
$$Y_3 = Y_2 + \varepsilon_3 = \varepsilon_1 + \varepsilon_2 + \varepsilon_3 \qquad\qquad \sigma_{Y_3}^2 = 3\sigma_\varepsilon^2$$
$$Y_4 = Y_3 + \varepsilon_4 = \varepsilon_1 + \varepsilon_2 + \varepsilon_3 + \varepsilon_4 \qquad\qquad \sigma_{Y_4}^2 = 4\sigma_\varepsilon^2$$
$$\text{etc.}$$

As can be seen, the variance of the tth value of Y is $t\sigma_\varepsilon^2$, therefore as the time period increases, so does the variance of Y.

11.10 (a) Use the random error from Prob. 11.3 to generate a unit-root series. (b) Graph the unit-root series and the original error term on the same axis. (c) Calculate an average for each series for $t = 1 - 5$, $t = 6 - 10$, $t = 11 - 15$, and $t = 16 - 20$.

(a) The results are listed in Table 11.9.   The method for generating Y is as follows:

$$Y_1 = \varepsilon_1 = 1.4884$$
$$Y_2 = Y_1 + \varepsilon_2 = 1.4884 + 0.2709 + 1.7593$$
$$Y_3 = Y_2 + \varepsilon_3 = 1.7593 - 0.2714 = 1.4879$$
$$Y_4 = Y_3 + \varepsilon_4 = 1.4879 - 2.3637 = -0.8758$$

etc.

Table 11.9   Unit-Root Series

t	ε	Y	$\bar{\varepsilon}$	$\bar{Y}$
1	1.4884	1.4884		
2	0.2709	1.7593		
3	−0.2714	1.4879		
4	−2.3637	−0.8758		
5	−1.7548	−2.6306		
			$\bar{\varepsilon}_{1-5} = -0.5261$	$\bar{Y}_{1-5} = 0.2458$
6	0.0142	−2.6164		
7	−0.3184	−2.9348		
8	0.6471	−2.2877		
9	0.7578	−1.5299		
10	0.7866	−0.7433		
			$\bar{\varepsilon}_{6-10} = 0.3775$	$\bar{Y}_{6-10} = -2.0224$
11	0.0231	−0.7202		
12	−0.2975	−1.0177		
13	2.0248	1.0071		
14	0.3581	1.3652		
15	−0.2191	1.1461		
			$\bar{\varepsilon}_{11-15} = 0.3779$	$\bar{Y}_{11-15} = 0.3561$
16	0.5701	1.7162		
17	−0.4038	1.3124		
18	−0.2615	1.0509		
19	0.2056	1.2565		
20	0.6881	1.9446		
			$\bar{\varepsilon}_{16-20} = 0.1597$	$\bar{Y}_{16-20} = 1.4561$

(b) Figure 11-3 graphs the two series.

(c) The averages are shown in Table 11.9. The average for the stationary series (ε) stays near zero for all subsets, while the averages for the unit-root series, Y, fluctuate to extreme negative values (−2.0224) and extreme positive values (1.4561), giving different inference for different subsets.

11.11 Table 11.10 reports the close of the NYSE (New York Stock Exchange) composite stockmarket index Y, and the population of Sri Lanka in thousands X for the years 1966 to 1992.   (a) Regress Y on X and test the coefficient on X at the 5% level of significance.   (b) Regress ΔY on ΔX and test the coefficient on X at the 5% level of significance.

(a) For the initial regression in levels, we obtain

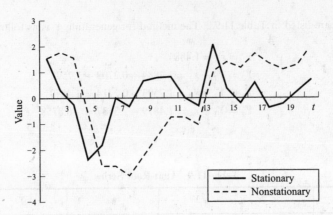

Fig. 11-3 Stationary (—) and Nonstationary (– – –) Series

Table 11.10 NYSE Closing Value and Population of Sri Lanka in Thousands, 1966–1992

Year	1966	1967	1968	1969	1970	1971	1972	1973	1974
Y	43.72	53.83	58.9	51.53	50.23	54.63	64.48	51.82	36.13
X	11440	11702	11992	12252	12516	12608	12861	13091	13284
Year	1975	1976	1977	1978	1979	1980	1981	1982	1983
Y	47.64	57.88	52.5	53.62	61.95	77.86	71.11	81.03	95.18
X	13496	13717	13942	14184	14471	14738	14988	15189	15417
Year	1984	1985	1986	1987	1988	1989	1990	1991	1992
Y	96.38	121.58	138.58	138.23	156.26	195.01	180.49	229.44	240.21
X	15599	15837	16117	16361	16587	16806	16993	17190	17405

Source: New York Stock Exchange (Index) and Penn-World Tables (Pop).

$$\hat{Y}_t = -313.01 + 0.03X_t \qquad R^2 = 0.75$$
$$(8.72)$$

There is a positive relationship between Y_t and X_t, which is significant at the 5% level (critical value = 2.06 with 25 df). Also, the R^2 is relatively high. We would conclude that the population of Sri Lanka is an important indicator of the NYSE.

(b) Taking the unit root into account, and regressing ΔY_t on ΔX_t, we get reliable results:

$$\Delta \hat{Y}_t = 7.14 + 0.0018 \, \Delta X_t \qquad R^2 < 0.01$$
$$(0.02)$$

The Sri Lankan population is no longer an indicator of the NYSE.

TESTING FOR UNIT ROOT

11.12 (a) Test Y_t from Prob. 11.11 for a unit root without a trend at the 5% level of significance.
(b) Test ΔY_t from Prob. 11.11 for a unit root without a trend at the 5% level of significance.

(a)
$$\Delta \hat{Y}_t = -1.20 + 0.10 Y_{t-1} \qquad R^2 = 0.12$$
$$(1.80)$$

Since $t_{\hat{b}_1} = 1.80 > -3.33$ (from App. 11), we fail to reject the null hypothesis that there is a unit root. The correct procedure is then to take first differences of Y before using it in a regression.

(b)
$$\Delta\Delta\hat{Y}_t = 8.55 - 1.14\,\Delta Y_{t-1} \qquad R^2 = 0.57$$
$$(-5.56)$$

Since $t_{\hat{b}_1} = -5.56 < -3.33$ (from App. 11), we reject the null hypothesis that there is a unit root. Therefore ΔY_t is a stationary series which can be used in a regression.

11.13 (a) Test X_t from Prob. 11.11 for a unit root without a trend at the 5% level of significance. (b) Test ΔX_t from Prob. 11.11 for a unit root without a trend at the 5% level of significance.

(a)
$$\Delta\hat{X}_t = 291.80 - 0.0043 X_{t-1} \qquad R^2 = 0.03$$
$$(-0.93)$$

Since $t_{\hat{b}_1} = -0.93 > -3.33$ (from App. 11), we fail to reject the null hypothesis that there is a unit root. The correct procedure is then to take first differences of X before using it in a regression.

(b)
$$\Delta\Delta\hat{X}_t = 206.94 - 0.91\,\Delta X_{t-1} \qquad R^2 = 0.46$$
$$(-4.42)$$

Since $t_{\hat{b}_1} = -4.42 < -3.33$ (from App. 11), we reject the null hypothesis that there is a unit root. Therefore ΔX_t is a stationary series which can be used in a regression.

11.14 (a) Test Y_t from Prob. 11.11 for a unit root using the F-test form of the ADF with a trend. (b) Test X_t from Prob. 11.11 for a unit root using the F-test form of the ADF with a trend and two lags of ΔX_t.

(a) Since the restriction for the null hypothesis involves testing if any coefficient is significant, the standard F test may be used with the Dickey-Fuller adjusted critical values (App. 11). We run the regression:

$$\Delta\hat{Y}_t = -5.27 - 0.10 Y_{t-1} + 1.77t \qquad R^2 = 0.28$$
$$(-0.91) \qquad (2.12) \qquad\qquad F = 4.09$$

Since $F = 4.09 < 7.24$, we cannot reject the null of unit root in favor of trend stationary.

(b) Recall from Chap. 7 the formula for the F test on a subset of variables is

$$F_{p,n-k} = \left(\frac{\dfrac{\sum e_{Ri}^2 - \sum e_i^2}{p}}{\left(\dfrac{\sum e_i^2}{n-k}\right)} \right)$$

where R indicates a restricted regression under the null hypothesis. The F test therefore requires two regressions to be run

Unrestricted:

$$\Delta\hat{X}_t = 6922.06 - 0.58 X_{t-1} + 134.17t + 0.33\,\Delta X_{t-1} + 0.28\,\Delta X_{t-2} \qquad R^2 = 0.29$$
$$(-2.77) \qquad (2.77) \qquad (1.54) \qquad (1.31) \qquad\qquad \text{ESS} = 26{,}483.44$$

Restricted:

$$\Delta\hat{X}_t = 219.57 + 0.05\,\Delta X_{t-1} - 0.02\,\Delta X_{t-2} \qquad R^2 < 0.01$$
$$(0.22) \qquad\quad (-0.09) \qquad\qquad \text{ESS} = 37{,}225.56$$

Calculating the F statistic, we obtain

$$F_{2,22} = \frac{\left(\dfrac{37{,}225.56 - 26{,}483.44}{2}\right)}{\left(\dfrac{26{,}483.44}{22}\right)} = 4.46$$

Since $F = 4.46 < 7.24$, we accept the null that X follows a unit-root process.

COINTEGRATION AND ERROR CORRECTION

11.15 (*a*) What is cointegration?   (*b*) How does cointegration affect the specification of a regression model?

(*a*) Two variables are cointegrated if they individually follow a unit root process, but jointly move together in the long run.   Individually, movements appear random and unpredictable, but the location of one can give information about the other.   If the prediction errors of Y regressed on X are stationary, there is evidence of cointegration.

(*b*) If cointegration exists, the long-run process should be used to explain the dependent variable.   If Y is above (resp. below) its long-run equilibrium, we would expect Y to decrease (resp. increase) in the next period.   Therefore an error-correction model includes deviations from the long-run relationship as an explanatory variable.

11.16 Show algebraically that estimating the model $Y_t = b_0 + b_1 X_t + b_2 X_{t-1} + b_3 Y_{t-1} + u_t$ when Y and X are cointegrated implies the use of an error-correction model.

Error correction stipulates that Y and X follow a long-run relationship:

$$Y = a_0 + a_1 X + \varepsilon$$

Taking the original model, $Y_t = b_0 + b_1 X_t + b_2 X_{t-1} + b_3 Y_{t-1} + u_t$, in the long run (as $t \to \infty$), we obtain

$$Y_\infty = b_0 + b_1 X_\infty + b_2 X_\infty + b_3 Y_\infty + u_t \quad \text{or} \quad (1 - b_3) Y_\infty = b_0 + (b_1 + b_2) X_\infty + u_t$$

Solving for Y, and dropping the subscript since it is contemporaneous, we have

$$Y = \frac{b_0}{(1 - b_3)} + \frac{(b_1 + b_2)}{(1 - b_3)} X + \varepsilon$$

Since Y and X follow the long-run relationship, we know that $b_0/(1 - b_3) = a_0$, and $(b_1 + b_2)/(1 - b_3) = a_1$. Since these parameters move in a constant ratio, we can solve for b_3 and b_2 in terms of b_0, b_1, a_0, and a_1.

$$b_3 = 1 - \frac{b_0}{a_0} \quad \text{and} \quad b_2 = a_1(1 - b_3) - b_1 = a_1 \frac{b_0}{a_0} - b_1$$

Substituting into the original model yields

$$Y_t = b_0 + b_1 X_t + \left(a_1 \frac{b_0}{a_0} - b_1 \right) X_{t-1} + \left(1 - \frac{b_0}{a_0} \right) Y_{t-1} + u_t$$

Grouping terms, we obtain

$$\Delta Y_t = b_0 + b_1 \, \Delta X_t - \frac{b_0}{a_0} (Y_{t-1} - a_1 X_{t-1}) + u_t$$

Since $Y_{t-1} - a_1 X_{t-1} = a_0 + \varepsilon_{t-1}$

$$\Delta Y_t = b_0 + b_1 \, \Delta X_t - \frac{b_0}{a_0} (a_0 + \varepsilon_{t-1}) + u_t$$

or

$$\Delta Y_t = b_1 \, \Delta X_t - \frac{b_0}{a_0} (\varepsilon_{t-1}) + u_t$$

this is the error-correction model (we usually include a constant even though it theoretically should be zero).

11.17 (*a*) Estimate a long-run relationship between Y and X from Prob. 11.11.   (*b*) Graph the residuals.   (*c*) Test for the presence of cointegration.

(*a*) A regression of Y and X is identical to that of Prob. 11.11:

$$\hat{Y}_t = -313.01 + 0.03X_t \qquad R^2 = 0.75$$
$$(8.72)$$

(b) Graphing the residuals from the regression in part *a* in Fig. 11-4, however, gives a picture that does not look stationary.

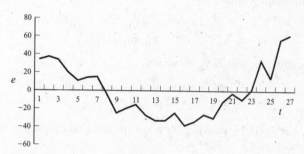

Fig. 11-4

(c) Testing the residuals for a unit root to find evidence of cointegration (or lack thereof) yields

$$\Delta\hat{e}_t = 0.91 - 0.08e_{t-1} \qquad R^2 = 0.02$$
$$(-0.67)$$

Since $t_{\hat{b}_1} = -0.67 > -3.33$ (from App. 11), we fail to reject the null hypothesis that there is a unit root. There is no evidence of cointegration, so error correction would not be appropriate.

11.18 Table 11.11 reports the Consumer Price Index for the Los Angeles area Y and the Chicago area X on a monthly basis from Jan. 1998 to Dec. 2000 (base year = 1982–1984).   (a) Test each variable for a unit root.   (b) Test for evidence of cointegration between Y and X.

Table 11.11   Consumer Price Index for Los Angeles and Chicago (Base Year = 82–84): Jan. 1998–Dec. 2000

Date	Jan-98	Feb-98	Mar-98	Apr-98	May-98	Jun-98	Jul-98	Aug-98	Sep-98	Oct-98	Nov-98	Dec-98
Y	161.0	161.1	161.4	161.8	162.3	162.2	162.1	162.6	162.6	163.2	163.4	163.5
X	162.8	163.1	164.1	164.8	165.6	166.0	166.5	165.4	165.3	165.7	165.4	165.1
Date	Jan-99	Feb-99	Mar-99	Apr-99	May-99	Jun-99	Jul-98	Aug-99	Sep-99	Oct-99	Nov-99	Dec-99
Y	164.2	164.6	165.0	166.6	166.2	165.4	165.8	166.3	167.2	167.2	167.1	167.3
X	166.1	166.4	167.0	167.6	168.2	168.9	169.4	169.3	169.7	169.7	169.3	169.2
Date	Jan-00	Feb-00	Mar-00	Apr-00	May-00	Jun-00	Jul-00	Aug-00	Sep-00	Oct-00	Nov-00	Dec-00
Y	167.9	169.3	170.7	170.6	171.1	171.0	171.7	172.2	173.3	173.8	173.5	173.5
X	170.2	171.4	172.2	171.9	173.7	176.0	174.6	173.7	174.8	175.4	176.0	175.8

Source: Bureau of Labor Statistics.

(a)
$$\Delta\hat{Y}_t = -0.73 + 0.01Y_{t-1} \qquad R^2 < 0.01$$
$$(0.29)$$

Since $t_{\hat{b}_1} = 0.29 > -3.33$ (from App. 11), we fail to reject the null hypothesis that there is a unit root for Y_t.

$$\Delta \hat{X}_t = 2.51 - 0.01 X_{t-1} \qquad R^2 < 0.01$$
$$(-0.37)$$

Since $t_{\hat{b}_1} = -0.37 > -3.33$ (from App. 11), we fail to reject the null hypothesis that there is a unit root for X_t.

(b) Since both Y and X are unit-root variables, we can proceed to test for cointegration. Estimating the long-run relationship yields

$$\hat{Y}_t = 10.45 + 0.95 X_t \qquad R^2 = 0.95$$
$$(26.69)$$

Testing the residual for unit root, we obtain

$$\Delta \hat{e}_t = 0.03 - 0.50 e_{t-1} \qquad R^2 = 0.26$$
$$(-3.38)$$

Since $t_{\hat{b}_1} = -3.38 < -3.33$, we reject the null hypothesis that there is a unit root for e_t. Therefore Y and X are cointegrated.

11.19 Estimate the error-correction model for the data in Prob. 11.18.

Since both variables are unit root and cointegrated, we run the model in differences with the inclusion of the lag residual of the long-run model:

$$\Delta \hat{Y}_t = 0.30 + 0.16 \Delta X_t + 0.12 e_{t-1} \qquad R^2 = 0.04$$
$$(1.11) \qquad (0.96)$$

CAUSALITY

11.20 How does Granger causality differ from other types of causality?

Granger causality is an econometric representation of the timing of causation. Unfortunately, Granger causality can never prove causality with certainty. There are several other factors that could mimic the results of Granger causality. X could Granger-cause Y because of a third factor causing both. This would not show up in the model. X could move before Y in anticipation of Y moving. X would Granger-cause Y, but it is the movement in Y which is the true cause. Also, the reactions of Y could be transitory, indicating that while X may Granger-cause Y, the effect does not last.

11.21 The data in Table 11.12 report housing starts Y in thousands and personal consumption X in billions of 1996 US dollars. We want to determine if housing starts is a leading indicator of consumption using Granger causality. What form should variables take in the regression (levels, differences, etc.)?

Since Granger causality is a time-series regression, its form will depend on the time-series properties of the variables, specifically if they possess a unit root, and if so, whether they are cointegrated. Testing for unit root in levels yields

$$\Delta \hat{Y}_t = 458.92 - 0.28 Y_{t-1} \qquad R^2 = 0.16$$
$$(-2.48)$$

Since $t_{\hat{b}_1} = -2.48 > -3.33$ (from App. 11), we fail to reject the null hypothesis that there is a unit root for Y_t. Since housing starts are a flow variable, it is not obvious that it should follow a unit root. In fact, the t statistic is close to the critical value. Unit-root testing suffers from being a low-power test in that it seldom rejects a unit root when it should. Since a unit root causes many statistical problems, however, we err on the side of correcting for the unit root when we do not have to.

$$\Delta \hat{X}_t = -94.06 + 0.02 X_{t-1} \qquad R^2 = 0.08$$
$$(1.72)$$

Table 11.12   Housing Starts in Thousands of Units and Real Personal Consumption in Billions of 1996 Dollars in the United States, Jan. 1997 to Dec. 1999

Year	1997	1997	1997	1997	1997	1997	1997	1997	1997	1997	1997	1997
Month	Jan.	Feb.	March	April	May	June	July	Aug.	Sept.	Oct.	Nov.	Dec.
Y	1355	1486	1457	1492	1442	1494	1437	1390	1546	1520	1510	1566
X	5342.1	5351.2	5358.7	5368.2	5361.5	5397.4	5454.0	5464.9	5467.3	5484.8	5506.5	5530.0
Year	1998	1998	1998	1998	1998	1998	1998	1998	1998	1998	1998	1998
Month	Jan.	Feb.	March	April	May	June	July	Aug.	Sept.	Oct.	Nov.	Dec.
Y	1525	1584	1567	1540	1536	1641	1598	1614	1582	1715	1660	1792
X	5540.8	5573.0	5603.5	5609.8	5658.4	5686.4	5685.9	5708.7	5738.4	5758.3	5771.5	5809.5
Year	1999	1999	1999	1999	1999	1999	1999	1999	1999	1999	1999	1999
Month	Jan.	Feb.	March	April	May	June	July	Aug.	Sept.	Oct.	Nov.	Dec.
Y	1804	1738	1737	1561	1649	1562	1704	1657	1628	1636	1663	1769
X	5817.9	5854.5	5908.4	5915.8	5928.4	5976.6	5987.1	6020.4	6033.9	6062.1	6090.8	6150.0

Source: St. Louis Federal Reserve (Bureau of Economic Analysis).

Since $t_{\hat{b}_1} = 1.72 > -3.33$ (from App. 11), we fail to reject the null hypothesis that there is a unit root for X_t. Testing for unit root in differences yields

$$\Delta\Delta\hat{Y}_t = 12.06 - 1.41\,\Delta Y_{t-1} \qquad R^2 = 0.71$$
$$(-8.92)$$

Since $t_{\hat{b}_1} = -8.92 < -3.33$ (from App. 11), we can reject the null hypothesis that there is a unit root for ΔY_t.

$$\Delta\Delta\hat{X}_t = 25.63 - 1.10\,\Delta X_{t-1} \qquad R^2 = 0.52$$
$$(-5.84)$$

Since $t_{\hat{b}_1} = -5.84 < -3.33$ (from App. 11), we fail to reject the null hypothesis that there is a unit root for ΔX_t. Since both Y and X are unit-root variables, we can proceed to test for cointegration. Estimating the long-run relationship yields

$$\hat{Y}_t = 3060.41 + 1.66X_t \qquad R^2 = 0.58$$
$$(6.84)$$

Testing the residual for a unit root yields

$$\Delta\hat{e}_t = 2.01 - 0.33e_{t-1} \qquad R^2 = 0.16$$
$$(-2.46)$$

Since $t_{\hat{b}_1} = -2.46 > -3.33$ we cannot reject the null hypothesis that there is a unit root for e_t. Therefore there is no evidence of cointegration. We can conclude that the correct model is to use both Y and X in first differences with no error correction.

11.22 Calculate the AIC for the Granger causality model from Prob. 11.21 for one to six lags with the first difference of consumption as the dependent variable. What is the optimal specification?

Since we are concerned only with the sum of squared errors (ESS), we omit reporting the regression coefficients.

With one lag each of the first difference of consumption and the first difference of housing starts ESS = 9297.932, $T = 34$, $j = 3$ (intercept and one lag of each):

$$\text{AIC} = \ln\left(\frac{\text{ESS}}{T}\right) + \frac{2j}{T} = \ln\left(\frac{9297.932}{34}\right) + \frac{2(3)}{34} = 5.79$$

With two lags each of the first difference of consumption and the first difference of housing starts ESS = 7797.001, $T = 33$, $j = 5$:

$$\text{AIC} = \ln\left(\frac{\text{ESS}}{T}\right) + \frac{2j}{T} = \ln\left(\frac{7797.001}{33}\right) + \frac{2(5)}{33} = 5.77$$

With three lags each of the first difference of consumption and the first difference of housing starts ESS = 7354.929, $T = 32$, $j = 7$:

$$\text{AIC} = \ln\left(\frac{\text{ESS}}{T}\right) + \frac{2j}{T} = \ln\left(\frac{7354.929}{32}\right) + \frac{2(7)}{32} = 5.87$$

With four lags each of the first difference of consumption and the first difference of housing starts ESS = 4617.587, $T = 31$, $j = 9$:

$$\text{AIC} = \ln\left(\frac{\text{ESS}}{T}\right) + \frac{2j}{T} = \ln\left(\frac{4617.587}{31}\right) + \frac{2(9)}{31} = 5.58$$

With five lags each of the first difference of consumption and the first difference of housing starts ESS = 3742.738, $T = 30$, $j = 11$:

$$\text{AIC} = \ln\left(\frac{\text{ESS}}{T}\right) + \frac{2j}{T} = \ln\left(\frac{3742.738}{30}\right) + \frac{2(11)}{30} = 5.56$$

With six lags each of the first difference of consumption and the first difference of housing starts ESS = 3085.670, $T = 29$, $j = 13$:

$$\text{AIC} = \ln\left(\frac{\text{ESS}}{T}\right) + \frac{2j}{T} = \ln\left(\frac{3085.670}{29}\right) + \frac{2(13)}{29} = 5.564$$

Since five lags has the lowest AIC, that is the optimal model.

11.23 Determine if housing starts Granger-cause personal consumption at the 5% level of significance using the data from Prob. 11.21 and the optimal model found in Probs. 11.21 and 11.22.

We run the model restricted and unrestricted, then use the F test to test whether housing starts are a statistically significant predictor of personal consumption.

Unrestricted:

$$\Delta\hat{X}_t = 42.06 - 0.01\Delta Y_{t-1} - 0.01\Delta Y_{t-2} + 0.02\Delta Y_{t-3} - 0.03\Delta Y_{t-4} + 0.09\Delta Y_{t-5} - 0.12\Delta X_{t-1}$$
$$(-0.22)(-0.23)(0.43)(-0.60)(1.97)(-0.45)$$
$$-0.57\Delta X_{t-2} + 0.05\Delta X_{t-3} - 0.13\Delta X_{t-4} + 0.02\Delta X_{t-5} R^2 = 0.52$$
$$(-2.71)(0.16)(-0.67)(0.09)\text{ESS} = 3742.74$$

Restricted:

$$\Delta\hat{X}_t = 58.68 - 0.42\Delta X_{t-1} - 0.57\Delta X_{t-2} - 0.18\Delta X_{t-3} - 0.16\Delta X_{t-4} - 0.11\Delta X_{t-5} R^2 = 0.28$$
$$(-1.78)(-2.71)(-0.78)(-0.83)(-0.58)\text{ESS} = 5648.53$$

$$F_{5,19} = \frac{\left(\dfrac{\Sigma e_{Ri}^2 - \Sigma e_i^2}{p}\right)}{\left(\dfrac{\Sigma e_i^2}{n-k}\right)} = \frac{\left(\dfrac{5648.53 - 3742.74}{5}\right)}{\left(\dfrac{3742.74}{19}\right)} = 1.93$$

The critical value for $F_{5,19}$ at the 5% level of significance is 2.74; since $F = 1.93 < 2.74$, we conclude that housing starts do not Granger-cause personal consumption.

Supplementary Problems

ARMA

11.24 Using the random variable from Table 11.13, and an AR(2) process for y_t with $\gamma_1 = 0.4$ and $\gamma_2 = -0.3$, (a) calculate y_4 (b) y_8 (c) y_{20} (d) y_{30}.
Ans. (a) 0.0855 (b) 0.3618 (c) 0.7625 (d) 0.5188

Table 11.13   Random-Error Terms

t	1	2	3	4	5	6	7	8	9	10
ε	0.1291	0.6910	0.1348	0.1510	0.3869	0.7318	0.4515	0.3334	0.8943	0.0773
t	11	12	13	14	15	16	17	18	19	20
ε	0.4303	0.5805	0.9250	0.0408	0.9621	0.6577	0.8292	0.5996	0.4197	0.8095
t	21	22	23	24	25	26	27	28	29	30
ε	0.4661	0.2208	0.2334	0.5894	0.8296	0.4352	0.1958	0.6074	0.7228	0.3146

11.25 Using the random variable from Table 11.13, and an MA(2) process for y_1 with $\theta_1 = 0.2$ and $\theta_2 = -0.5$, (a) calculate y_4, (b) y_8, (c) y_{20}, (d) y_{30}.
Ans. (a) 0.4695 (b) 0.6090 (c) 1.0254 (d) 0.4737

IDENTIFYING ARMA

11.26 Table 11.14 reports the average temperature in New York's Central Park from 1969 to 1999. Calculate the autocorrelation function of average temperature up to six lags.
Ans. $\mathrm{ACF}_1 = -0.0051$, $\mathrm{ACF}_2 = -0.0013$, $\mathrm{ACF}_3 = -0.2007$, $\mathrm{ACF}_4 = 0.2448$, $\mathrm{ACF}_5 = -0.1598$, $\mathrm{ACF}_6 = 0.1023$

Table 11.14   Average Temperature T in Central Park: New York, 1969–1999

Year	1969	1970	1971	1972	1973	1974	1975	1976	1977	1978	1979
T, °C	12.71	12.33	12.58	12.24	13.34	12.59	12.91	12.15	12.13	11.5	13.06
Year	1980	1981	1982	1983	1984	1985	1986	1987	1988	1989	1990
T, °C	13.16	12.70	12.41	13.68	12.63	13.53	12.68	12.83	12.84	12.67	13.23
Year	1991	1992	1993	1994	1995	1996	1997	1998	1999		
T, °C	14.04	12.27	13.07	12.66	13.39	11.66	12.51	13.73	13.75		

Source: NASA Goddard Institute for Space Studies.

11.27 (a) Calculate the Q statistic for the autocorrelations in Prob. 11.26. (b) Are there statistically significant correlations at the 5% level of significance?
Ans. (a) 4.22 (b) No

NONSTATIONARY SERIES

11.28 (a) Calculate the t statistic for the ADF test of unit root without a trend and no lags of ΔY_t for the temperatures in Table 11.14. (b) Do the temperatures possess a unit root?
Ans. (a) −5.09 (b) No

11.29 (a) Calculate the F statistic for the ADF test of unit root with a trend and no lags of ΔY_t for the temperatures in Table 11.14. (b) Do the temperatures possess a unit root?
Ans. (a) 15.85 (b) No

COINTEGRATION AND ERROR CORRECTION

11.30 Table 11.15 reports the value of the Dow Jones Industrial Average (DJIA) Y, the S&P 500 Stock Index X, and the Toronto Stock Exchange 300 Index Z, from Jan. 2 to 30, 2001. (a) Does the DJIA have a unit root? (b) Does the S&P 500 have a unit root? (c) Are Y and X cointegrated?
Ans. (a) Yes (b) Yes (c) No

Table 11.15   DJIA, S&P 500 Index, and TSE 300 Index: Jan. 2–30, 2001

Date	2-Jan-01	3-Jan-01	4-Jan-01	5-Jan-01	8-Jan-01	9-Jan-01	10-Jan-01	11-Jan-01	12-Jan-01	16-Jan-01
Y	10,646.15	10,881.2	10,945.75	10,912.41	10,662.01	10,621.35	10,572.55	10,604.27	10,609.55	10,525.38
X	1283.27	1373.73	1347.56	1333.34	1298.35	1295.86	1300.8	1313.27	1326.82	1318.55
Z	8611.5	8937.8	8905.7	8690.2	8671.7	8572	8600.8	8805.4	8716.4	8744.0

Date	17-Jan-01	18-Jan-01	19-Jan-01	22-Jan-01	23-Jan-01	24-Jan-01	25-Jan-01	26-Jan-01	29-Jan-01	30-Jan-01
Y	10,652.66	10,584.34	10,678.28	10,587.59	10,578.24	10,649.81	10,646.97	10,729.52	10,659.98	10,702.19
X	1326.65	1329.47	1347.97	1342.54	1342.9	2360.4	1364.3	1357.51	1354.95	1364.17
Z	8879.4	8899.1	9161.1	9121	9268.8	9306.2	9183.4	9158.2	9302.2	9348.4

Source: *quote.yahoo.com*.

11.31 Using the data in Table 11.15 (a) Does the Toronto Stock Exchange have a unit root? (b) Are X and Z cointegrated?
Ans. (a) Yes (b) Yes

CAUSALITY

11.32 Table 11.16 reports monthly first differences of an industrial production index for the United States Y and the S&P 500 Stock Market Index X from February 1998 to December 2000. (a) Using one lag of Y and X, does X Granger-cause Y? (b) If so, what is the short-run magnitude of the causality? (c) What is the long-run magnitude?
Ans. (a) Yes (b) −0.0166 (c) −0.0118

Table 11.16   Industrial Production Index and S&P 500 Index: United States, Feb. 1998–Dec. 2000

Date		Feb-98	Mar-98	Apr-98	May-98	Jun-98	Jul-98	Aug-98	Sep-98	Oct-98	Nov-98	Dec-98
Y		1.87	1.26	−2.00	0.68	2.62	−4.56	7.95	0.53	−0.12	−3.48	−1.75
X		69.05	52.41	10.00	−20.93	43.02	−13.16	−163.39	59.73	81.66	64.96	65.59

Date	Jan-99	Feb-99	Mar-99	Apr-99	May-99	Jun-99	Jul-99	Aug-99	Sep-99	Oct-99	Nov-99	Dec-99
Y	1.13	1.88	2.12	−2.09	0.41	4.46	−4.99	7.12	0.75	0.61	−2.49	−1.39
X	50.41	−41.31	48.04	48.81	−33.34	70.87	−43.99	−8.30	−37.70	80.22	26.13	80.18

Date	Jan-00	Feb-00	Mar-00	Apr-00	May-00	Jun-00	Jul-00	Aug-00	Sep-00	Oct-00	Nov-00	Dec-00
Y	1.37	2.16	2.03	−1.34	0.20	5.43	−6.76	7.66	1.65	−1.57	−3.01	−2.19
X	−74.79	−28.04	132.16	−46.14	−31.83	34.00	−23.77	86.85	−81.17	−7.10	−114.45	5.32

Source: Federal Reserve Board of Governors (Industrial Production) and *quote.yahoo.com* (S&P 500).

11.33 Using the data from Table 11.16,   (a) What is the F statistic used to test if X Granger-causes Y with six lags?   (b) Does X Granger-cause Y with six lags?   (c) How would one know the correct number of lags to use?

    Ans.   (a) 2.60   (b) No   (c) Calculate the AIC for different number of lags and use model with lowest AIC

CHAPTER 12

Computer Applications in Econometrics

12.1 DATA FORMATS

If data are found from an existing source (rather than collected by the researcher), they often come in a text format. Text format is flexible since any statistical package and brand of computer can read it. There are two main types of text formats:

1. *Delimited format* (also called *free format*)—each variable is separated by a character, usually a space, tab, or comma.
2. *Fixed format*—each variable occupies a specific column or group of columns in the text file.

To determine the format, order of the variables, and any codes (e.g., missing value code) one must consult a codebook which accompanies the data set.

EXAMPLE 1. We report the data from Chap. 2, Example 1 as a text file in several formats.

1, 6	1 6	1 6
2, 7	2 7	2 7
3, 6	3 6	3 6
4, 8	4 8	4 8
5, 5	5 5	5 5
6, 7	6 7	6 7
7, 6	7 6	7 6
8, 9	8 9	8 9
9, 10	9 10	9 10
10, 6	10 6	10 6
Comma-delimited	Space-delimited	Fixed format: test no. in
order: test no., grade	order: test no., grade	columns 1–2, grade in columns 4–5

Below we explore three specific statistics packages. Our aim is to give a general understanding of the programming language of each package, as well as procedures to carry out the calculations from this text. As it is impossible to cover every statement and procedure of the software, we have chosen windows-based programs which include a detailed help file for further reference.

12.2 MICROSOFT EXCEL

Excel is a spreadsheet package which includes functions for most common statistical calculations. Excel uses a graphical interface, which means that the user enters data and function in certain locations on the spreadsheet (called *cells*). Cell location in Excel is defined by the row number and column letter of each cell. Data may be read from external files (see Example 2) or typed directly into the cells by clicking on the cell and typing the text or number desired. Functions are designated by an equal sign (=) and perform many statistical calculations. To identify the values for the calculation, either individually enter each cell (A1, A2) or use a colon to indicate a range of cells (A1:A10, all cells from A1 to A10). Below are some commonly used Excel functions:

Description	Excel function
Add, subtract, multiply, divide, exponent	+, -, *, /, ^
Square root	= sqrt(A1)
Summation	= sum(A1:A10)
Mean	= average(A1:A10)
Median	= median(A1:A10)
Mode	= mode(A1:A10)
Population variance	= varp(A1:A10)
Sample variance	= var(A1:A10)
Population standard deviation	= stdevp(A1:A10)
Sample standard deviation	= stdev(A1:A10)
Covariance	= covar(A1:A10,B1:B10)
Random number between 0 and 1	= rand()
Prob < A1 under standard normal distribution	= normsdist(A1)
Prob <-A1 and > A1 under t distribution (20 df, 2-tail test)	= tdist(A1,20,2)

All functions may be accessed through the toolbar Insert-Function, which includes descriptions of the function. Graphing is done through the toolbar Insert-Chart. More advanced calculations (histogram, t test, ANOVA, regression) are found in the toolbar Tools-Data Analysis. Note that if the Data Analysis option is not present under tools, then the Analysis Tool Pack has not been installed. To add the option either go to Microsoft Office Setup or Tools-Ad-Ins and install Analysis Tool Pack.

EXAMPLE 2. We saved the data from the comma-delimited version of Example 1 to a text file. Using Excel, we can open the data directly into a worksheet with the following steps:

1. File-Open, in the Open dialog box set "Files of type" to "All Files (*.*)," select the desired file, in this case example.txt.

2. The Text Import Wizard dialog box appears since the selected file is not an Excel file.

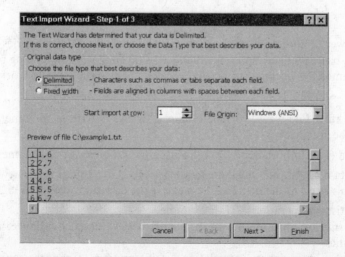

We have the option of specifying "Delimited" or "Fixed width" (fixed format). If "Fixed width" is selected, the next box allows the selection of columns. Since our data are delimited, we choose "Delimited" and click "Next."

The next box allows the selection of the delimiter. Our data are comma-delimited, so we check the box next to "Comma." For most data purposes, this is enough for Excel to import the data, so we click "Finish."

Our data are now in Excel and may be used in calculations, and saved as an Excel spreadsheet.

12.3 EVIEWS

Eviews is a powerful statistical package designed especially for time-series regression analysis. Eviews is a windows-based statistical package that works through windows dialog boxes. All regression options are programmed by checking the desired options. The basic steps to work with data in Eviews are

1. Open a workfile (File-New-Workfile). Since Eviews is written for time series, start and end dates must be specified.

2. Read in data (File-Import-Read Text, Lotus, Excel). Give variable names, delimiters, sample.

3. Redefine data if necessary (Quick-Generate Series). Give equation for new variable using usual math symbols (e.g., to define x2 as 2 times x1, the equation would be "x2 = 2*x1").

4. Perform statistical operations. For example,
   Descriptive statistic—histogram, mean, standard deviation, covariance, ACF, ADF (Quick-Series Statistics)
   Joint statistics—covariance, correlation, cointegration, Granger causality (Quick-Group Statistics)
   Estimation—regression, ARMA corrections (Quick-Estimate Equation)

EXAMPLE 3.   Using the text file example1.txt from above, we can import the data into Eviews:

1. To start a new workspace, we click File-New-Workfile.   A dialog box queries the period length and dates. Since our data do not constitute a time series, we enter 1 as start date and 10 as end date to clear enough space for 10 observations.

2. We click File-Import-Read Text, Lotus, Excel, to read data from an external text file.   The ASCII Text Import dialog box appears:

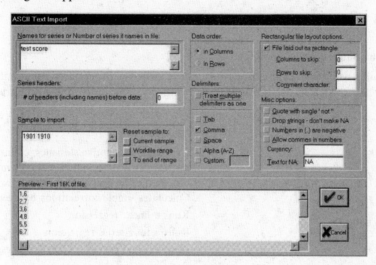

We list the variable names in the order in which they appear in the data set.   Data are arranged in columns, comma-delimited, so those options are checked.   Checking the box for rectangular file layout indicates that there is one observation per row.   Clicking "OK" reads the data into Eviews and sets up an entry in the workfile for each variable.   The workfile may be saved at this point.

12.4   SAS

The current version of SAS (we are using V 8.0) operates in Windows, but is programmed by entering statements rather than checking options.   There are three main windows in SAS: the *Program Editor* where statements are written; the *Log*, where comments are stored when a program is submitted for processing (processing time, error messages, etc.), and the *Output* window, where results are written on successful processing of a program.   The *Explorer* window, which accesses SAS data sets, and the *Results* window, which catalogs previous results, are useful for the organization of large projects.

SAS programming involves two distinct parts:

1. The data step where the data are read and the variables are defined.   Its basic structure is:

Program	Description
`libname lname 'c:\';`	Gives path where SAS data set will be stored. This can be omitted if the data set will be used once (i.e., temporary data set). *lname* refers to the user-defined name given to the library. All names in SAS must begin with a letter and be no more than 8 characters.
`data lname.dname;`	Names data set *dname* to be stored in library *lname*.
`infile 'path:\file.ext' delimiter='','';`	Gives location of text file containing data. The delimiter option may be omitted if the data are space-delimited or in fixed format.
`input var1 var2;`	Reads in variables in order of columns. If data are in fixed format, list variables followed by the column numbers where the data fall (e.g., *var1* 1–2).

After the data step, new variables can be calculated through equations (as with Eviews "Generate"). The usual math notation is used for add, subtract, multiply, and divide (+, -, *, /). Exponents are achieved by two stars (**). Data manipulations must come in the data step. If a procedure has been run, a new data step must be started in order to create new variables. Previous data sets can be called into a data step with the "Set" command. For example

```
data recall
set lname.dname;
```

calls back the data set read in above.

2. The procedures where the estimation routines are called. Procedures are identified by "proc" followed by the specific procedure name and options. Some commonly used procedures are listed here:

Procedure	Description
proc means;	Calculates descriptive statistics, count, mean, standard deviation, minimum, maximum
proc freq;	Calculates descriptive statistics of discrete variables
proc corr;	Calculates simple correlations between variables
proc reg;	Runs a linear regression
proc autoreg;	Runs a time-series regression
proc arima;	Identifies and corrects ARMA processes
proc probit;	Runs a binary choice regression
proc syslin;	Estimates simultaneous equations
proc print;	Prints the data set to the Output window
proc plot;	Plots a graph
proc iml;	Matrix language; performs matrix mathematics

All lines of a SAS program are followed by a semicolon. Sections of the program to be processed are followed by the "run;" command; "quit;" designates the end of the program. The program is run by clicking Run-Submit, or clicking the 🏃 button.

EXAMPLE 4. Using the text file example1.txt from above, we can import the data into SAS through the data step. The data step is as follows:

```
data example;
infile ''c:\example1.txt'' delimiter = '','';
input test score;
run;
quit;
```

The SAS Log window reports the following information:

```
1    data example;
2            infile ''c:\example1.txt'' delimiter = '','';
3            input test score;
4            run;
```

NOTE: The infile ''c:\example1.txt'' is:
      File Name=c:\example1.txt,
      RECFM=V, LRECL=256

NOTE: 10 records were read from the infile ''c:\example1.txt''.
      The minimum record length was 3.
      The maximum record length was 4.
NOTE: The data set WORK.EXAMPLE has 10 observations and 2 variables.
NOTE: DATA statement used:
      real time   1.25 seconds

```
5            quit;
```

The Log window tells us that the file was found that 10 records (observations) and 2 variables were read. It also reports the processing time of 1.25 s.

Solved Problems

DATA FORMATS

12.1 (*a*) Why are computers important in statistics and econometrics?   (*b*) What are common sources of computer-readable data?

   (*a*) Much of statistical theory relies on the large-sample properties of estimators.   As the data set gets larger, standard errors get smaller; therefore confidence intervals get narrower and more precise.   The minimum acceptable number of observations for most practical purposes is 30.   As data sets get larger, however, calculations get more time-consuming.   Without computers, even simple calculations involving large data sets would not be feasible.   More complex calculations, such as probit or simultaneous equations, are too computationally demanding even with relatively small data sets.   Reading text files on the computer also eliminates typing errors from data entry.   What must be remembered is that while the computer is a tool for processing calculations quickly, the researcher still must verify that the model has been specified correctly.

   (*b*) Government agencies have large amounts of public, computer-readable data (Census, Bureau of Labor Statistics, Federal Reserve, etc.).   Other sources are college and university research departments, Internet search engines, nonprofit agencies, and political lobbying groups.   Financial data may be obtained through securities ratings companies and for-profit information services, but usually at a substantial cost.   Appendix 12 lists all Internet data sources used in this text.

12.2 (*a*) What is the difference between delimited and fixed-format data?   (*b*) What are some possible problems with delimited data?

   (*a*) Delimited data have some type of character separating the different variables.   In fixed-format data sets, data are arranged so that each variable occupies specific columns of the text file.

   (*b*) Tab delimiters can be a problem since some statistical packages do not read tabs well (SAS).   Tabs can especially be problems with non–Microsoft Windows programs such as mainframes and DOS.   Space delimiters can cause a problem with text variables that contain spaces within them.   Consider reading in data of countries for the list "United States of America   Hong Kong   Italy   Germany."   Reading this as space-delimited would yield eight variables; the first variable would be "United," the second "States," the third "of," and so on.   Comma-delimited data would solve this problem since "United States of America, Hong Kong, Italy, Germany" would be read correctly.

12.3 Identify the format of the following population estimates (in millions) for July, 1999 from the U.S. Census Bureau:

(*a*)		(*b*)	(*c*)
New Mexico	1.7	New Mexico, 1.7	New Mexico; 1.7
New York	18.2	New York, 18.2	New York; 18.2
North Carolina	7.7	North Carolina, 7.7	North Carolina; 7.7
North Dakota	0.6	North Dakota, 0.6	North Dakota; 0.6
Ohio	11.3	Ohio, 11.3	Ohio; 11.3

   (*a*) Fixed format, state in columns 1 to 14, population in columns 16 to 19.

   (*b*) Comma-delimited

   (*c*) Semicolon-delimited

MICROSOFT EXCEL

12.4 Using the data from Example 1, (*a*) Use the data analysis tools to graph the histogram and ogive of test scores. (*b*) Calculate, a mean, median, mode, sample variance, sample standard deviation, and coefficient of variation to statistially describe the data. (*c*) Use Excel functions to standardize each test score.

(*a*) For a histogram in Excel, choose Tools-Data Analysis. In the resulting dialog box, select "Histogram" and click "OK." We then choose the options we want for our histogram in the following box:

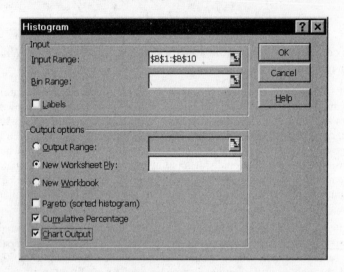

Our data are in column B, from row 1 to row 10. The default is a frequency distribution, checking "Chart Output" draws the histogram, and checking "Cumulative Percentage" plots the ogive. Custom class intervals may be typed into Excel and indicated as the "Bin Range." The results are as follows:

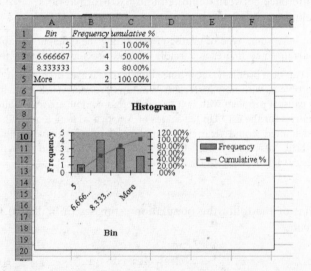

Parts *b* and *c* are reported in the image below as both numerical results and Excel formulas. For *b* the descriptive statistics can all be performed through functions. The coefficient of variation is simply the standard deviation divided by its mean (dividing by the mean gives a relative measure of variation without units). For part *c*, note that when formulas are copied and pasted, the cell references adjust to the new location. In standardizing, we want to subtract the same mean and divide by the same standard deviation for all calculations. Including a dollar sign ($) before the column and row reference keeps it from changing when pasted to a new location.

	A	B	C	D
1	1	6	(c)	-0.6396
2	2	7		0
3	3	6		-0.6396
4	4	8		0.639602
5	5	5		-1.2792
6	6	7		0
7	7	6		-0.6396
8	8	9		1.279204
9	9	10		1.918806
10	10	6		-0.6396
11				
12	(b)			
13	mean	7		
14	median	6.5		
15	mode	6		
16	sample var	2.444444		
17	sample std. dev.	1.563472		
18	coefficient of var.	0.223353		

	A	B	C	D
1	1	6	(c)	=(B1-$B$13)/$B$17
2	2	7		=(B2-$B$13)/$B$17
3	3	6		=(B3-$B$13)/$B$17
4	4	8		=(B4-$B$13)/$B$17
5	5	5		=(B5-$B$13)/$B$17
6	6	7		=(B6-$B$13)/$B$17
7	7	6		=(B7-$B$13)/$B$17
8	8	9		=(B8-$B$13)/$B$17
9	9	10		=(B9-$B$13)/$B$17
10	10	6		=(B10-$B$13)/$B$17
11				
12	(b)			
13	mean	=AVERAGE(B1:B10)		
14	median	=MEDIAN(B1:B10)		
15	mode	=MODE(B1:B10)		
16	sample var	=VAR(B1:B10)		
17	sample std. dev.	=STDEV(B1:B10)		
18	coefficient of var.	=B17/B13		

12.5 For the data in Chap. 5, Example 9 (*a*) perform a *t* test of the null hypothesis that wrapping 1 has average sales equal to 85. (*b*) Peform a *t* test of the null hypothesis that wrapping 1 and wrapping 2 have the same average sales. (*c*) Perform an ANOVA test of the null hypothesis that all three wrappings have the same average sales.

(*a*) We calculate the *t* statistic using the Excel Formulas. Since the probability in the tails of the *t* distribution is greater than 0.05, we accept the null that the average sales are 85 at the 5% level of significance.

	A	B	C
1	Wrapping 1	Wrapping 2	Wrapping 3
2	87	78	90
3	83	81	91
4	79	79	84
5	81	82	82
6	80	80	88
7			
8		82	mean
9	3.1622777	sample st. dev.	
10	-2.12132	t	
11	0.1011915	prob. of t under Null(p-value)	

	A	B	C
1	Wrapping 1	Wrapping 2	Wrapping 3
2	87	78	90
3	83	81	91
4	79	79	84
5	81	82	82
6	80	80	88
7			
8	=AVERAGE(A2:A6)	mean	
9	=STDEV(A2:A6)	sample st. dev.	
10	=(A8-85)/(A9/SQRT(5))	t	
11	=TDIST(-A10,4,2)	prob. of t under Null(p-value)	

(*b*) The two-sample *t* test is found in Tools-Data Analysis. There are several options. Since it is specified in Chap. 5, Example 9 that the data have equal variances, we select "*t* test: Two Sample Assuming Equal Variances," and click "OK." The following dialog box appears:

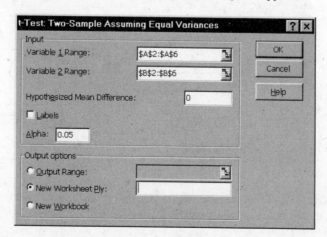

We enter the range for wrapping 1 as variable 1 and from wrapping 2 as variable 2. The hypothesized mean difference is 0 since our null states that the means are equal. Alpha is the desired level of significance. The result fails to reject the null that both means are the same at the 5% level of significance.

	A	B	C
1	t-Test: Two-Sample Assuming Equal Variances		
2			
3		Variable 1	Variable 2
4	Mean	82	80
5	Variance	10	2.5
6	Observations	5	5
7	Pooled Variance	6.25	
8	Hypothesized Mean Difference	0	
9	df	8	
10	t Stat	1.264911	
11	P(T<=t) one-tail	0.120752	
12	t Critical one-tail	1.859548	
13	P(T<=t) two-tail	0.241504	
14	t Critical two-tail	2.306006	

(c) The ANOVA test is also found through Tools-Data Analysis. We choose "ANOVA Single Factor."

We enter the entire range of all three variables and select "Grouped By: Columns" since the variables are in separate columns. Again we set the level of significance to 5%.

	A	B	C	D	E	F	G
1	Anova: Single Factor						
2							
3	SUMMARY						
4	Groups	Count	Sum	Average	Variance		
5	Column 1	5	410	82	10		
6	Column 2	5	400	80	2.5		
7	Column 3	5	435	87	15		
8							
9							
10	ANOVA						
11	Source of Variation	SS	df	MS	F	P-value	F crit
12	Between Groups	130	2	65	7.090909	0.00927	3.88529
13	Within Groups	110	12	9.166667			
14							
15	Total	240	14				

Since the calculated F value exceeds the critical value ("F crit" in the table), we reject the null hypothesis that all three wrappings have the same average sales.

12.6   In Example 1 of Chap. 6, Table 6.1 reports corn per acre Y and fertilizer used X from 1971 to 1980.   (*a*) Calculate the covariance between X and Y.   (*b*) Use Excel to plot X and Y.   (*c*) Fit a regression line to the graph.

(*a*)   As seen below, the covariance between the X and Y is positive.

	A	B	C	D
1	year	n	corn	fertilizer
2	1971	1	40	6
3	1972	2	44	10
4	1973	3	46	12
5	1974	4	48	14
6	1975	5	52	16
7	1976	6	58	18
8	1977	7	60	22
9	1978	8	68	24
10	1979	9	74	26
11	1980	10	80	32
12				
13			covariance	95.6

	A	B	C	D
1	year	n	corn	fertilizer
2	1971	1	40	6
3	1972	2	44	10
4	1973	3	46	12
5	1974	4	48	14
6	1975	5	52	16
7	1976	6	58	18
8	1977	7	60	22
9	1978	8	68	24
10	1979	9	74	26
11	1980	10	80	32
12				
13			covariance	=COVAR(C2:C11,D2:D11)

(*b*)   To plot the two variables, we highlight both variables and choose Insert-Chart.   We click XY (scatter) plot and click next.   The series can be named in the "Series" tab, we also switch the X and Y variable so that Y is on the vertical axis.   In the next window the chart and axes can be named.   In the next box the location can be determined, and we can click "Finish."   The following graph is created:

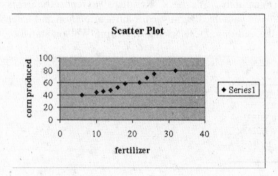

(*c*)   To fit a regression line to the plot, click the right mouse button over the plot, and select "Add Trend-line" (this may take some practice aiming).   We select to add a linear trendline; under the "Options" tab we can select to have the regression equation and R^2 reported.

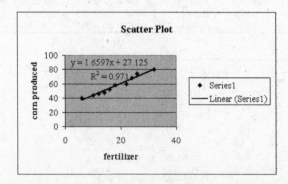

12.7 Example 1 of Chap. 7 extends the corn production table to add insecticide use. Run a multiple regression of Y on X_1 and X_2, reporting the residual error terms.

Regression estimation is under Tools-Data Analysis; we select "Regression."

In the dialog box, we give the loction of the Y and X variables (this can be done easily by clicking in the desired box and highlighting the variable on the worksheet). It is important that all the independent variables are in a continuous range of columns. We check "Residuals" to report the errors of the regression. Note that checking "Residual Plots" is a valuable diagnostic for autocorrelation and heteroscedasticity. The residuals can be used to calculate the additional tests such as a Durbin-Watson statistic.

	A	B	C	D	E	F	G
1	SUMMARY OUTPUT						
3	*Regression Statistics*						
4	Multiple R	0.995808106					
5	R Square	0.991633784					
6	Adjusted R Square	0.989243437					
7	Standard Error	1.39746693					
8	Observations	10					
10	ANOVA						
11		*df*	*SS*	*MS*	*F*	*Significance F*	
12	Regression	2	1620.329603	810.1648	414.84923	5.35614E-08	
13	Residual	7	13.67039674	1.9529138			
14	Total	9	1634				
16		*Coefficients*	*Standard Error*	*t Stat*	*P-value*	*Lower 95%*	*Upper 95%*
17	Intercept	31.98067141	1.631795719	19.598453	2.248E-07	28.12209044	35.83925238
18	X Variable 1	0.650050865	0.250161261	2.5985273	0.0355012	0.058513903	1.241587826
19	X Variable 2	1.109867752	0.267433637	4.1500679	0.0042947	0.47748814	1.742247364
21	RESIDUAL OUTPUT						
23	*Observation*	*Predicted Y*	*Residuals*				
24	1	40.32044761	-0.320447609				
25	2	42.92065107	1.079348932				
26	3	45.33062055	0.669379451				
27	4	48.85045778	-0.850457782				
28	5	52.37029502	-0.370295015				
29	6	57	1				
30	7	61.81993896	-1.819938962				
31	8	69.7792472	-1.779247202				
32	9	72.18921668	1.810783316				
33	10	79.41912513	0.580874873				

EVIEWS

12.8   Save the variables from Prob. 12.7 in an Excel worksheet, and import the values into Eviews.

    Since Eviews can read Excel worksheets, we save the data in Excel format.   To make reading the data easier, we eliminate all functions and labels.   Below is the Excel worksheet and Eviews import options to read the data.

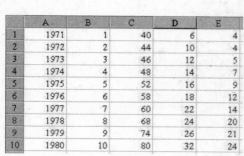

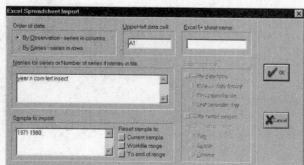

12.9   Using the Eviews workfile from Problem 12.8   (*a*) generate a variable for the proportion of fertilizer per bushel of corn.   (*b*) Calculate descriptive statistics for the fertilizer ratio.   (*c*) Graph the correlogram for the fertilizer ratio.

    (*a*)   To generate a new variable, we go to Quick-Generate Series.   We get the dialog box below.

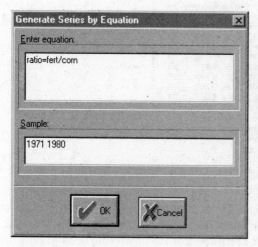

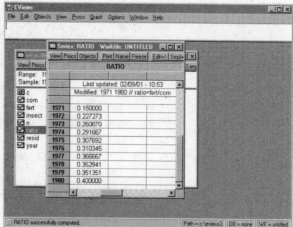

    We name the new variable "ratio" and define it by the equation " = fert/corn," and click "OK." Clicking on the ratio variable in the workfile shows the results of the circulation.

    (*b*)   For descriptive statistics of a series, choose Quick-Series Statistics-Histogram and Stats.   Enter the desired series for the resulting information.   (Descriptive statistics of the entire data set can be found in Quick-Group Statistics.)

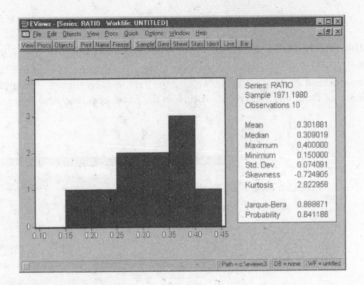

(c)   The correlogram is found in Quick-Series Statistics-Correlogram.   After specifying the series name, correlations in levels, and eight lags, we get the following output.   Note the high initial correlation which fades out, and the large spike at one lag for partial correlations indicates AR(1).

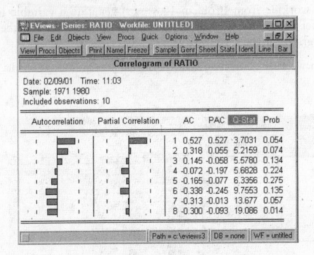

12.10   Using the data from the Eview workfile in Prob. 12.8   (a) Estimate the regression of corn on fertilizer and pesticides.   (b) Is there evidence of autocorrelation in part a? If so, correct for autocorrelation.   (c) Estimate the regression of the fertilizer/corn ratio on only a constant.   (d) Is there evidence of autocorrelation in part c?   If so, correct for autocorrelation.

(a)   To run a regression, select Quick-Estimate equation.   To specify the equation in the dialog box, list the variables to be used in the regression with the dependent variable first; "c" includes a constant (intercept), and then the dependent variables.   The "Method" setting allows for different estimation techniques.   For OLS, the default setting is correct.   "Sample" allows the user to estimate the regression on a subset of the data set.   The default setting is to estimate for the entire data set.   The specification of the regression equation and the output are listed in the following dialog box.

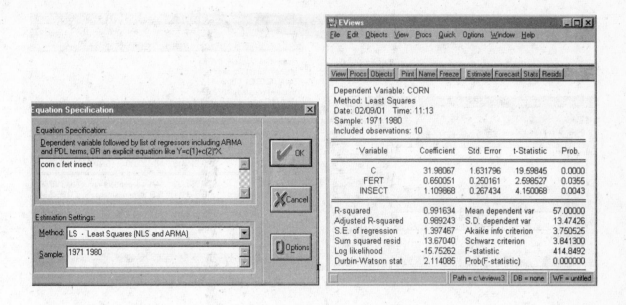

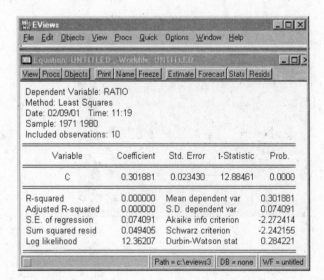

(b) Eviews automatically calculates many diagnostic statistics, including the R^2, the F statistic, AIC, log likelihood, and the Durbin-Watson statistic. Since the Durbin-Watson statistic is near 2, there is no evidence of first-order autocorrelation.

(c) In our estimate of the regression for "ratio," however, the Durbin-Watson statistic is near zero in the output below, indicating autocorrelation.

(d) To correct for autocorrelation, the same procedure is used as for the standard regression, except that "ar(1)" is included in the regression equation. This same method can be used to correct for any ARMA process by including "ar(p)" for autoregression processes, and "ma(q)" for moving average processes (where p and q are the appropriate numbers). Lags can also be inserted quickly by using $(-L)$ where L is the desired lag length. For example, to insert one lag of ratio as an alternate control for autocorrelation, the equation specification would be "ratio c ratio(−1)." From the resulting output, we can see that first-order autocorrelation is no longer present.

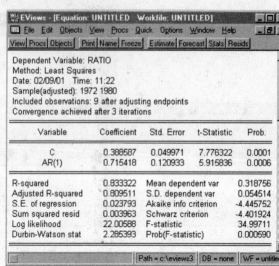

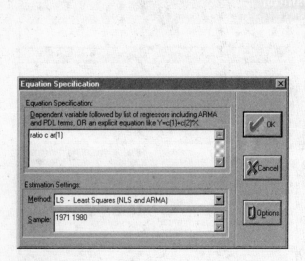

12.11 From the data in Example 2 of Chap. 11, use Eviews to   (*a*) run an ADF test to test the null hypothesis of a unit root in Y.   (*b*) Run an ADF test to test the null hypothesis of a unit root in ΔY.

   (*a*)   The ADF test is found in the Quick-Series Statistics-Unit Root Test.   The resulting dialog box is as follows:

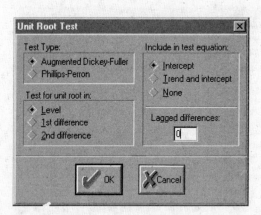

Eview allows flexibility in the unit root test, allowing choice of intercept; trend, or neither, levels or difference; and different lags of the differenced terms to control for autocorrelation.   We choose the test in levels, with an intercept and no lags.

The output reports the regression as well as the critical values.   Since the ADF statistic is greater than the critical value, we accept the null of unit root.

   (*b*)   Running the ADF test in first differences allows us to reject the null of unit root at the 10% significance level, but not at the 5% significance level.

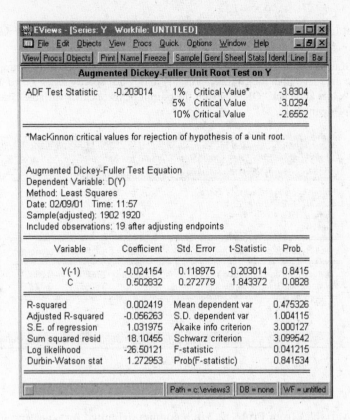

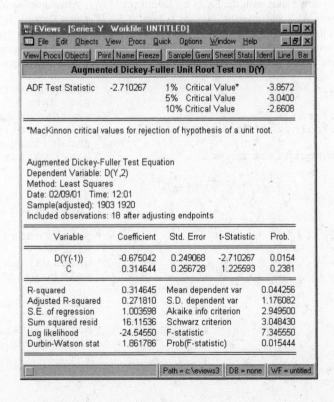

12.12 From the data in Table 11.16 for Prob. 11.33, use Eviews to test if   (*a*) *X* Granger-causes *Y* with six lags and   (*b*) *Y* Granger-causes *X* with six lags.

(*a*), (*b*) Granger causality is found in the toolbar Quick-Group Statistics-Granger Causality Test.   We have input the data set and in the Granger causality dialog box specify series *Y* and *X*, and click "OK."   We then specify six lags and click "OK."   From the output below, neither variable Granger-causes the other at the 5% level of significance.

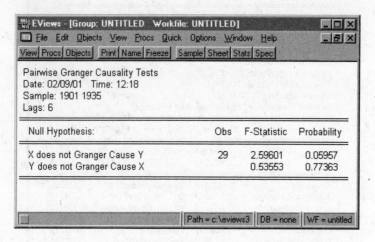

SAS

12.13 (*a*) Save the variables from Prob. 12.7 in a comma-delimited text file, and import the values into SAS.   (*b*) Create a variable for the fertilizer:corn ratio.   (*c*) Print the ratio variable to the output window.   (*d*) Calculate descriptive statistics for the ratio variable.   (*e*) Calculate the correlogram for the fertilizer:corn ratio.

(*a*), (*b*), (*c*), (*d*) We start with the Excel file from Prob. 12.7.   After deleting all but the variable values, we click "File-Save As," and save the data set as type "CSV (Comma Delimited)" for easy accessibility by SAS.   The SAS program to accomplish parts *a* through *d* is presented below.   Note that in SAS "/*" and "*/" enclose comments which are not read by SAS.   It is important to annotate programs and give variables descriptive names so the program is easily debugged, if necessary, and others can read your program.

```
libname main 'c:\';              /* designates the directory to
                                    save data */
data main.corn;                  /* starts the data step */
infile 'c:\corn.csv' delimiter='','';  /* gives location of text file
                                    and delimiter */
input year n corn fert insect;   /* names variables and gives
                                    order */

ratio=fert/corn;                 /* defines ratio as division of
                                    fert and corn */

proc print;                      /* prints data to output
                                    window */
var ratio;                       /* names variables for print,
                                    omit to print all variables */

proc means;                      /* calculates descriptive
                                    stats */
var ratio;                       /* names variables, omit for
                                    stats of all variables */

run;
quit;
```

The output window reports the results. From proc print:

Obs	year	n	corn	fert	insect	ratio
1	1971	1	40	6	4	0.15000
2	1972	2	44	10	4	0.22727
3	1973	3	46	12	5	0.26087
4	1974	4	48	14	7	0.29167
5	1975	5	52	16	9	0.30769
6	1976	6	58	18	12	0.31034
7	1977	7	60	22	14	0.36667
8	1978	8	68	24	20	0.35294
9	1979	9	74	26	21	0.35135
10	1980	10	80	32	24	0.40000

From proc means:

The MEANS Procedure

Variable	N	Mean	Std Dev	Minimum	Maximum
year	10	1975.50	3.0276504	1971.00	1980.00
n	10	5.5000000	3.0276504	1.0000000	10.0000000
corn	10	57.0000000	13.4742553	40.0000000	80.0000000
fert	10	18.0000000	8.0000000	6.0000000	32.0000000
insect	10	12.0000000	7.4833148	4.0000000	24.0000000
ratio	10	0.3018805	0.0740907	0.1500000	0.4000000

(e) To diagnose ARMA processes in SAS, there is "proc arima" which has two stages: identify (designated by "i") and estimate (designated by "e"). Calling back up the data set from the previous parts and continuing yields

```
libname main 'c:\';    /* names library and gives location to find data */
data arma;             /* begins data step and names temporary data set */
set main.corn;         /* reads previously saved data */

proc arima;            /* procedure to calculate correlogram */
i var=ratio;           /* selects variable to identify */

run;
quit;
```

This produces the following output:

The ARIMA Procedure

Name of Variable = ratio

Mean of Working Series	0.301881
Standard Deviation	0.070289
Number of Observations	10

Autocorrelations

Lag	Covariance	Correlation	-1 9 8 7 6 5 4 3 2 1 0 1 2 3 4 5 6 7 8 9 1	Std Error
0	0.0049405	1.00000	| |***************|	0
1	0.0026036	0.52700	| . |*********** . |	0.316228
2	0.0015690	0.31758	| . |****** . |	0.394393

"." marks two standard errors

Inverse Autocorrelations

Lag	Correlation	-1 9 8 7 6 5 4 3 2 1 0 1 2 3 4 5 6 7 8 9 1
1	-0.37607	| . ********| . |
2	-0.04410	| . *| . |

```
                                Partial Autocorrelations

        Lag        Correlation    -1 9 8 7 6 5 4 3 2 1 0 1 2 3 4 5 6 7 8 9 1

          1          0.52700       |    .    |***********    .    |
          2          0.05517       |    .    |*              .    |
```

If we diagnosed an AR(1) process, we could add the line "e p=(1);" to the arima procedure after the identify line.   More complex processes can be estimated similarly.   For example, "e p=(1) (8);" estimates an autoregressive process at the first and eighth lags, and "e q=(1 8);" estimates moving average at the first and eighth lags.

12.14 Using the permanent SAS data set from Prob. 12.13.   (a) Estimate the regression of corn on fertilizer and pesticides.   (b) Is there evidence of autocorrelation in part a? If so, correct for autocorrelation.   (c) Estimate the regression of the fertilizer:corn ratio on only a constant.   (d) Is there evidence of autocorrelation in part c? If so, correct for autocorrelation.

(a), (c) The Durbin-Watson statistic can be calculated in the basic regression procedure, "proc reg," but can also be calculated in "proc autoreg" with the added benefit of a p value which eliminates the need for supplementary critical value tables, and can be used for longer lags of autoregression.   We will use both procedures.

```
libname main 'c:\';              /* names library and gives location to
                                    find data */
data dw;                         /* begins data step and names temporary
                                    data set */
set main.corn;                   /* reads previously saved data */

proc reg;                        /* starts regression procedure */
model corn=fert insect /dw;      /* specifies the regression model, SAS
                                    automatically includes constant, /dw
                                    is omitted for no Durbin-Watson */

proc autoreg;                    /* starts autoregression procedure */
model ratio= /dw=1 dwprob;       /* specifies the regression model, /dw=1
                                    calculates Durbin-Watson start for
                                    1 lag, dwprob calculates
                                    significance */

run;
quit;
```

The resulting output is

```
                            The REG Procedure
                             Model: MODEL1
                         Dependent Variable: corn

                          Analysis of Variance

                           Sum of        Mean
 Source            DF     Squares        Square       F Value     Pr > F

 Model              2    1620.32960     810.16480      414.85     < .0001
 Error              7      13.67040       1.95291
 Corrected Total    9    1634.00000

         Root MSE              1.39747     R-Square     0.9916
         Dependent Mean       57.00000     Adj R-Sq     0.9892
         Coeff Var             2.45170
```

Parameter Estimates

Variable	DF	Parameter Estimate	Standard Error	t Value	Pr > |t|
Intercept	1	31.98067	1.63180	19.60	<.0001
fert	1	0.65005	0.25016	2.60	0.0355
insect	1	1.10987	0.26743	4.15	0.0043

The REG Procedure
Model : MODEL1
Dependent Variable: corn

Durbin-Watson D	2.114
Number of Observations	10
1st Order Autocorrelation	-0.073

The AUTOREG Procedure

Dependent Variable: ratio

Ordinary Least Squares Estimates

SSE	0.04940489	DFE	9
MSE	0.00549	Root MSE	0.07409
SBC	-22.421554	AIC	-22.724139
Regress R-Square	0.0000	Total R-Square	0.0000
Durbin-Watson	0.2842	Pr < DW	<.0001
Pr > DW	1.0000		

NOTE: Pr<DW is the p-value for testing positive autocorrelation, and Pr>DW is the p-value for testing negative autocorrelation.

Variable	DF	Estimate	Standard Error	t Value	Approx Pr > |t|
Intercept	1	0.3019	0.0234	12.88	<.0001

(b), (d) The Durbin-Watson statistic for the model in part *a* does not indicate autocorrelation, but the model in part *d* shows statistically significant autocorrelation since *d* is near 0 and Pr < DW is less than a 5% level of significance (0.05).   To correct for autocorrelation, we also use "proc autoreg."

```
libname main 'c:\';
data dw;
set main.corn;

proc autoreg;
model ratio= /dw=1 dwprob nlag=1;    /* nlag=1 corrects for AR(1) */

run;
quit;
```

This gives the following output:

The AUTOREG Procedure

Dependent Variable: ratio

Ordinary Least Squares Estimates

SSE	0.04940489	DFE	9
MSE	0.00549	Root MSE	0.07409
SBC	-22.421554	AIC	-22.724139
Regress R-Square	0.0000	Total R-Square	0.0000
Durbin-Watson	0.2842	Pr < DW	<.0001
Pr > DW	1.0000		

NOTE: Pr<DW is the p-value for testing positive autocorrelation, and Pr>DW is the p-value for testing negative autocorrelation.

Variable	DF	Estimate	Standard Error	t Value	Approx Pr > |t|
Intercept	1	0.3019	0.0234	12.88	<.0001

Estimates of Autocorrelations

Lag	Covariance	Correlation	-1 9 8 7 6 5 4 3 2 1 0 1 2 3 4 5 6 7 8 9 1
0	0.00494	1.000000	| |***************|
1	0.00260	0.527000	| |*********** |

Preliminary MSE   0.00357

Estimates of Autoregressive Parameters

Lag	Coefficient	Standard Error	t Value
1	-0.527000	0.300473	-1.75

Yule-Walker Estimates

SSE	0.02653769	DFE	8
MSE	0.00332	Root MSE	0.05760
SBC	-26.008448	AIC	-26.613619
Regress R-Square	0.0000	Total R-Square	0.4629
Durbin-Watson	0.9705	Pr < DW	0.0334
Pr > DW	0.9666		

NOTE: Pr<DW is the p-value for testing positive autocorrelation, and Pr>DW is the p-value for testing negative autocorrelation.

The AUTOREG Procedure

Variable	DF	Estimate	Standard Error	t Value	Approx Pr > |t|
Intercept	1	0.2970	0.0348	8.53	<.0001

The correction calculates the magnitude of autocorrelation and estimates the corrected regression. The Durbin-Watson statistics indicates that autocorrelation is still present at the 5% level of significance, but not at the 1% level. Note that the results differ from Eview since SAS uses a different estimation method by default.

12.15 Estimate the binary choice model from Example 5 in Chap. 8 using   (a) probit and   (b) logit.

(a) (b) Since the Logit specification is an option "proc probit," we will put both parts in one program. We will also show the method of manually inputing data through the "cards" statement to bypass creating a separate text file.   The probit procedure in SAS also requires that the data be sorted with successes first for the estimation.   This can be done with "proc sort."

```
data country;
input open gdpcap;
cards;
0        569
0        408
0        2240
0        1869
1        16471
0        1282
1        2102
0        1104
0        914
1        5746
1        2173
0        978
0        762
1        12653
1        3068
1        3075
1        547
1        5185
1        7082
0        1162
;

proc sort;                              /* calls sort procedure */
by descending open;                     /* sorts data set by open variable,
                                           descending option puts larger
                                           values first */

proc probit order=data;                 /* probit procedure, order=data
                                           specifies that successes are
                                           first in data */
class open;                             /* dependent variable */
model open=gdpcap;                      /* regression model */

proc probit order=data;
class open;
model open=gdpcap /d=logistic;          /* regression model, /d option
                                           specifies distribution */

run;
quit;
```

The output is

                        Probit Procedure

                     Class Level Information

                  Name   Levels   Values

                  open   2   1   0

Model Information

```
Data Set                          WORK.COUNTRY
Dependent Variable                        open
Number of Observations                      20
Name of Distribution                    NORMAL
Log Likelihood                     -6.86471345
```

Response Profile

```
Level           Count

  1              10
  0              10
```

Algorithm converged.

Analysis of Parameter Estimates

Variable	DF	Estimate	Standard Error	Chi-Square	Pr > ChiSq	Label
Intercept	1	-1.99418	0.82471	5.8470	0.0156	Intercept
gdpcap	1	0.0010035	0.0004712	4.5347	0.0332	

Probit Model in Terms of Tolerance Distribution

```
        MU                        SIGMA

   1987.23361                  966.514769
```

Probit Procedure

Estimated Covariance Matrix
for Tolerance Parameters

	MU	SIGMA
MU	188389.39327	96239.205174
SIGMA	96239.205174	218986.43870

Probit Procedure

Class Level Information

Name	Levels	Values
open	2	1 0

Model Information

```
Data Set                          WORK.COUNTRY
Dependent Variable                        open
Number of Observations                      20
Name of Distribution                  LOGISTIC
Log Likelihood                    -6.766465426
```

Response Profile

```
Level           Count

  1              10
  0              10
```

Algorithm converged.

Analysis of Parameter Estimates

Variable	DF	Estimate	Standard Error	Chi-Square	Pr > ChiSq	Label
Intercept	1	-3.60499	1.68107	4.5987	0.0320	Intercept
gdpcap	1	0.0017958	0.0008999	3.9817	0.0460	

Probit Model in Terms of Tolerance Distribution

MU	SIGMA
2007.49509	556.864971

Probit Procedure

Estimated Covariance Matrix for Tolerance Parameters

	MU	SIGMA
MU	166670.35772	41952.902987
SIGMA	41952.902987	77881.332977

Note that both distributions give similar results.

12.16 Using the data from Chap. 10, Table 10.1, estimate the simultaneous equations model for Money Supply on GDP by two-stage least squares (2SLS) using investment and government expenditure as instrumental variables (Example 6).

```
data simul;
infile 'c:\table101.csv' delimiter='','';
input year m y i g;

proc syslin 2sls;              /* simultaneous equations procedure, 2sls
                                  indicates two-stage least squares */
endogenous m y;                /* designates endogenous variables */
instruments i g;               /* designates instrumental variables */
money: model m=y;              /* model to be estimated */

run;
quit;
```

This gives the output

The SYSLIN Procedure
Two-Stage Least Squares Estimation

Model	MONEY
Dependent Variable	m

Analysis of Variance

Source	DF	Sum of Squares	Mean Square	F Value	Pr > F
Model	1	783204.1	783204.1	92.50	<.0001
Error	16	135469.4	8466.839		
Corrected Total	17	931628.7			

```
Root MSE              92.01543     R-Square      0.85254
Dependent Mean       874.72667     Adj R-Sq      0.84332
Coeff Var             10.51933
```

```
                        Parameter Estimates
```

Variable	DF	Parameter Estimate	Standard Error	t Value	Pr > |t|
Intercept	1	166.5660	76.75781	2.17	0.0454
y	1	0.115286	0.011987	9.62	<.0001

12.17 From the data in Table 11.16 for Prob. 11.33, use SAS to test if (a) X Granger-causes Y with six lags and   (b) Y Granger-causes X with six lags.

(a), (b) In SAS, the F test can be calculated by adding a "test" line to "proc reg."

```
data granger;
infile 'c:\granger.csv' delimiter='','';
input y x;
                                          /* create lagged variables */
y1=lag1(y);
y2=lag2(y);
y3=lag3(y);
y4=lag4(y);
y5=lag5(y);
y6=lag6(y);

x1=lag1(x);
x2=lag2(x);
x3=lag3(x);
x4=lag4(x);
x5=lag5(x);
x6=lag6(x);

proc reg;
model y=y1 y2 y3 y4 y5 y6 x1 x2 x3 x4 x5 x6;   /* model with 6 lags of each */
grangxy: test x1, x2, x3, x4, x5, x6;          /* test null that all are zero
                                                  with F test */

proc reg;
model x=y1 y2 y3 y4 y5 y6 x1 x2 x3 x4 x5 x6;
grangyx: test y1, y2, y3, y4, y5, y6;
run;
quit;
```

This gives the following output:

```
                   The REG Procedure
                     Model: MODEL1
                 Dependent Variable: y

                   Analysis of Variance
```

Source	DF	Sum of Squares	Mean Square	F Value	Pr > F
Model	12	244.13732	20.34478	2.98	0.0219
Error	16	109.18963	6.82435		
Corrected Total	28	353.32694			

Root MSE	2.61235	R-Square	0.6910	
Dependent Mean	0.56138	Adj R-Sq	0.4592	
Coeff Var	465.34419			

Parameter Estimates

Variable	DF	Parameter Estimate	Standard Error	t Value	Pr > |t|
Intercept	1	2.70383	1.78699	1.51	0.1498
y1	1	-0.73704	0.22650	-3.25	0.0050
y2	1	-0.82864	0.36461	-2.27	0.0372
y3	1	-1.16165	0.42922	-2.71	0.0156
y4	1	-0.67208	0.46783	-1.44	0.1701
y5	1	0.26792	0.44364	0.60	0.5544
y6	1	0.09995	0.27288	0.37	0.7190
x1	1	-0.01778	0.00800	-2.22	0.0410
x2	1	-0.01157	0.01166	-0.99	0.3360
x3	1	-0.01493	0.01499	-1.00	0.3341
x4	1	-0.02471	0.01592	-1.55	0.1403
x5	1	0.01126	0.01750	0.64	0.5288
x6	1	0.03078	0.01391	2.21	0.0417

The REG Procedure
Model: MODEL1

Test GRANGXY Results for Dependent Variable y

Source	DF	Mean Square	F Value	Pr > F
Numerator	6	17.71606	2.60	0.0596
Denominator	16	6.82435		

The REG Procedure
Model: MODEL1
Dependent Variable: x

Analysis of Variance

Source	DF	Sum of Squares	Mean Square	F Value	Pr > F
Model	12	28986	2415.49262	0.37	0.9544
Error	16	103157	6447.29086		
Corrected Total	28	132143			

Root MSE	80.29502	R-Square	0.2194	
Dependent Mean	6.88310	Adj R-Sq	-0.3661	
Coeff Var	1166.55262			

Parameter Estimates

Variable	DF	Parameter Estimate	Standard Error	t Value	Pr > |t|
Intercept	1	68.79389	54.92640	1.25	0.2284
y1	1	-4.02016	6.96180	-0.58	0.5717
y2	1	-6.15257	11.20694	-0.55	0.5906
y3	1	-12.96359	13.19274	-0.98	0.3404
y4	1	-10.12374	14.37952	-0.70	0.4915

y5	1	-14.33754	13.63601	-1.05	0.3087
y6	1	-8.95082	8.38735	-1.07	0.3017
x1	1	-0.17522	0.24590	-0.71	0.4864
x2	1	-0.35289	0.35841	-0.98	0.3395
x3	1	-0.34052	0.46075	-0.74	0.4706
x4	1	-0.55908	0.48943	-1.14	0.2701
x5	1	-0.33701	0.53778	-0.63	0.5397
x6	1	-0.37859	0.42743	-0.89	0.3889

The REG Procedure
Model: MODEL1

Test GRANGYX Results for Dependent Variable x

Source	DF	Mean Square	F Value	Pr > F
Numerator	6	3452.68866	0.54	0.7736
Denominator	16	6447.29086		

Again, neither variable Granger-causes the other at the 5% level of significance.

Supplementary Problems

DATA FORMATS

12.18 Using the data from the Federal Reserve Board of Governors (the Website is listed in App. 12), what two data formats would be able to read the text file of the interest rate data?
Ans. Space-delimited and fixed format.

12.19 Can all space-delimited data be read in fixed format?
Ans. No, often space-delimited data do not line up into columns if observations are of differing lengths.

MICROSOFT EXCEL

12.20 In Problem 12.6, a simple regression line was fit to agricultural data using Excel. From the output (*a*) what was $\hat{b}_0$? (*b*) what was $\hat{b}_1$? (*c*) What was the R^2?
Ans. (*a*) 27.125   (*b*) 1.6597   (*c*) 0.971

12.21 In Prob. 12.7, a multiple regression was estimated using Excel. From the output (*a*) what was the sum of squared errors? (*b*) What was the standard error of $\hat{b}_0$? (*c*) What was the R^2?
Ans. (*a*) 13.6704   (*b*) 0.2674   (*c*) 0.9916

EVIEWS

12.22 Using the output from Eviews in Prob. 12.9(*b*) (*a*) What would the *t* statistic be to test the null hypothesis that the population mean of the fertilizer ratio is 0.25? (*b*) Is this statistically significant at the 5% level?
Ans. (*a*) 2.21   (*b*) No

12.23 What is the critical value for the Granger causality *F* statistic calculated in Prob. 12.12 (*a*) At the 5% level of significance? (*b*) At the 1% level of significance?
Ans. (*a*) 2.74   (*b*) 4.20

SAS

12.24 From the estimation in Prob. 12.15   (a) What is the log-likelihood value for the logit regression?   (b) What is the t statistic for $\hat{b}_1$ in the logit regression?
Ans.   (a) -6.7665 (b) $t = 0.0018/0.0009 = 2$

12.25 In Prob. 12.17, we see X Granger-causes Y at the 10% level of significance.   From the output   (a) What is the short-run effect of X on Y?   (b) What is the long-run effect of X on Y?
Ans.   (a) -0.02695 (b) -0.00668

Econometrics Examination

1. Table 1 gives the quantity supplied of a commodity Y at various prices X, holding everything else constant. (a) Estimate the regression equation of Y on X. (b) Test for the statistical significance of the parameter estimates at the 5% level of significance. (c) Find R^2 and report all previous results in standard summary form. (d) Predict Y and calculate a 95% confidence or prediction interval for $X = 10$.

Table 1. Quantity Supplied at Various Prices

n	1	2	3	4	5	6	7	8
Y	12	14	10	13	17	12	11	15
X	5	11	7	8	11	7	6	9

2. Suppose that from 24 yearly observations on the quantity demand of a commodity in kilograms per year Y, its price in dollars X_1, consumer's income in thousands of dollars X_2, and the price of a substitute commodity in dollars X_3, the following estimated regression is obtained, where the numbers in parentheses represent standard errors:

$$\hat{Y} = 13 - 7X_1 + 2.4X_2 - 4X_3$$
$$(2) \quad (0.8) \quad (18)$$

(a) Indicate whether the signs of the parameters conform to those predicted by demand theory. (b) Are the estimated slope parameters significant at the 5% level? (c) Find R^2, if $\sum y^2 = 40$, $\sum yx_1 = 10$, and $\sum yx_2 = 45$ (where small letters indicate deviations from the mean). (d) Find $\bar{R}^2$. (e) Is R^2 significantly different from zero at the 5% level? (f) Find the standard error of the regression. (g) Find the coefficient of price and income elasticity of demand at the means, given $\bar{Y} = 32$, $\bar{X}_1 = 8$, and $\bar{X}_2 = 16$.

3. When the level of business expenditures for new plants and equipment of nonmanufacturing firms in the United States Y_t from 1960 to 1979 is regressed on the GNP X_{1t}, and the consumer price index, X_{2t}, the following results are obtained:

$$\hat{Y}_t = 31.75 + 0.08 X_{1t} - 0.58X_{2t} \qquad R^2 = 0.98$$
$$(6.08) \qquad (-3.08) \qquad\qquad d = 0.77$$

(a) How do you know that autocorrelation is present? What is meant by *autocorrelation*? Why is autocorrelation a problem? (b) How can you estimate ρ, the coefficient of autocorrelation? (c) How can the value of ρ be used to transform the variables in order to correct for autocorrelation? How do you find the first value of the transformed variables? (d) Is there any evidence of remaining autocorrelation from the following results obtained by running the regression on the transformed variables (indicated by an asterisk)?

$$Y_t^* = 3.79 + 0.04X_{1t}^* - 0.05X_{2t}^* \qquad R^2 = 0.96$$
$$(8.10) \qquad (-0.72) \qquad\quad d = 0.89$$

What could be the cause of any remaining autocorrelation? How could this be corrected?

4. The following two equations represent a simple macroeconomic model:
$$R_t = a_0 + a_1 M_t + a_2 Y_t + u_{1t}$$
$$Y_t = b_0 + b_1 R_t + u_{2t}$$

where R is the interest rate, M is the money supply, and Y is income. (a) Why is this a simultaneous-equations model? Which are the endogenous and exogenous variables? Why would the estimation of the R and Y equations by OLS give biased and inconsistent parameter estimates? (b) Find the reduced form of the model. (c) Is this model underidentified, over-identified, or just identified? Why? What are the values of the structural coefficients? What is an appropriate estimation technique for the model? Explain this technique. (d) If the first, or R, equation included Y_{t-1} as an additional explanatory variable, would this model be identified, overidentified, or underidentified? What are the values of the structural slope coefficients? What would be an appropriate estimation technique? Explain this technique.

5. The ARIMA procedure in SAS gives the following output for a data set of 220 time-series observations. (a) What type of time-series process do the data seem to follow? (b) Calculate the Box-Pierce statistic up to 20 lags. (c) Is there evidence of statistically significant time-series correlations at the 5% level of significance? (d) How would one choose the exact order or correlation to correct for?

The ARIMA Procedure

Name of Variable = y

Mean of Working Series   0.033797
Standard Deviation   2.122958
Number of Observations   220

Autocorrelations

Lag	Covariance	Correlation	-1 9 8 7 6 5 4 3 2 1 0 1 2 3 4 5 6 7 8 9 1	Std Error
0	4.506949	1.00000	| |******************** |	0
1	3.709889	0.82315	| . |**************** |	0.067420
2	2.908734	0.64539	| . |************ |	0.103466
3	2.245384	0.49820	| . |********** |	0.120382
4	1.652113	0.36657	| . |******* |	0.129415
5	1.098705	0.24378	| . |***** |	0.134052
6	0.521525	0.11572	| . |** . |	0.136052
7	-0.133209	-.02956	| . * | . |	0.136498
8	-0.868708	-.19275	| . **** | . |	0.136528
9	-1.567477	-.34779	| ******* | . |	0.137759
10	-2.185962	-.48502	|********** | . |	0.141694
11	-2.185497	-.48492	|********** | . |	0.149049
12	-2.009321	-.44583	|********* | . |	0.156056
13	-1.979412	-.43919	|********* | . |	0.161742
14	-1.759277	-.39035	|******** | . |	0.167074
15	-1.434070	-.31819	| .******* | . |	0.171170
16	-1.137798	-.25245	| . ***** | . |	0.173837
17	-0.872123	-.19351	| . **** | . |	0.175496
18	-0.670881	-.14885	| . *** | . |	0.176463
19	-0.314030	-.06968	| . * | . |	0.177033
20	-0.0008474	-.00019	| . | . |	0.177158

"." marks two standard errors

Partial Autocorrelations

Lag	Correlation	-1 9 8 7 6 5 4 3 2 1 0 1 2 3 4 5 6 7 8 9 1
1	0.82315	| . |***************** |
2	-0.09982	| . ** | . |
3	-0.01224	| . | . |
4	-0.05172	| . * | . |
5	-0.06475	| . * | . |
6	-0.11227	| . ** | . |
7	-0.16538	| *** | . |
8	-0.20610	| **** | . |
9	-0.17777	| **** | . |
10	-0.18760	| **** | . |
11	0.22327	| . |**** |
12	0.03572	| . |* . |
13	-0.11763	| .** | . |
14	0.10343	| . |**. |
15	0.04442	| . |* . |
16	-0.06357	| . * | . |
17	-0.10960	| .** | . |
18	-0.18580	| **** | . |
19	0.02378	| . | . |
20	-0.08972	| .** | . |

Answers

1. (*a*) See Table 2.

Table 2. Worksheet

n	Y_i	X_i	y_i	x_i	x_iy_i	x_i^2	$\hat{Y}_i$	e_i	e_i^2	X_i^2	y_i^2
1	12	5	−1	−3	3	9	10.54	1.46	2.1316	25	1
2	14	11	1	3	3	9	15.46	−1.46	2.1316	121	1
3	10	7	−3	−1	3	1	12.18	−2.18	4.7524	49	9
4	13	8	0	0	0	0	13.00	0.00	0.0000	64	0
5	17	11	4	3	12	9	15.46	1.54	2.3716	121	16
6	12	7	−1	−1	1	1	12.18	−0.18	0.0324	49	1
7	11	6	−2	−2	4	4	11.36	−0.36	0.1296	36	4
8	15	9	2	1	2	1	13.82	1.18	1.3924	81	4
$n=8$	$\sum Y_i = 104$ $\bar{Y}=13$	$\sum X_i = 64$ $\bar{X}=8$	$\sum y_i = 0$	$\sum x_i = 0$	$\sum x_iy_i = 28$	$\sum x_i^2 = 34$		$\sum e_i = 0$	$\sum e_i^2 = 12.9416$	$\sum X_i^2 = 546$	$\sum y_i^2 = 36$

$$\hat{b}_1 = \frac{\sum x_i y_i}{\sum x_i^2} = \frac{28}{34} \cong 0.82 \qquad \text{(from the first 7 columns of Table 2).}$$

$$\hat{b}_0 = \overline{Y} - \hat{b}_1 \overline{X} \cong 13 - (0.82)(8) \cong 6.44$$

$$\hat{Y}_i = 6.44 + 0.82\, X_i$$

(b)

$$s_{\hat{b}_0}^2 = \frac{\sum e_i^2}{(n-k)} \frac{\sum X_i^2}{n \sum x_i^2} = \frac{(12.9416)(546)}{(8-2)(8)(34)} \cong 4.33 \quad \text{and} \quad s_{\hat{b}_0} \cong 2.08$$

$$s_{\hat{b}_1}^2 = \frac{\sum e_i^2}{(n-k) \sum x_i^2} = \frac{12.9416}{(8-2)(34)} \cong 0.06 \quad \text{and} \quad s_{\hat{b}_1} \cong 0.25$$

$$t_0 = \frac{\hat{b}_0}{s_{\hat{b}_0}} = \frac{6.44}{2.08} \cong 3.10 \qquad \text{and is significant at the 5\% level}$$

$$t_1 = \frac{\hat{b}_1}{s_{\hat{b}_1}} = \frac{0.82}{0.25} \cong 3.28 \qquad \text{and is also significant at the 5\% level}$$

(c)

$$R^2 = 1 - \frac{\sum e_i^2}{\sum y_i^2} = 1 - \frac{12.9416}{36} \cong 0.6405, \text{ or } 64.05\%$$

$$\hat{Y}_i = 6.44 + 0.82 X_i \qquad R^2 \cong 64.05$$

$$(3.10) \quad (3.28)$$

(d)

$$\hat{Y}_F = 6.44 + 0.82(10) = 14.64$$

$$s_F^2 = \frac{\sum e_i^2}{(n-2)} \left[1 + \frac{1}{n} + \frac{(X_F - \overline{X})^2}{\sum x_i^2} \right] = \frac{12.9416}{6} \left[1 + \frac{1}{8} + \frac{(10-8)^2}{34} \right]$$

$$s_F^2 = 2.67 \quad \text{and} \quad s_F \cong 1.63$$

Therefore, the 95% confidence or prediction interval for Y_F is given by $Y_F = 14.64 \pm 2.45(1.63)$, where $t_{0.025} = \pm 2.45$, with $n - k = 8 - 2 = 6$ df, so that we are 95% confident that $10.65 \leq Y_F \leq 18.63$.

2. (a) Consumer demand theory postulates that the quantity demanded of a commodity is inversely related to its price but directly related to consumers' income (if the commodity is a normal good) and to the price of substitute commodities. Thus the signs of $\hat{b}_1$ and $\hat{b}_2$ conform, but the sign of $\hat{b}_3$ does not conform to that predicted by demand theory.

(b) $t_1 = -7/2 = -3.5$, $t_2 = 2.4/0.8 = 3$, and $t_3 = 4/18 \cong 0.22$. Therefore, $\hat{b}_1$ and $\hat{b}_2$ are statistically significant at the 5% level, but $\hat{b}_3$ is not.

(c)

$$R^2 = \frac{\hat{b}_1 \sum y x_1 + \hat{b}_2 \sum y x_2}{\sum y^2} = \frac{-7(10) + 2.4(45)}{40} = \frac{-70 + 108}{40} = 0.9500, \text{ or } 95\%$$

(d)

$$\overline{R}^2 = 1 - (1 - R^2)\frac{n-1}{n-4} = 1 - (1 - 0.95)\frac{23}{20} = 1 - (0.05)(1.15) = 0.9425, \text{ or } 94.25\%$$

(e) Since

$$F_{3,20} = \frac{R^2/k - 1}{(1 - R^2)/n - k} = \frac{0.95/4 - 1}{(1 - 0.95)/24 - 4} \cong \frac{0.3167}{0.0025} = 126.68$$

R^2 is significantly different from zero at the 5% level.

(f) Since $R^2 = 1 - (\sum e^2 / \sum y^2)$, it follows that $\sum e^2 = (1 - R^2) \sum y^2 = (1 - 0.95)(40) = 2$. Thus

$$s = \sqrt{\frac{\sum e^2}{n - k}} = \sqrt{2/20} \cong 0.32$$

(g) $\eta_{x_1} = \hat{b}_1(\overline{X}_1/\overline{Y}) = -7(8/32) = -1.75$. $\eta_{x_2} = \hat{b}_2(\overline{X}_2/\overline{Y}) = 2.4(16/32) = 1.2$.

3. (*a*) Evidence of the presence of autocorrelation is given by the very low value of the Durbin-Watson statistic d. *Autocorrelation* refers to the case in which the error term in one time period is associated with the error term in any other period. The most common form of autocorrelation in time-series data is positive first-order autocorrelation. With autocorrelation, the OLS parameters are still unbiased and consistent, but the standard errors of the estimated regression parameters are biased, leading to incorrect statistical tests and biased confidence intervals.

(*b*) An estimate of the coefficient of autocorrelation ρ can be obtained from the coefficient of Y_{t-1} in the following regression:

$$\hat{Y}_t = \hat{b}_0 + \hat{\rho}Y_{t-1} + \hat{b}_1 X_{1t} - \hat{b}_1\rho X_{1t} + \hat{b}_2 X_{2t} - \hat{b}_2\rho X_{t-1}$$

(*c*) The value of the transformed variables to correct for autocorrelation can be found as follows (where the asterisk refers to the transformed variables):

$$Y_t^* = Y_t - \hat{\rho}Y_{t-1} \qquad X_{1t}^* = X_{1t} - \hat{\rho}X_{1t-1} \qquad X_{2t}^* = X_{2t} - \hat{\rho}X_{2t-1}$$

$$Y_1^* = Y_1\sqrt{1-\hat{\rho}^2} \qquad X_{11}^* = X_1\sqrt{1-\hat{\rho}^2} \qquad X_{21} = X_2\sqrt{1-\hat{\rho}^2}$$

(*d*) Since d remains very low, evidence of autocorrelation remains even after the adjustment. In this case, autocorrelation is very likely due to the fact that some important explanatory variables were not included in the regression, to improper functional form, or more generally to biased model specification. Therefore, before transforming the variables in an attempt to overcome autocorrelation, it is crucial to include all the variables, use the functional form suggested by investment theory, and generally avoid an incorrect model specification.

4. (*a*) This two-equation model is simultaneous because R and Y are jointly determined; that is, $R = f(Y)$ and $Y = f(R)$. The endogenous variables of the model are R and Y, while M is exogenous or determined outside the model. The estimation of the R function by OLS gives biased and inconsistent parameter estimates because Y_t is correlated with u_{1t}. Similarly, estimating the second, or Y, equation by OLS also gives biased and inconsistent parameter estimates because R and u_2 are correlated.

(*b*) Substituting the value of Y given by the second equation into the first equation, we get

$$R_t = a_0 + a_1 M_t + a_2(b_0 + b_1 R_t + u_{2t}) + u_{1t}$$

$$R_t - a_2 b_1 R_1 = a_0 + a_2 b_0 + a_1 M_t + a_2 u_{2t} + u_{1t}$$

$$R_t = \frac{a_0 + a_2 b_0}{1 - a_2 b_1} + \frac{a_1}{1 - a_2 b_1}M_t + \frac{a_2 u_{2t} + u_{1t}}{1 - a_2 b_1} \qquad \text{or} \qquad R_t = \pi_0 + \pi_1 M_t + \upsilon_{1t}$$

Substituting the value of R_t given by the first equation into the second equation, we get

$$Y_t = b_0 + b_1(a_0 + a_1 M_t + a_2 Y_t + u_{1t}) + u_{2t}$$

$$Y_t - a_2 b_1 Y_t = a_0 b_1 + b_0 + a_1 b_1 M_t + b_1 u_{1t} + u_{2t}$$

$$Y_t = \frac{a_0 b_1 + b_0}{1 - a_2 b_1} + \frac{a_1 b_1}{1 - a_2 b_1}M_t + \frac{b_1 u_{1t} + u_{2t}}{1 - a_2 b_1} \qquad \text{or} \qquad Y_t = \pi_2 + \pi_3 M_t + \upsilon_{2t}$$

(*c*) Since the first, or R, equation does not exclude any exogenous variable, it is unidentified. Since the number of excluded exogenous variables from the second, or Y, equation (which is one, i.e., the M variable) equals the number of endogenous variables (i.e, R and Y) minus 1, the second, or Y, equation is exactly identified. $b_1 = \pi_3/\pi_1$ and $b_0 = \pi_2 - b_1\pi_0$. The values of a_1 and a_2 cannot be found because the R equation is underidentified. An appropriate technique for estimating the exactly identified Y equation is indirect least squares (ILS). This involves OLS estimation of the R_t reduced-form equation and then use of $\hat{R}_t$ to estimate the Y structural equation. When this is done, $\hat{b}_1$ is consistent.

(*d*) If the first, or R, equation included the additional Y_{t-1} variable, the first equation would continue to be underidentified, but the second equation would now be overidentified. Two different values of b_1 can be calculated from the reduced-form coefficients, but it would be impossible to calculate any of the structural slope coefficients of the unidentified R equation. An appropriate technique for estimating the overidentified Y equation is two-stage least squares (2SLS). This involves first regressing R_t on M_t and Y_{t-1}, and then using $\hat{R}_t$ to estimate the Y structural equation. When this is done, $\hat{b}_1$ is consistent.

5. (*a*) The large correlations at the first and tenth lag indicate the presence of time-series correlations. The spike at one lag fades away slowly, and the partial correlation at one lag leaves quickly, indicating AR(1). The tenth lag is more troublesome since it exhibits features of AR in the correlations, but the partial correlation is not clear. The combination of the two effects makes diagnosis more difficult.

(*b*) The Box-Pierce statistic is

$$Q = T \sum \mathrm{ACF}_s^2 = 220(2.9523) = 649.56$$

(*c*) The critical value of the chi-square distribution with 20 df is 31.41 at the 5% level of significance. Since $Q = 649.56 > 31.41$, we reject the null of no correlations. Therefore the correlations are statistically significant.

(*d*) One could try possible specifications and take the one with the lowest AIC. For our case, we try AR(1,10), AR(1) and MA(10), and MA(1) and MA(10) since we have an idea of the lag lengths, but not the process. We do this by adding the following procedure in our SAS program:

```
proc arima;
i var=y;
e p=(1) (10);      /* AR(1) and AR(10) */
e p=(1) q=(10);    /* AR(1) and MA(10) */
e q=(1 10);        /* MA(1) and MA(10) */
```

The resulting AIC is 670.97, 644.38, and 786.79, respectively, telling us that the second model of AR(1) and MA(10) is the best specification.

Binomial Distribution

n	x	.01	.05	.10	.15	.20	.25	p .30	.35	.40	.45	.50
1	0	.9900	.9500	.9000	.8500	.8000	.7500	.7000	.6500	.6000	.5500	.5000
	1	.0100	.0500	.1000	.1500	.2000	.2500	.3000	.3500	.4000	.4500	.5000
2	0	.9801	.9025	.8100	.7225	.6400	.5625	.4900	.4225	.3600	.3025	.2500
	1	.0198	.0950	.1800	.2550	.3200	.3750	.4200	.4550	.4800	.4950	.5000
	2	.0001	.0025	.1100	.0225	.0400	.0625	.0900	.1225	.1600	.2025	.2500
3	0	.9703	.8574	.7290	.6141	.5120	.4219	.3430	.2746	.2160	.1664	.1250
	1	.0294	.1354	.2430	.3251	.3840	.4219	.4410	.4436	.4320	.4084	.3750
	2	.0003	.0071	.0.270	.0574	.0960	.1406	.1890	.2289	.2880	.3341	.3750
	3	.0000	.0001	.0010	.0034	.0080	.0156	.0270	.0429	.0640	.0911	.1250
4	0	.9606	.8145	.6561	.5220	.4096	.3164	.2401	.1785	.1296	.0915	.0625
	1	.0388	.1715	.2916	.3685	.4096	.4219	.4116	.3845	.3456	.2995	.2500
	2	.0006	.0135	.0486	.0975	.1536	.2109	.2646	.3105	.3456	.3675	.3750
	3	.0000	.0005	.0036	.0115	.0256	.0469	.0756	.1115	.1536	.2005	.2500
	4	.0000	.0000	.0001	.0005	.0016	.0039	.0081	.0150	.0256	.0410	.0625
5	0	.9510	.7738	.5905	.4437	.3277	.2373	.1681	.1160	.0778	.0503	.0312
	1	.0480	.2036	.3280	.3915	.4096	.3955	.3602	.3124	.2592	.2059	.1562
	2	.0010	.0214	.0729	.1382	.2048	.2637	.3087	.3364	.3456	.3369	.3125
	3	.0000	.0011	.0081	.0244	.0512	.0879	.1323	.1811	.2304	.2757	.3125
	4	.0000	.0000	.0004	.0022	.0064	.0146	.0284	.0488	.0768	.1128	.1562
	5	.0000	.0000	.0000	.0001	.0003	.0010	.0024	.0053	.0102	.0185	.0312
6	0	.9415	.7351	.5314	.3771	.2621	.1780	.1176	.0754	.0467	.0277	.0156
	1	.0571	.2321	.3543	.3993	.3932	.3560	.3025	.2437	.1866	.1359	.0938
	2	.0014	.0305	.0984	.1762	.2458	.2966	.3241	.3280	.3110	.2780	.2344
	3	.0000	.0021	.0146	.0415	.0819	.1318	.1852	.2355	.2765	.3032	.3125
	4	.0000	.0001	.0012	.0055	.0154	.0330	.0595	.0951	.1382	.1861	.2344
	5	.0000	.0000	.0001	.0004	.0015	.0044	.0102	.0205	.0369	.0609	.0938
	6	.0000	.0000	.0000	.0000	.0001	.0002	.0007	.0018	.0041	.0083	.0156
7	0	.9321	.6983	.4783	.3206	.2097	.1335	.0824	.0490	.0280	.0152	.0078
	1	.0659	.2573	.3720	.3960	.3670	.3115	.2471	.1848	.1306	.0872	.0547
	2	.0020	.0406	.1240	.2097	.2753	.3115	.3177	.2985	.2613	.2140	.1641

n	x	.01	.05	.10	.15	.20	.25	p .30	.35	.40	.45	.50
	3	.0000	.0036	.0230	.0617	.1147	.1730	.2269	.2679	.2903	.2918	.2734
	4	.0000	.0002	.0026	.0109	.0287	.0577	.0972	.1442	.1935	.2388	.2734
	5	.0000	.0000	.0002	.0012	.0043	.0115	.0250	.0466	.0774	.1172	.1641
	6	.0000	.0000	.0000	.0001	.0004	.0013	.0036	.0084	.0172	.0320	.0547
	7	.0000	.0000	.0000	.0000	.0000	.0001	.0002	.0006	.0016	.0037	.0078
8	0	.9227	.6634	.4305	.2725	.1678	.1002	.0576	.0319	.0168	.0084	.0039
	1	.0746	.2793	.3826	.3847	.3355	.2670	.1977	.1373	.0896	.0548	.0312
	2	.0026	.0515	.1488	.2376	.2936	.3115	.2065	.2587	.2090	.1569	.1094
	3	.0001	.0054	.0331	.0839	.1468	.2076	.2541	.2786	.2787	.2568	.2188
	4	.0000	.0004	.0046	.0185	.0459	.0865	.1361	.1875	.2322	.2627	.2734
	5	.0000	.0000	.0004	.0026	.0092	.0231	.0467	.0808	.1239	.1719	.2188
	6	.0000	.0000	.0000	.0002	.0011	.0038	.0100	.0217	.0413	.0403	.1094
	7	.0000	.0000	.0000	.0000	.0001	.0004	.0012	.0033	.0079	.0164	.0312
	8	.0000	.0000	.0000	.0000	.0000	.0000	.0001	.0002	.0007	.0017	.0039
9	0	.9135	.6302	.3874	.2316	.1342	.0751	.0404	.0207	.0101	.0046	.0020
	1	.0830	.2985	.3874	.3679	.3020	.2253	.1556	.1004	.0605	.0339	.0176
	2	.0034	.0629	.1722	.2597	.3020	.3003	.2668	.2162	.1612	.1110	.0703
	3	.0001	.0077	.0446	.1069	.1762	.2336	.2668	.2716	.2508	.2119	.1641
	4	.0000	.0006	.0074	.0283	.0661	.1168	.1715	.2194	.2508	.2600	.2461
	5	.0000	.0000	.0008	.0050	.0165	.0389	.0735	.1181	.1672	.2128	.2461
	6	.0000	.0000	.0001	.0006	.0028	.0087	.0210	.0424	.0743	.1160	.1641
	7	.0000	.0000	.0000	.0000	.0003	.0012	.0039	.0098	.0212	.0407	.0703
	8	.0000	.0000	.0000	.0000	.0000	.0001	.0004	.0013	.0035	.0083	.0176
	9	.0000	.0000	.0000	.0000	.0000	.0000	.0000	.0001	.0003	.0008	.0020
10	0	.9044	.5987	.3487	.1969	.1074	.0563	.0282	.0135	.0060	.0025	.0010
	1	.0914	.3151	.3874	.3474	.2684	.1877	.1211	.0725	.0403	.0207	.0098
	2	.0042	.0746	.1937	.2759	.3020	.2816	.2335	.1757	.1209	.0763	.0439
	3	.0001	.0105	.0574	.1298	.2013	.2503	.2668	.2522	.2150	.1665	.1172
	4	.0000	.0010	.0112	.0401	.0881	.1460	.2001	.2377	.2508	.2384	.2051
	5	.0000	.0001	.0015	.0085	.0264	.0584	.1029	.1536	.2007	.2340	.2461
	6	.0000	.0000	.0001	.0012	.0055	.0162	.0368	.0689	.1115	.1596	.2051
	7	.0000	.0000	.0000	.0001	.0008	.0031	.0090	.0212	.0425	.0746	.1172
	8	.0000	.0000	.0000	.0000	.0001	.0004	.0014	.0043	.0106	.0229	.0439
	9	.0000	.0000	.0000	.0000	.0000	.0000	.0001	.0005	.0016	.0042	.0098
	10	.0000	.0000	.0000	.0000	.0000	.0000	.0000	.0000	.0001	.0003	.0010
11	0	.8953	.5688	.3138	.1673	.0859	.0422	.0198	.0088	.0036	.0014	.0005
	1	.0995	.3293	.3835	.3248	.2363	.1549	.0932	.0518	.0266	.0125	.0054
	2	.0050	.0867	.2131	.2866	.2953	.2581	.1998	.1395	.0887	.0513	.0269
	3	.0002	.0137	.0710	.1517	.2215	.2581	.2568	.2254	.1774	.1259	.0806
	4	.0000	.0014	.0158	.0536	.1107	.1721	.2201	.2428	.2365	.2060	.1611
	5	.0000	.0001	.0025	.0132	.0388	.0803	.1321	.1830	.2207	.2360	.2256
	6	.0000	.0000	.0003	.0023	.0097	.0268	.0566	.0985	.1471	.1931	.2256
	7	.0000	.0000	.0000	.0003	.0017	.0064	.0173	.0379	.0701	.1128	.1611
	8	.0000	.0000	.0000	.0000	.0002	.0011	.0037	.0102	.0234	.0462	.0806
	9	.0000	.0000	.0000	.0000	.0000	.0001	.0005	.0018	.0052	.0126	.0269
	10	.0000	.0000	.0000	.0000	.0000	.0000	.0000	.0002	.0007	.0021	.0054
	11	.0000	.0000	.0000	.0000	.0000	.0000	.0000	.0000	.0000	.0002	.0005
12	0	.8864	.5404	.2824	.1422	.0687	.0317	.0138	.0057	.0022	.0008	.0002
	1	.1074	.3413	.3766	.3012	.2062	.1267	.0712	.0368	.0174	.0075	.0029
	2	.0060	.0988	.2301	.2924	.2835	.2323	.1678	.1088	.0639	.0339	.0161
	3	.0002	.0173	.0852	.1720	.2362	.2581	.2397	.1954	.1419	.0923	.0537
	4	.0000	.0021	.0213	.0683	.1329	.1936	.2311	.2367	.2128	.1700	.1208
	5	.0000	.0002	.0038	.0193	.0532	.1032	.1585	.2039	.2270	.2225	.1934
	6	.0000	.0000	.0005	.0040	.0155	.0401	.0792	.1281	.1766	.2124	.2256

n	x	.01	.05	.10	.15	.20	.25	p .30	.35	.40	.45	.50
	7	.0000	.0000	.0000	.0006	.0033	.0115	.0291	.0591	.1009	.1489	.1934
	8	.0000	.0000	.0000	.0001	.0005	.0024	.0078	.0199	.0420	.0762	.1208
	9	.0000	.0000	.0000	.0000	.0001	.0004	.0015	.0048	.0125	.0277	.0537
	10	.0000	.0000	.0000	.0000	.0000	.0000	.0002	.0008	.0025	.0068	.0161
	11	.0000	.0000	.0000	.0000	.0000	.0000	.0000	.0001	.0003	.0010	.0029
	12	.0000	.0000	.0000	.0000	.0000	.0000	.0000	.0000	.0000	.0001	.0002
13	0	.8775	.5133	.2542	.1209	.0550	.0238	.0097	.0037	.0013	.0004	.0001
	1	.1152	.3512	.3672	.2774	.1787	.1029	.0540	.0259	.0113	.0045	.0016
	2	.0070	.1109	.2448	.2937	.2680	.2059	.1388	.0836	.0453	.0220	.0095
	3	.0003	.0214	.0997	.1900	.2457	.2517	.2181	.1651	.1107	.0660	.0349
	4	.0000	.0028	.0277	.0838	.1535	.2097	.2337	.2222	.1845	.1350	.0873
	5	.0000	.0003	.0055	.0266	.0691	.1258	.1803	.2154	.2214	.1989	.1571
	6	.0000	.0000	.0008	.0063	.0230	.0559	.1030	.1546	.1968	.2169	.2095
	7	.0000	.0000	.0001	.0011	.0058	.0186	.0442	.0833	.1312	.1775	.2095
	8	.0000	.0000	.0001	.0001	.0011	.0047	.0142	.0336	.0656	.1089	.1571
	9	.0000	.0000	.0000	.0000	.0001	.0009	.0034	.0101	.0243	.0495	.0873
	10	.0000	.0000	.0000	.0000	.0000	.0001	.0006	.0022	.0065	.0162	.0349
	11	.0000	.0000	.0000	.0000	.0000	.0000	.0001	.0003	.0012	.0036	.0095
	12	.0000	.0000	.0000	.0000	.0000	.0000	.0000	.0000	.0001	.0005	.0016
	13	.0000	.0000	.0000	.0000	.0000	.0000	.0000	.0000	.0000	.0000	.0001
14	0	.8687	.4877	.2288	.1028	.0440	.0178	.0068	.0024	.0008	.0002	.0001
	1	.1229	.3593	.3559	.2539	.1539	.0832	.0467	.0181	.0073	.0027	.0009
	2	.0081	.1229	.2570	.2912	.2501	.1802	.1134	.0634	.0317	.0141	.0056
	3	.0003	.0259	.1142	.2056	.2501	.2402	.1943	.1366	.0845	.0462	.0222
	4	.0000	.0037	.0349	.0998	.1720	.2202	.2290	.2022	.1549	.1040	.0611
	5	.0000	.0004	.0078	.0352	.0860	.1468	.1963	.2178	.2066	.1701	.1222
	6	.0000	.0000	.0013	.0093	.0322	.0734	.1262	.1759	.2066	.2088	.1833
	7	.0000	.0000	.0002	.0019	.0092	.0280	.0618	.1082	.1574	.1952	.2095
	8	.0000	.0000	.0000	.0003	.0020	.0082	.0232	.0510	.0918	.1398	.1833
	9	.0000	.0000	.0000	.0000	.0003	.0018	.0066	.0183	.0408	.0762	.1222
	10	.0000	.0000	.0000	.0000	.0000	.0003	.0014	.0049	.0136	.0312	.0611
	11	.0000	.0000	.0000	.0000	.0000	.0000	.0002	.0010	.0033	.0093	.0222
	12	.0000	.0000	.0000	.0000	.0000	.0000	.0000	.0001	.0005	.0019	.0056
	13	.0000	.0000	.0000	.0000	.0000	.0000	.0000	.0000	.0001	.0002	.0009
	14	.0000	.0000	.0000	.0000	.0000	.0000	.0000	.0000	.0000	.0000	.0001
15	0	.8601	.4633	.2059	.0874	.0352	.0134	.0047	.0016	.0005	.0001	.0000
	1	.1303	.3658	.3432	.2312	.1319	.0668	.0305	.0126	.0047	.0016	.0005
	2	.0092	.1348	.2669	.2856	.2309	.1559	.0916	.0476	.0219	.0090	.0032
	3	.0004	.0307	.1285	.2184	.2501	.2252	.1700	.1110	.0634	.0318	.0139
	4	.0000	.0049	.0428	.1156	.1876	.2252	.2186	.1792	.1268	.0780	.0417
	5	.0000	.0006	.0105	.0449	.1032	.1651	.2061	.2123	.1859	.1404	.0916
	6	.0000	.0000	.0019	.0132	.0430	.0917	.1472	.1906	.2066	.1914	.1527
	7	.0000	.0000	.0003	.0030	.0138	.0393	.0811	.1319	.1771	.2013	.1964
	8	.0000	.0000	.0000	.0005	.0035	.0131	.0348	.0710	.1181	.1647	.1964
	9	.0000	.0000	.0000	.0001	.0007	.0034	.0116	.0298	.0612	.1048	.1527
	10	.0000	.0000	.0000	.0000	.0001	.0007	.0030	.0096	.0245	.0515	.0916
	11	.0000	.0000	.0000	.0000	.0000	.0001	.0006	.0024	.0074	.0191	.0417
	12	.0000	.0000	.0000	.0000	.0000	.0000	.0001	.0004	.0016	.0052	.0139
	13	.0000	.0000	.0000	.0000	.0000	.0000	.0000	.0001	.0003	.0010	.0032
	14	.0000	.0000	.0000	.0000	.0000	.0000	.0000	.0000	.0000	.0001	.0005
	15	.0000	.0000	.0000	.0000	.0000	.0000	.0000	.0000	.0000	.0000	.0000
16	0	.8515	.4401	.1853	.0743	.0281	.0100	.0033	.0010	.0003	.0001	.0000
	1	.1376	.3706	.3294	.2097	.1126	.0535	.0228	.0087	.0030	.0009	.0002
	2	.0104	.1463	.2745	.2775	.2111	.1336	.0732	.0353	.0150	.0056	.0018

n	x	.01	.05	.10	.15	.20	.25	p .30	.35	.40	.45	.50
	3	.0005	.0359	.1423	.2285	.2463	.2079	.1465	.0888	.0468	.0215	.0085
	4	.0000	.0061	.0514	.1311	.2001	.2252	.2040	.1553	.1014	.0572	.0278
	5	.0000	.0008	.0137	.0555	.1201	.1802	.2099	.2008	.1623	.1123	.0667
	6	.0000	.0001	.0028	.0180	.0550	.1101	.1649	.1982	.1983	.1684	.1222
	7	.0000	.0000	.0004	.0045	.0197	.0524	.1010	.1524	.1889	.1969	.1746
	8	.0000	.0000	.0001	.0009	.0055	.0197	.0487	.0923	.1417	.1812	.1964
	9	.0000	.0000	.0000	.0001	.0012	.0058	.0185	.0442	.0840	.1318	.1746
	10	.0000	.0000	.0000	.0000	.0002	.0014	.0056	.0167	.0392	.0755	.1222
	11	.0000	.0000	.0000	.0000	.0000	.0002	.0013	.0049	.0142	.0337	.0667
	12	.0000	.0000	.0000	.0000	.0000	.0000	.0002	.0011	.0040	.0115	.0278
	13	.0000	.0000	.0000	.0000	.0000	.0000	.0000	.0002	.0008	.0029	.0085
	14	.0000	.0000	.0000	.0000	.0000	.0000	.0000	.0000	.0001	.0005	.0018
	15	.0000	.0000	.0000	.0000	.0000	.0000	.0000	.0000	.0000	.0001	.0002
	16	.0000	.0000	.0000	.0000	.0000	.0000	.0000	.0000	.0000	.0000	.0000
17	0	.8429	.4181	.1668	.0631	.0225	.0075	.0023	.0007	.0002	.0000	.0000
	1	.1447	.3741	.3150	.1893	.0957	.0426	.0169	.0060	.0019	.0005	.0001
	2	.0117	.1575	.2800	.2673	.1914	.1136	.0581	.0260	.0102	.0035	.0010
	3	.0006	.0415	.1556	.2359	.2393	.1893	.1245	.0701	.0341	.0144	.0052
	4	.0000	.0076	.0605	.1457	.2093	.2209	.1868	.1320	.0796	.0411	.0182
	5	.0000	.0010	.0175	.0668	.1361	.1914	.2081	.1849	.1379	.0875	.0472
	6	.0000	.0001	.0039	.0236	.0680	.1276	.1784	.1991	.1839	.1432	.1944
	7	.0000	.0000	.0007	.0065	.0267	.0668	.1201	.1685	.1927	.1841	.1484
	8	.0000	.0000	.0001	.0014	.0084	.0279	.0644	.1134	.1606	.1883	.1855
	9	.0000	.0000	.0000	.0003	.0021	.0093	.0276	.0611	.1070	.1540	.1855
	10	.0000	.0000	.0000	.0000	.0004	.0025	.0095	.0263	.0571	.1008	.1484
	11	.0000	.0000	.0000	.0000	.0001	.0005	.0026	.0090	.0242	.0525	.0944
	12	.0000	.0000	.0000	.0000	.0000	.0001	.0006	.0024	.0081	.0215	.0472
	13	.0000	.0000	.0000	.0000	.0000	.0000	.0001	.0005	.0021	.0068	.0182
	14	.0000	.0000	.0000	.0000	.0000	.0000	.0000	.0001	.0004	.0016	.0052
	15	.0000	.0000	.0000	.0000	.0000	.0000	.0000	.0000	.0001	.0003	.0010
	16	.0000	.0000	.0000	.0000	.0000	.0000	.0000	.0000	.0000	.0000	.0001
	17	.0000	.0000	.0000	.0000	.0000	.0000	.0000	.0000	.0000	.0000	.0000
18	0	.8345	.3972	.1501	.0536	.0180	.0056	.0016	.0004	.0001	.0000	.0000
	1	.1517	.3763	.3002	.1704	.0811	.0338	.0126	.0042	.0012	.0003	.0001
	2	.0130	.1683	.2835	.2556	.1723	.0958	.0458	.0190	.0069	.0022	.0006
	3	.0007	.0473	.1680	.2406	.2297	.1704	.1046	.0547	.0246	.0095	.0031
	4	.0000	.0093	.0700	.1592	.2153	.2130	.1681	.1104	.0614	.0291	.0117
	5	.0000	.0014	.0218	.0787	.1507	.1988	.2017	.1664	.1146	.0666	.0327
	6	.0000	.0002	.0052	.0301	.0816	.1436	.1873	.1941	.1655	.1181	.0708
	7	.0000	.0000	.0010	.0091	.0350	.0820	.1376	.1792	.1892	.1657	.1214
	8	.0000	.0000	.0002	.0022	.0120	.0376	.0811	.1327	.1734	.1864	.1669
	9	.0000	.0000	.0000	.0004	.0033	.0139	.0386	.0794	.1284	.1694	.1855
	10	.0000	.0000	.0000	.0001	.0008	.0042	.0149	.0385	.0771	.1248	.1669
	11	.0000	.0000	.0000	.0000	.0001	.0010	.0046	.0151	.0374	.0742	.1214
	12	.0000	.0000	.0000	.0000	.0000	.0002	.0012	.0047	.0145	.0354	.0708
	13	.0000	.0000	.0000	.0000	.0000	.0000	.0002	.0012	.0045	.0134	.0327
	14	.0000	.0000	.0000	.0000	.0000	.0000	.0000	.0002	.0011	.0039	.0117
	15	.0000	.0000	.0000	.0000	.0000	.0000	.0000	.0000	.0002	.0009	.0031
	16	.0000	.0000	.0000	.0000	.0000	.0000	.0000	.0000	.0000	.0001	.0006
	17	.0000	.0000	.0000	.0000	.0000	.0000	.0000	.0000	.0000	.0000	.0001
	18	.0000	.0000	.0000	.0000	.0000	.0000	.0000	.0000	.0000	.0000	.0000
19	0	.8262	.3774	.1351	.0456	.0144	.0042	.0011	.0003	.0001	.0000	.0000
	1	.1586	.3774	.2852	.1529	.0685	.0268	.0093	.0029	.0008	.0002	.0000
	2	.0144	.1787	.2852	.2428	.1540	.0803	.0358	.0138	.0046	.0013	.0003

n	x	.01	.05	.10	.15	.20	.25	p	.30	.35	.40	.45	.50
	3	.0008	.0533	.1796	.2428	.2182	.1517		.0869	.0422	.0175	.0062	.0018
	4	.0000	.0112	.0798	.1714	.2182	.2023		.1491	.0909	.0467	.0203	.0074
	5	.0000	.0018	.0266	.0907	.1636	.2023		.1916	.1468	.0933	.0497	.0222
	6	.0000	.0002	.0069	.0374	.0955	.1574		.1916	.1844	.1451	.0949	.0518
	7	.0000	.0000	.0014	.0122	.0443	.0974		.1525	.1844	.1797	.1443	.0961
	8	.0000	.0000	.0002	.0032	.0166	.0487		.0981	.1489	.1797	.1771	.1442
	9	.0000	.0000	.0000	.0007	.0051	.0198		.0514	.0980	.1464	.1771	.1762
	10	.0000	.0000	.0000	.0001	.0013	.0066		.0220	.0528	.0976	.1449	.1762
	11	.0000	.0000	.0000	.0000	.0003	.0018		.0077	.0233	.0532	.0970	.1442
	12	.0000	.0000	.0000	.0000	.0000	.0004		.0022	.0083	.0237	.0529	.0961
	13	.0000	.0000	.0000	.0000	.0000	.0001		.0005	.0024	.0085	.0233	.0518
	14	.0000	.0000	.0000	.0000	.0000	.0000		.0001	.0006	.0024	.0082	.0222
	15	.0000	.0000	.0000	.0000	.0000	.0000		.0000	.0001	.0005	.0022	.0074
	16	.0000	.0000	.0000	.0000	.0000	.0000		.0000	.0000	.0001	.0005	.0018
	17	.0000	.0000	.0000	.0000	.0000	.0000		.0000	.0000	.0000	.0001	.0003
	18	.0000	.0000	.0000	.0000	.0000	.0000		.0000	.0000	.0000	.0000	.0000
	19	.0000	.0000	.0000	.0000	.0000	.0000		.0000	.0000	.0000	.0000	.0000
20	0	.8179	.3585	.1216	.0388	.0115	.0032		.0008	.0002	.0000	.0000	.0000
	1	.1652	.3774	.2702	.1368	.0576	.0211		.0068	.0020	.0005	.0001	.0000
	2	.0159	.1887	.2852	.2293	.1369	.0669		.0278	.0100	.0031	.0008	.0002
	3	.0010	.0596	.1901	.2428	.2054	.1339		.0716	.0323	.0123	.0040	.0011
	4	.0000	.0133	.0898	.1821	.2182	.1897		.1304	.0738	.0350	.0139	.0046
	5	.0000	.0022	.0319	.1028	.1746	.2023		.1789	.1272	.0746	.0365	.0148
	6	.0000	.0003	.0089	.0454	.1091	.1686		.1916	.1712	.1244	.0746	.0370
	7	.0000	.0000	.0020	.0160	.0545	.1124		.1643	.1844	.1659	.1221	.0739
	8	.0000	.0000	.0004	.0046	.0222	.0609		.1144	.1614	.1797	.1623	.1201
	9	.0000	.0000	.0001	.0011	.0074	.0271		.0654	.1158	.1597	.1771	.1602
	10	.0000	.0000	.0000	.0002	.0020	.0099		.0308	.0686	.1171	.1593	.1762
	11	.0000	.0000	.0000	.0000	.0005	.0030		.0120	.0336	.0710	.1185	.1602
	12	.0000	.0000	.0000	.0000	.0001	.0008		.0039	.0136	.0355	.0727	.1201
	13	.0000	.0000	.0000	.0000	.0000	.0002		.0010	.0045	.0146	.0366	.0739
	14	.0000	.0000	.0000	.0000	.0000	.0000		.0002	.0012	.0049	.0150	.0370
	15	.0000	.0000	.0000	.0000	.0000	.0000		.0000	.0000	.0013	.0049	.0148
	16	.0000	.0000	.0000	.0000	.0000	.0000		.0000	.0000	.0003	.0013	.0046
	17	.0000	.0000	.0000	.0000	.0000	.0000		.0000	.0000	.0000	.0002	.0011
	18	.0000	.0000	.0000	.0000	.0000	.0000		.0000	.0000	.0000	.0000	.0002
	19	.0000	.0000	.0000	.0000	.0000	.0000		.0000	.0000	.0000	.0000	.0000
	20	.0000	.0000	.0000	.0000	.0000	.0000		.0000	.0000	.0000	.0000	.0000
25	0	.7778	.2774	.0718	.0172	.0038	.0008		.0001	.0000	.0000	.0000	.0000
	1	.1964	.3650	.1994	.0759	.0236	.0063		.0014	.0003	.0000	.0000	.0000
	2	.0238	.2305	.2659	.1607	.0708	.0251		.0074	.0018	.0004	.0001	.0000
	3	.0018	.0930	.2265	.2174	.1358	.0641		.0243	.0076	.0019	.0004	.0001
	4	.0001	.0269	.1384	.2110	.1867	.1175		.0572	.0224	.0071	.0018	.0004
	5	.0000	.0060	.0646	.1564	.1960	.1645		.1030	.0506	.0199	.0063	.0016
	6	.0000	.0010	.0239	.0920	.1633	.1828		.1472	.0908	.0442	.0172	.0053
	7	.0000	.0001	.0072	.0441	.1108	.1654		.1712	.1327	.0800	.0381	.0143
	8	.0000	.0000	.0018	.0175	.0623	.1241		.1651	.1607	.1200	.0701	.0322
	9	.0000	.0000	.0004	.0058	.0294	.0781		.1336	.1635	.1511	.1084	.0609
	10	.0000	.0000	.0000	.0016	.0118	.0417		.0916	.1409	.1612	.1419	.0974
	11	.0000	.0000	.0000	.0004	.0040	.0189		.0536	.1034	.1465	.1583	.1328
	12	.0000	.0000	.0000	.0000	.0012	.0074		.0268	.0650	.1140	.1511	.1550
	13	.0000	.0000	.0000	.0000	.0003	.0025		.0115	.0350	.0760	.1236	.1550
	14	.0000	.0000	.0000	.0000	.0000	.0007		.0042	.0161	.0434	.0867	.1328
	15	.0000	.0000	.0000	.0000	.0000	.0002		.0013	.0064	.0212	.0520	.0974

n	x	.01	.05	.10	.15	.20	.25	p .30	.35	.40	.45	.50
	16	.0000	.0000	.0000	.0000	.0000	.0000	.0004	.0021	.0088	.0266	.0609
	17	.0000	.0000	.0000	.0000	.0000	.0000	.0001	.0006	.0031	.0115	.0322
	18	.0000	.0000	.0000	.0000	.0000	.0000	.0000	.0001	.0009	.0042	.0143
	19	.0000	.0000	.0000	.0000	.0000	.0000	.0000	.0000	.0002	.0013	.0053
	20	.0000	.0000	.0000	.0000	.0000	.0000	.0000	.0000	.0000	.0001	.0016
	21	0000	.0000	.0000	.0000	.0000	.0000	.0000	.0000	.0000	.0000	.0004
	22	.0000	.0000	.0000	.0000	.0000	.0000	.0000	.0000	.0000	.0000	.0001
30	0	.7397	.2146	.0424	.0076	.0012	.0002	.0000	.0000	.0000	.0000	.0000
	1	.2242	.3389	.1413	.0404	.0093	.0018	.0003	.0000	.0000	.0000	.0000
	2	.0328	.2586	.2277	.1034	.0337	.0086	.0018	.0003	.0000	.0000	.0000
	3	.0031	.1270	.2361	.1703	.0785	.0269	.0072	.0015	.0003	.0000	.0000
	4	.0002	.0451	.1771	.2028	.1325	.0604	.0208	.0056	.0012	.0002	.0000
	5	.0000	.0124	.1023	.1861	.1723	.1047	.0464	.0157	.0041	.0008	.0001
	6	.0000	.0027	.0474	.1368	.1795	.1455	.0829	.0353	.0115	.0029	.0006
	7	.0000	.0005	.0180	.0828	.1538	.1662	.1219	.0652	.0263	.0081	.0019
	8	.0000	.0001	.0058	.0420	.1106	.1593	.1501	.1009	.0505	.0191	.0055
	9	.0000	.0000	.0016	.0181	.0676	.1298	.1573	.1328	.0823	.0382	.0133
	10	.0000	.0000	.0004	.0067	.0355	.0909	.1416	.1502	.1152	.0656	.0280
	11	.0000	.0000	.0001	.0022	.0161	.0551	.1103	.1471	.1396	.0976	.0509
	12	.0000	.0000	.0000	.0006	.0064	.0291	.0749	.1254	.1474	.1265	.0806
	13	.0000	.0000	.0000	.0001	.0022	.0134	.0444	.0935	.1360	.1433	.1115
	14	.0000	.0000	.0000	.0000	.0007	.0054	.0231	.0611	.1101	.1424	.1354
	15	.0000	.0000	.0000	.0000	.0002	.0019	.0106	.0351	.0783	.1242	.1445
	16	.0000	.0000	.0000	.0000	.0000	.0006	.0042	.0177	.0489	.0953	.1354
	17	.0000	.0000	.0000	.0000	.0000	.0002	.0015	.0079	.0269	.0642	.1115
	18	.0000	.0000	.0000	.0000	.0000	.0000	.0005	.0031	.0129	.0379	.0806
	19	.0000	.0000	.0000	.0000	.0000	.0000	.0001	.0010	.0054	.0196	.0509
	20	.0000	.0000	.0000	.0000	.0000	.0000	.0000	.0003	.0020	.0088	.0280
	21	.0000	.0000	.0000	.0000	.0000	.0000	.0000	.0001	.0006	.0034	.0133
	22	.0000	.0000	.0000	.0000	.0000	.0000	.0000	.0000	.0002	.0012	.0055
	23	.0000	.0000	.0000	.0000	.0000	.0000	.0000	.0000	.0000	.0003	.0019
	24	.0000	.0000	.0000	.0000	.0000	.0000	.0000	.0000	.0000	.0001	.0006
	25	.0000	.0000	.0000	.0000	.0000	.0000	.0000	.0000	.0000	.0000	.0001

*Example: $P(X = 3, n = 5, p = 0.30) = 0.1323$.

Poisson Distribution

Values of $e^{-\lambda}$

λ	$e^{-\lambda}$	λ	$e^{-\lambda}$
0.0	1.00000	2.5	.08208
0.1	.90484	2.6	.07427
0.2	.81873	2.7	.06721
0.3	.74082	2.8	.06081
0.4	.67032	2.9	.05502
0.5	.60653	3.0	.04979
0.6	.54881	3.2	.04076
0.7	.49659	3.4	.03337
0.8	.44933	3.6	.02732
0.9	.40657	3.8	.02237
1.0	.36788	4.0	.01832
1.1	.33287	4.2	.01500
1.2	.30119	4.4	.01228
1.3	.27253	4.6	.01005
1.4	.24660	4.8	.00823
1.5	.22313	5.0	.00674
1.6	.20190	5.5	.00409
1.7	.18268	6.0	.00248
1.8	.16530	6.5	.00150
1.9	.14957	7.0	.00091
2.0	.13534	7.5	.00055
2.1	.12246	8.0	.00034
2.2	.00180	8.5	.00020
2.3	.10026	9.0	.00012
2.4	.09072	10.0	.00005

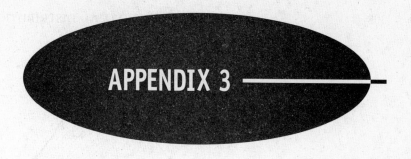

Standard Normal Distribution

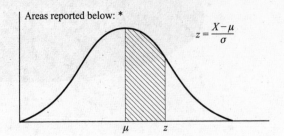

Areas reported below: *

$$z = \frac{X - \mu}{\sigma}$$

Proportions of Area for the Standard Normal Distribution

z	.00	.01	.02	.03	.04	.05	.06	.07	.08	.09
0.0	.0000	.0040	.0080	.0120	.0160	.0199	.0239	.0279	.0319	.0359
0.1	.0398	.0438	.0478	.0517	.0557	.0596	.0636	.0675	.0714	.0753
0.2	.0793	.0832	.0871	.0910	.0948	.0987	.1026	.1064	.1103	.1141
0.3	.1179	.1217	.1255	.1293	.1331	.1368	.1406	.1443	.1480	.1517
0.4	.1554	.1591	.1628	.1664	.1700	.1736	.1772	.1808	.1844	.1879
0.5	.1915	.1950	.1985	.2019	.2054	.2088	.2123	.2157	.2190	.2224
0.6	.2257	.2291	.2324	.2357	.2389	.2422	.2454	.2486	.2518	.2549
0.7	.2580	.2612	.2642	.2673	.2704	.2734	.2764	.2794	.2823	.2852
0.8	.2881	.2910	.2939	.2967	.2995	.3023	.3051	.3078	.3106	.3133
0.9	.3159	.3186	.3212	.3238	.3264	.3289	.3315	.3340	.3365	.3389
1.0	.3413	.3438	.3461	.3485	.3508	.3531	.3554	.3577	.3599	.3621
1.1	.3643	.3665	.3686	.3708	.3729	.3749	.3770	.3790	.3810	.3830
1.2	.3849	.3869	.3888	.3907	.3925	.3944	.3962	.3980	.3997	.4014
1.3	.4032	.4049	.4066	.4082	.4099	.4115	.4131	.4147	.4162	.4177
1.4	.4192	.4207	.4222	.4236	.4251	.4265	.4279	.4292	.4306	.4319
1.5	.4332	.4345	.4357	.4370	.4382	.4394	.4406	.4418	.4429	.4441
1.6	.4452	.4463	.4474	.4484	.4495	.4505	.4515	.4525	.4535	.4545
1.7	.4554	.4564	.4573	.4582	.4591	.4599	.4608	.4616	.4625	.4633
1.8	.4641	.4649	.4656	.4664	.4671	.4678	.4686	.4693	.4699	.4706
1.9	.4713	.4719	.4726	.4732	.4738	.4744	.4750	.4756	.4761	.4767
2.0	.4772	.4778	.4783	.4788	.4793	.4798	.4803	.4808	.4812	.4817
2.1	.4821	.4826	.4830	.4834	.4838	.4842	.4846	.4850	.4854	.4857
2.2	.4861	.4864	.4868	.4871	.4875	.4878	.4881	.4884	.4887	.4890
2.3	.4893	.4896	.4898	.4901	.4904	.4906	.4909	.4911	.4913	.4916
2.4	.4918	.4920	.4922	.4925	.4927	.4929	.4931	.4932	.4934	.4936
2.5	.4938	.4940	.4941	.4943	.4945	.4946	.4948	.4949	.4951	.4952
2.6	.4953	.4955	.4956	.4957	.4959	.4960	.4961	.4962	.4963	.4964
2.7	.4965	.4966	.4967	.4968	.4969	.4970	.4971	.4972	.4973	.4974
2.8	.4974	.4975	.4976	.4977	.4977	.4978	.4979	.4979	.4980	.4981
2.9	.4981	.4982	.4983	.4983	.4984	.4984	.4985	.4985	.4986	.4986
3.0	.4987									
3.5	.4997									
4.0	.4999									

*Example: For z = 1.96, shaded area is 0.4750 out of the total area of 1.0000.

Table of Random Numbers

10097	85017	84532	13618	23157	86952	02438	76520	91499	38631	79430	64241	97959	67422	69992	68479
37542	16719	82789	69041	05545	44109	05403	64894	80336	49172	16332	44670	35089	17691	89246	26940
08422	65842	27672	82186	14871	22115	86529	19645	44104	89232	57327	34679	62235	79655	81336	85157
99019	76875	20684	39187	38976	94324	43204	09376	12550	02844	15026	32439	58537	48274	81330	11100
12807	93640	39160	41453	97312	41548	93137	80157	63606	40387	65406	37920	08709	60623	02237	16505
66065	99478	70086	71265	11742	18226	29004	34072	61196	80240	44177	51171	08723	39323	05798	26457
31060	65119	26486	47353	43361	99436	42753	45571	15474	44910	99321	72173	56239	04595	10836	95270
85269	70322	21592	48233	93806	32584	21828	02051	94557	33663	86347	00926	44915	34823	51770	67897
63573	58133	41278	11697	49540	61777	67954	05325	42481	86430	19102	37420	41976	76559	24358	97344
73796	44655	81255	31133	36768	60452	38537	03529	23523	31379	68588	81675	15694	43438	36879	73208
98520	02295	13487	98662	07092	44673	61303	14905	04493	98086	32533	17767	14523	52494	24826	75246
11805	85035	54881	35587	43310	48897	48493	39808	00549	33185	04805	05431	94598	97654	16232	64051
83452	01197	86935	28021	61570	23350	65710	06288	35963	80951	68953	99634	81949	15307	00406	26898
88685	97907	19078	40646	31352	48625	44369	86507	59808	79752	02529	40200	73742	08391	49140	45427
99594	63268	96905	28797	57048	46359	74294	87517	46058	18633	99970	67348	49329	95236	32537	01390
65481	52841	59684	67411	09243	56092	84369	17468	32179	74029	74717	17674	90446	00597	45240	87379
80124	53722	71399	10916	07959	21225	13018	17727	69234	54178	10805	35635	45266	61406	41941	20117
74350	11434	51908	62171	93732	26958	02400	77402	19565	11664	77602	99817	28573	41430	96382	01758
69916	62375	99292	21177	72721	66995	07289	66252	45155	48324	32135	26803	16213	14938	71961	19476
09893	28337	20923	87929	61020	62841	31374	14225	94864	69074	45753	20505	78317	31994	98145	36168

Student's
t Distribution

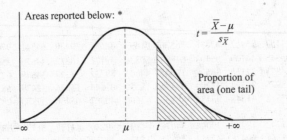

Areas reported below: *

$$t = \frac{\bar{X} - \mu}{s_{\bar{X}}}$$

Proportion of area (one tail)

$-\infty \qquad \mu \quad t \qquad +\infty$

Proportions of Area for the t Distributions

df	0.10	0.05	0.025	0.01	0.005	df	0.10	0.05	0.025	0.01	0.005
1	3.078	6.314	12.706	31.821	63.657	18	1.330	1.734	2.101	2.552	2.878
2	1.886	2.920	4.303	6.965	9.925	19	1.328	1.729	2.093	2.539	2.861
3	1.638	2.353	3.182	4.541	5.841	20	1.325	1.725	2.086	2.528	2.845
4	1.533	2.132	2.776	3.747	4.604	21	1.323	1.721	2.080	2.518	2.831
5	1.476	2.015	2.571	3.365	4.032	22	1.321	1.717	2.074	2.508	2.819
6	1.440	1.943	2.447	3.143	3.707	23	1.319	1.714	2.069	2.500	2.807
7	1.415	1.895	2.365	2.998	3.499	24	1.318	1.711	2.064	2.492	2.797
8	1.397	1.860	2.306	2.896	3.355	25	1.316	1.708	2.060	2.485	2.787
9	1.383	1.833	2.262	2.821	3.250	26	1.315	1.706	2.056	2.479	2.779
10	1.372	1.812	2.228	2.764	3.169	27	1.314	1.703	2.052	2.473	2.771
11	1.363	1.796	2.201	2.718	3.106	28	1.313	1.701	2.048	2.467	2.763
12	1.356	1.782	2.179	2.681	3.055	29	1.311	1.699	2.045	2.462	2.756
13	1.350	1.771	2.160	2.650	3.012	30	1.310	1.697	2.042	2.457	2.750
14	1.345	1.761	2.145	2.624	2.977	40	1.303	1.684	2.021	2.423	2.704
15	1.341	1.753	2.131	2.602	2.947	60	1.296	1.671	2.000	2.390	2.660
16	1.337	1.746	2.120	2.583	2.921	120	1.289	1.658	1.980	2.358	2.617
17	1.333	1.740	2.110	2.567	2.898	∞	1.282	1.645	1.960	2.326	2.576

*Example: For the shaded area to represent 0.05 of the total area of 1.0, value of t with 10 degrees of freedom is 1.812

Source: From Table III of Fisher and Yates, *Statistical Tables for Biological, Agricultural and Medical Research*, 6th ed., 1974, published by Longman Group Ltd., London (previously published by Oliver & Boyd, Edinburgh), by permission of the authors and publishers.

Chi-Square Distribution

Areas reported below: *

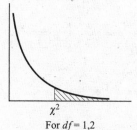

For df = 1,2   For df ≥ 3

Proportions of Area for the χ^2 Distributions

df	\multicolumn{11}{c}{Proportion of Area}
	0.995	0.990	0.975	0.950	0.900	0.500	0.100	0.050	0.025	0.010	0.005
1	0.00004	0.00016	0.00098	0.00393	0.0158	0.455	2.71	3.84	5.02	6.63	7.88
2	0.0100	0.0201	0.0506	0.103	0.211	1.386	4.61	5.99	7.38	9.21	10.60
3	0.072	0.115	0.216	0.352	0.584	2.366	6.25	7.81	9.35	11.34	12.84
4	0.207	0.297	0.484	0.711	1.064	3.357	7.78	9.49	11.14	13.28	14.86
5	0.412	0.554	0.831	1.145	1.61	4.251	9.24	11.07	12.83	15.09	16.75
6	0.676	0.872	1.24	1.64	2.20	5.35	10.64	12.59	14.45	16.81	18.55
7	0.989	1.24	1.69	2.17	2.83	6.35	12.02	14.07	16.01	18.48	20.28
8	1.34	1.65	2.18	2.73	3.49	7.34	13.36	15.51	17.53	20.09	21.96
9	1.73	2.09	2.70	3.33	4.17	8.34	14.68	16.92	19.02	21.67	23.59
10	2.16	2.56	3.25	3.94	4.87	9.34	15.99	18.31	20.48	23.21	25.19
11	2.60	3.05	3.82	4.57	5.58	10.34	17.28	19.68	21.92	24.73	26.76
12	3.07	3.57	4.40	5.23	6.30	11.34	18.55	21.03	23.34	26.22	28.30
13	3.57	4.11	5.01	5.89	7.04	12.34	19.81	22.36	24.74	27.69	29.82
14	4.07	4.66	5.63	6.57	7.79	13.34	21.06	23.68	26.12	29.14	31.32
15	4.60	5.23	6.26	7.26	8.55	14.34	22.31	25.00	27.49	30.58	32.80
16	5.14	5.81	6.91	7.96	9.31	15.34	23.54	26.30	28.85	32.00	34.27
17	5.70	6.41	7.56	8.67	10.09	16.34	24.77	27.59	30.19	33.41	35.72
18	6.26	7.01	8.23	9.39	10.86	17.34	25.99	28.87	31.53	34.81	37.16
19	6.84	7.63	8.91	10.12	11.65	18.34	27.20	30.14	32.85	36.19	38.58
20	7.43	8.26	9.59	10.85	12.44	19.34	28.41	31.41	34.17	37.57	40.00

df					Proportion of Area						
	0.995	0.990	0.975	0.950	0.900	0.500	0.100	0.050	0.025	0.010	0.005
21	8.03	8.90	10.28	11.59	13.24	20.34	29.62	32.67	35.48	38.93	41.40
22	8.64	9.54	10.98	12.34	14.04	21.34	30.81	33.92	36.78	40.29	42.80
23	9.26	10.20	11.69	13.09	14.85	22.34	32.01	35.17	38.08	41.64	44.18
24	9.89	10.86	12.40	13.85	15.66	23.34	33.20	36.42	39.36	42.98	45.56
25	10.52	11.52	13.12	14.61	16.47	24.34	34.38	37.65	40.65	44.31	46.93
26	11.16	12.20	13.84	15.38	17.29	25.34	35.56	38.89	41.92	45.64	48.29
27	11.81	12.83	14.57	16.15	18.11	26.34	36.74	40.11	43.19	46.96	49.64
28	12.46	13.56	15.31	16.93	18.94	27.34	37.92	41.34	44.46	48.28	50.99
29	13.12	14.26	16.05	17.71	19.77	28.34	39.09	42.56	45.72	49.59	52.34
30	13.79	14.95	16.79	18.49	20.60	29.34	40.26	43.77	46.98	50.89	53.67
40	20.71	22.16	24.43	26.51	29.05	39.34	51.81	55.76	59.34	63.69	66.77
50	27.99	29.71	32.36	34.76	37.69	49.33	63.17	67.50	71.42	76.15	79.49
60	35.53	37.43	40.48	43.19	46.46	59.33	74.40	79.08	83.30	88.38	91.95
70	43.28	45.44	48.76	51.74	55.33	69.33	85.53	90.53	95.02	100.4	104.2
80	51.17	53.54	51.17	60.39	64.28	79.33	98.58	101.9	106.6	112.3	116.3
90	59.20	61.75	65.65	69.13	73.29	89.33	107.6	113.1	118.1	124.1	128.3
100	67.33	70.06	74.22	77.93	82.36	99.33	118.5	124.3	129.6	135.8	140.2

*Example: For the shaded area to represent 0.05 of the total area of 1.0 under the density function, the value of x^2 is 18.31 when df = 10.

Source: From Table IV of Fisher and Yates, *Statistical Tables for Biological, Agricultural and Medical Research*, 6th ed., 1974, published by Longman Group Ltd., London (previously published by Oliver & Boyd, Edinburgh), by permission of the authors and publishers.

F Distribution

Values of F Exceeded with Probabilities of 5 and 1 Percent

Each cell lists the 5% value (top) / 1% value (bottom).

df (denominator)	1	2	3	4	5	6	7	8	9	10	11	12	14	16	20	24	30	40	50	75	100	200	500	∞
1	161 / 4,052	200 / 4,999	216 / 5,403	225 / 5,625	230 / 5,764	234 / 5,859	237 / 5,928	239 / 5,981	241 / 6,022	242 / 6,056	243 / 6,082	244 / 6,106	245 / 6,142	246 / 6,169	248 / 6,208	249 / 6,234	250 / 6,261	251 / 6,286	252 / 6,302	253 / 6,323	253 / 6,334	254 / 6,352	254 / 6,361	254 / 6,366
2	18.51 / 98.49	19.00 / 99.00	19.16 / 99.17	19.25 / 99.25	19.30 / 99.30	19.33 / 99.33	19.36 / 99.36	19.37 / 99.37	19.38 / 99.39	19.39 / 99.40	19.40 / 99.41	19.41 / 99.42	19.42 / 99.43	19.43 / 99.44	19.44 / 99.45	19.45 / 99.46	19.46 / 99.47	19.47 / 99.48	19.47 / 99.48	19.48 / 99.49	19.49 / 99.49	19.49 / 99.49	19.50 / 99.50	19.50 / 99.50
3	10.13 / 34.12	9.55 / 30.82	9.28 / 29.46	9.12 / 28.71	9.01 / 28.24	8.94 / 27.91	8.88 / 27.67	8.84 / 27.49	8.81 / 27.34	8.78 / 27.23	8.76 / 27.13	8.74 / 27.05	8.71 / 26.92	8.69 / 26.83	8.66 / 26.69	8.64 / 26.60	8.62 / 26.50	8.60 / 26.41	8.58 / 26.35	8.57 / 26.27	8.56 / 26.23	8.54 / 26.18	8.54 / 26.14	8.53 / 26.12
4	7.71 / 21.20	6.94 / 18.00	6.59 / 16.69	6.39 / 15.98	6.26 / 15.52	6.16 / 15.21	6.09 / 14.98	6.04 / 14.80	6.00 / 14.66	5.96 / 14.54	5.93 / 14.45	5.91 / 14.37	5.87 / 14.24	5.84 / 14.15	5.80 / 14.02	5.77 / 13.93	5.74 / 13.83	5.71 / 13.74	5.70 / 13.69	5.68 / 13.61	5.66 / 13.57	5.65 / 13.52	5.64 / 13.48	5.63 / 13.46
5	6.61 / 16.26	5.79 / 13.27	5.41 / 12.06	5.19 / 11.39	5.05 / 10.97	4.95 / 10.67	4.88 / 10.45	4.82 / 10.29	4.78 / 10.15	4.74 / 10.05	4.70 / 9.96	4.68 / 9.89	4.64 / 9.77	4.60 / 9.68	4.56 / 9.55	4.53 / 9.47	4.50 / 9.38	4.46 / 9.29	4.44 / 9.24	4.42 / 9.17	4.40 / 9.13	4.38 / 9.07	4.37 / 9.04	4.36 / 9.02
6	5.99 / 13.74	5.14 / 10.92	4.76 / 9.78	4.53 / 9.15	4.39 / 8.75	4.28 / 8.47	4.21 / 8.26	4.15 / 8.10	4.10 / 7.98	4.06 / 7.87	4.03 / 7.79	4.00 / 7.72	3.96 / 7.60	3.92 / 7.52	3.87 / 7.39	3.84 / 7.31	3.81 / 7.23	3.77 / 7.14	3.75 / 7.09	3.72 / 7.02	3.71 / 6.99	3.69 / 6.94	3.68 / 6.90	3.67 / 6.88
7	5.59 / 12.25	4.74 / 9.55	4.34 / 8.45	4.12 / 7.85	3.97 / 7.46	3.87 / 7.19	3.79 / 7.00	3.73 / 6.84	3.68 / 6.71	3.63 / 6.62	3.60 / 6.54	3.57 / 6.47	3.52 / 6.35	3.49 / 6.27	3.44 / 6.15	3.41 / 6.07	3.38 / 5.98	3.34 / 5.90	3.32 / 5.85	3.29 / 5.78	3.28 / 5.75	3.25 / 5.70	3.24 / 5.67	3.23 / 5.65
8	5.32 / 11.26	4.46 / 8.65	4.07 / 7.59	3.84 / 7.01	3.69 / 6.63	3.58 / 6.37	3.50 / 6.19	3.44 / 6.03	3.39 / 5.91	3.34 / 5.82	3.31 / 5.74	3.28 / 5.67	3.23 / 5.56	3.20 / 5.48	3.15 / 5.36	3.12 / 5.28	3.08 / 5.20	3.05 / 5.11	3.03 / 5.06	3.00 / 5.00	2.98 / 4.96	2.96 / 4.91	2.94 / 4.88	2.93 / 4.86
9	5.12 / 10.56	4.26 / 8.02	3.86 / 6.99	3.63 / 6.42	3.48 / 6.06	3.37 / 5.80	3.29 / 5.62	3.23 / 5.47	3.18 / 5.35	3.13 / 5.26	3.10 / 5.18	3.07 / 5.11	3.02 / 5.00	2.98 / 4.92	2.93 / 4.80	2.90 / 4.73	2.86 / 4.64	2.82 / 4.56	2.80 / 4.51	2.77 / 4.45	2.76 / 4.41	2.73 / 4.36	2.72 / 4.33	2.71 / 4.31
10	4.96 / 10.04	4.10 / 7.56	3.71 / 6.55	3.48 / 5.99	3.33 / 5.64	3.22 / 5.39	3.14 / 5.21	3.07 / 5.06	3.02 / 4.95	2.97 / 4.85	2.94 / 4.78	2.91 / 4.71	2.86 / 4.60	2.82 / 4.52	2.77 / 4.41	2.74 / 4.33	2.70 / 4.25	2.67 / 4.17	2.64 / 4.12	2.61 / 4.05	2.59 / 4.01	2.56 / 3.96	2.55 / 3.93	2.54 / 3.91
11	4.84 / 9.65	3.98 / 7.20	3.59 / 6.22	3.36 / 5.67	3.20 / 5.32	3.09 / 5.07	3.01 / 4.88	2.95 / 4.74	2.90 / 4.63	2.86 / 4.54	2.82 / 4.46	2.79 / 4.40	2.74 / 4.29	2.70 / 4.21	2.65 / 4.10	2.61 / 4.02	2.57 / 3.94	2.53 / 3.86	2.50 / 3.80	2.47 / 3.74	2.45 / 3.70	2.42 / 3.66	2.41 / 3.62	2.40 / 3.60
12	4.75 / 9.33	3.88 / 6.93	3.49 / 5.95	3.26 / 5.41	3.11 / 5.06	3.00 / 4.82	2.92 / 4.65	2.85 / 4.50	2.80 / 4.39	2.76 / 4.30	2.72 / 4.22	2.69 / 4.16	2.64 / 4.05	2.60 / 3.98	2.54 / 3.86	2.50 / 3.78	2.46 / 3.70	2.42 / 3.61	2.40 / 3.56	2.36 / 3.49	2.35 / 3.46	2.32 / 3.41	2.31 / 3.38	2.30 / 3.36
13	4.67 / 9.07	3.80 / 6.70	3.41 / 5.74	3.18 / 5.20	3.02 / 4.86	2.92 / 4.62	2.84 / 4.44	2.77 / 4.30	2.72 / 4.19	2.67 / 4.10	2.63 / 4.02	2.60 / 3.96	2.55 / 3.85	2.51 / 3.78	2.46 / 3.67	2.42 / 3.59	2.38 / 3.51	2.34 / 3.42	2.32 / 3.37	2.28 / 3.30	2.26 / 3.27	2.24 / 3.21	2.22 / 3.18	2.21 / 3.16
14	4.60 / 8.86	3.74 / 6.51	3.34 / 5.56	3.11 / 5.03	2.96 / 4.69	2.85 / 4.46	2.77 / 4.28	2.70 / 4.14	2.65 / 4.03	2.60 / 3.94	2.56 / 3.86	2.53 / 3.80	2.48 / 3.70	2.44 / 3.62	2.39 / 3.51	2.35 / 3.43	2.31 / 3.34	2.27 / 3.26	2.24 / 3.21	2.21 / 3.14	2.19 / 3.11	2.16 / 3.06	2.14 / 3.02	2.13 / 3.00
15	4.54 / 8.68	3.68 / 6.36	3.29 / 5.42	3.06 / 4.89	2.90 / 4.56	2.79 / 4.32	2.70 / 4.14	2.64 / 4.00	2.59 / 3.89	2.55 / 3.80	2.51 / 3.73	2.48 / 3.67	2.43 / 3.56	2.39 / 3.48	2.33 / 3.36	2.29 / 3.29	2.25 / 3.20	2.21 / 3.12	2.18 / 3.07	2.15 / 3.00	2.12 / 2.97	2.10 / 2.92	2.08 / 2.89	2.07 / 2.87
16	4.49 / 8.53	3.63 / 6.23	3.24 / 5.29	3.01 / 4.77	2.85 / 4.44	2.74 / 4.20	2.66 / 4.03	2.59 / 3.89	2.54 / 3.78	2.49 / 3.69	2.45 / 3.61	2.42 / 3.55	2.37 / 3.45	2.33 / 3.37	2.28 / 3.25	2.24 / 3.18	2.20 / 3.10	2.16 / 3.01	2.13 / 2.96	2.09 / 2.98	2.07 / 2.86	2.04 / 2.80	2.02 / 2.77	2.01 / 2.75

(Continued)

(*Appendix 7 continued*)

df (numerator)

df (denom.)	1	2	3	4	5	6	7	8	9	10	11	12	14	16	20	24	30	40	50	75	100	200	500	∞
17	4.45 **8.40**	3.59 **6.11**	3.20 **5.18**	2.96 **4.67**	2.81 **4.34**	2.70 **4.10**	2.62 **3.93**	2.55 **3.79**	2.50 **3.68**	2.45 **3.59**	2.41 **3.52**	2.38 **3.45**	2.33 **3.35**	2.29 **3.27**	2.23 **3.16**	2.19 **3.08**	2.15 **3.00**	2.11 **2.92**	2.08 **2.86**	2.04 **2.79**	2.02 **2.76**	1.99 **2.70**	1.97 **2.67**	1.96 **2.65**
18	4.41 **8.28**	3.55 **6.01**	3.16 **5.09**	2.93 **4.58**	2.77 **4.25**	2.66 **4.01**	2.58 **3.85**	2.51 **3.71**	2.46 **3.60**	2.41 **3.51**	2.37 **3.44**	2.34 **3.37**	2.29 **3.27**	2.25 **3.19**	2.19 **3.07**	2.15 **3.00**	2.11 **2.91**	2.07 **2.83**	2.04 **2.78**	2.00 **2.71**	1.98 **2.68**	1.95 **2.62**	1.93 **2.59**	1.92 **2.57**
19	4.38 **8.18**	3.52 **5.93**	3.13 **5.01**	2.90 **4.50**	2.74 **4.17**	2.63 **3.94**	2.55 **3.77**	2.48 **3.63**	2.43 **3.52**	2.38 **3.43**	2.34 **3.36**	2.31 **3.30**	2.26 **3.19**	2.21 **3.12**	2.15 **3.00**	2.11 **2.92**	2.07 **2.84**	2.02 **2.76**	2.00 **2.70**	1.96 **2.63**	1.94 **2.60**	1.91 **2.54**	1.90 **2.51**	1.88 **2.49**
20	4.35 **8.10**	3.49 **5.85**	3.10 **4.94**	2.87 **4.43**	2.71 **4.10**	2.60 **3.87**	2.52 **3.71**	2.45 **3.56**	2.40 **3.45**	2.35 **3.37**	2.31 **3.30**	2.28 **3.23**	2.23 **3.13**	2.18 **3.05**	2.12 **2.94**	2.08 **2.86**	2.04 **2.77**	1.99 **2.69**	1.96 **2.63**	1.92 **2.56**	1.90 **2.53**	1.87 **2.47**	1.85 **2.44**	1.84 **2.42**
21	4.32 **8.02**	3.47 **5.78**	3.07 **4.87**	2.84 **4.37**	2.68 **4.04**	2.57 **3.81**	2.49 **3.65**	2.42 **3.51**	2.37 **3.40**	2.32 **3.31**	2.28 **3.24**	2.25 **3.17**	2.20 **3.07**	2.15 **2.99**	2.09 **2.88**	2.05 **2.80**	2.00 **2.72**	1.96 **2.63**	1.93 **2.58**	1.89 **2.51**	1.87 **2.47**	1.84 **2.42**	1.82 **2.38**	1.81 **2.36**
22	4.30 **7.94**	3.44 **5.72**	3.05 **4.82**	2.82 **4.31**	2.66 **3.99**	2.55 **3.76**	2.47 **3.59**	2.40 **3.45**	2.35 **3.35**	2.30 **3.26**	2.26 **3.18**	2.23 **3.12**	2.18 **3.02**	2.13 **2.94**	2.07 **2.83**	2.03 **2.75**	1.98 **2.67**	1.93 **2.58**	1.91 **2.53**	1.87 **2.46**	1.84 **2.42**	1.81 **2.37**	1.80 **2.33**	1.78 **2.31**
23	4.28 **7.88**	3.42 **5.66**	3.03 **4.76**	2.80 **4.26**	2.64 **3.94**	2.53 **3.71**	2.45 **3.54**	2.38 **3.41**	2.32 **3.30**	2.28 **3.21**	2.24 **3.14**	2.20 **3.07**	2.14 **2.97**	2.10 **2.89**	2.04 **2.78**	2.00 **2.70**	1.96 **2.62**	1.91 **2.53**	1.88 **2.48**	1.84 **2.41**	1.82 **2.37**	1.79 **2.32**	1.77 **2.28**	1.76 **2.26**
24	4.26 **7.82**	3.40 **5.61**	3.01 **4.72**	2.78 **4.22**	2.62 **3.90**	2.51 **3.67**	2.43 **3.50**	2.36 **3.36**	2.30 **3.25**	2.26 **3.17**	2.22 **3.09**	2.18 **3.03**	2.13 **2.93**	2.09 **2.85**	2.02 **2.74**	1.98 **2.66**	1.94 **2.58**	1.89 **2.49**	1.86 **2.44**	1.82 **2.36**	1.80 **2.33**	1.76 **2.27**	1.74 **2.23**	1.73 **2.21**
25	4.24 **7.77**	3.38 **5.57**	2.99 **4.68**	2.76 **4.18**	2.60 **3.86**	2.49 **3.63**	2.41 **3.46**	2.34 **3.32**	2.28 **3.21**	2.24 **3.13**	2.20 **3.05**	2.16 **2.99**	2.11 **2.89**	2.06 **2.81**	2.00 **2.70**	1.96 **2.62**	1.92 **2.54**	1.87 **2.45**	1.84 **2.40**	1.80 **2.32**	1.77 **2.29**	1.74 **2.23**	1.72 **2.19**	1.71 **2.17**
26	4.22 **7.72**	3.37 **5.53**	2.98 **4.64**	2.74 **4.14**	2.59 **3.82**	2.47 **3.59**	2.39 **3.42**	2.32 **3.29**	2.27 **3.17**	2.22 **3.09**	2.18 **3.02**	2.15 **2.96**	2.10 **2.86**	2.05 **2.77**	1.99 **2.66**	1.95 **2.58**	1.90 **2.50**	1.85 **2.41**	1.82 **2.36**	1.78 **2.28**	1.76 **2.25**	1.72 **2.19**	1.70 **2.15**	1.69 **2.13**
27	4.21 **7.68**	3.35 **5.49**	2.96 **4.60**	2.73 **4.11**	2.57 **3.79**	2.46 **3.56**	2.37 **3.39**	2.30 **3.26**	2.25 **3.14**	2.20 **3.06**	2.16 **2.98**	2.13 **2.93**	2.08 **2.83**	2.03 **2.74**	1.97 **2.63**	1.93 **2.55**	1.88 **2.47**	1.84 **2.38**	1.80 **2.33**	1.76 **2.25**	1.74 **2.21**	1.71 **2.16**	1.68 **2.12**	1.67 **2.10**
28	4.20 **7.64**	3.34 **5.45**	2.95 **4.57**	2.71 **4.07**	2.56 **3.76**	2.44 **3.53**	2.36 **3.36**	2.29 **3.23**	2.24 **3.11**	2.19 **3.03**	2.15 **2.95**	2.12 **2.90**	2.06 **2.80**	2.02 **2.71**	1.96 **2.60**	1.91 **2.52**	1.87 **2.44**	1.81 **2.35**	1.78 **2.30**	1.75 **2.22**	1.72 **2.18**	1.69 **2.13**	1.67 **2.09**	1.65 **2.06**
29	4.18 **7.60**	3.33 **5.42**	2.93 **4.54**	2.70 **4.04**	2.54 **3.73**	2.43 **3.50**	2.35 **3.33**	2.28 **3.20**	2.22 **3.08**	2.18 **3.00**	2.14 **2.92**	2.10 **2.87**	2.05 **2.77**	2.00 **2.68**	1.94 **2.57**	1.90 **2.49**	1.85 **2.41**	1.80 **2.32**	1.77 **2.27**	1.73 **2.19**	1.71 **2.15**	1.69 **2.10**	1.65 **2.06**	1.64 **2.03**
30	4.17 **7.56**	3.32 **5.39**	2.92 **4.51**	2.69 **4.02**	2.53 **3.70**	2.42 **3.47**	2.34 **3.30**	2.27 **3.17**	2.21 **3.06**	2.16 **2.98**	2.12 **2.90**	2.09 **2.84**	2.04 **2.74**	1.99 **2.66**	1.93 **2.55**	1.89 **2.47**	1.84 **2.38**	1.79 **2.29**	1.76 **2.24**	1.72 **2.16**	1.69 **2.13**	1.66 **2.07**	1.64 **2.03**	1.62 **2.01**
32	4.15 **7.50**	3.30 **5.34**	2.90 **4.46**	2.67 **3.97**	2.51 **3.66**	2.40 **3.42**	2.32 **3.25**	2.25 **3.12**	2.19 **3.01**	2.14 **2.94**	2.10 **2.86**	2.07 **2.80**	2.02 **2.70**	1.97 **2.62**	1.91 **2.51**	1.86 **2.42**	1.82 **2.34**	1.76 **2.25**	1.74 **2.20**	1.69 **2.12**	1.67 **2.08**	1.64 **2.02**	1.61 **1.98**	1.59 **1.96**
34	4.13 **7.44**	3.28 **5.29**	2.88 **4.42**	2.65 **3.93**	2.49 **3.61**	2.38 **3.38**	2.30 **3.21**	2.23 **3.08**	2.17 **2.97**	2.12 **2.89**	2.08 **2.82**	2.05 **2.76**	2.00 **2.66**	1.95 **2.58**	1.89 **2.47**	1.84 **2.38**	1.80 **2.30**	1.74 **2.21**	1.71 **2.15**	1.67 **2.08**	1.64 **2.04**	1.61 **1.98**	1.59 **1.94**	1.57 **1.91**
36	4.11 **7.39**	3.26 **5.25**	2.86 **4.38**	2.63 **3.89**	2.48 **3.58**	2.36 **3.35**	2.28 **3.18**	2.21 **3.04**	2.15 **2.94**	2.10 **2.86**	2.06 **2.78**	2.03 **2.72**	1.98 **2.62**	1.93 **2.54**	1.87 **2.43**	1.82 **2.35**	1.78 **2.26**	1.72 **2.17**	1.69 **2.12**	1.65 **2.04**	1.62 **2.00**	1.59 **1.94**	1.56 **1.90**	1.55 **1.87**

(*Continued*)

(Appendix 7 continued)

df (numerator)

df (denom.)	1	2	3	4	5	6	7	8	9	10	11	12	14	16	20	24	30	40	50	75	100	200	500	∞
38	4.10 / 7.35	3.25 / 5.21	2.85 / 4.34	2.62 / 3.86	2.46 / 3.54	2.35 / 3.32	2.26 / 3.15	2.19 / 3.02	2.14 / 2.91	2.09 / 2.82	2.05 / 2.75	2.02 / 2.69	1.96 / 2.59	1.92 / 2.51	1.85 / 2.40	1.80 / 2.32	1.76 / 2.22	1.71 / 2.14	1.67 / 2.08	1.63 / 2.00	1.60 / 1.97	1.57 / 1.90	1.54 / 1.86	1.53 / 1.84
40	4.07 / 7.31	3.23 / 5.18	2.84 / 4.31	2.61 / 3.83	2.45 / 3.51	2.34 / 3.29	2.25 / 3.12	2.18 / 2.99	2.12 / 2.88	2.07 / 2.80	2.04 / 2.73	2.00 / 2.66	1.95 / 2.56	1.90 / 2.49	1.84 / 2.37	1.79 / 2.29	1.74 / 2.20	1.69 / 2.11	1.66 / 2.05	1.61 / 1.97	1.59 / 1.94	1.55 / 1.88	1.53 / 1.84	1.51 / 1.81
42	4.07 / 7.27	3.22 / 5.15	2.83 / 4.29	2.59 / 3.80	2.44 / 3.49	2.32 / 3.26	2.24 / 3.10	2.17 / 2.96	2.11 / 2.86	2.06 / 2.77	2.02 / 2.70	1.99 / 2.64	1.94 / 2.54	1.89 / 2.46	1.82 / 2.35	1.78 / 2.26	1.73 / 2.17	1.68 / 2.08	1.64 / 2.02	1.60 / 1.94	1.57 / 1.91	1.54 / 1.85	1.51 / 1.80	1.49 / 1.78
44	4.06 / 7.24	3.21 / 5.12	2.82 / 4.26	2.58 / 3.78	2.43 / 3.46	2.31 / 3.24	2.23 / 3.07	2.16 / 2.94	2.10 / 2.84	2.05 / 2.75	2.01 / 2.68	1.98 / 2.62	1.92 / 2.52	1.88 / 2.44	1.81 / 2.32	1.76 / 2.24	1.72 / 2.15	1.66 / 2.06	1.63 / 2.00	1.58 / 1.92	1.56 / 1.88	1.52 / 1.82	1.50 / 1.78	1.48 / 1.75
46	4.05 / 7.21	3.20 / 5.10	2.81 / 4.24	2.57 / 3.76	2.42 / 3.44	2.30 / 3.22	2.22 / 3.05	2.14 / 2.92	2.09 / 2.82	2.04 / 2.73	2.00 / 2.66	1.97 / 2.60	1.91 / 2.50	1.87 / 2.42	1.80 / 2.30	1.75 / 2.22	1.71 / 2.13	1.65 / 2.04	1.62 / 1.98	1.57 / 1.90	1.54 / 1.86	1.51 / 1.80	1.48 / 1.76	1.46 / 1.72
48	4.04 / 7.19	3.19 / 5.08	2.80 / 4.22	2.56 / 3.74	2.41 / 3.42	2.30 / 3.20	2.21 / 3.04	2.14 / 2.90	2.08 / 2.80	2.03 / 2.71	1.99 / 2.64	1.96 / 2.58	1.90 / 2.48	1.86 / 2.40	1.79 / 2.28	1.74 / 2.20	1.70 / 2.11	1.64 / 2.02	1.61 / 1.96	1.56 / 1.88	1.53 / 1.84	1.50 / 1.78	1.47 / 1.73	1.45 / 1.70
50	4.03 / 7.17	3.18 / 5.06	2.79 / 4.20	2.56 / 3.72	2.40 / 3.41	2.29 / 3.18	2.20 / 3.02	2.13 / 2.88	2.07 / 2.78	2.02 / 2.70	1.98 / 2.62	1.95 / 2.56	1.90 / 2.46	1.85 / 2.39	1.78 / 2.26	1.74 / 2.18	1.69 / 2.10	1.63 / 2.00	1.60 / 1.94	1.55 / 1.86	1.52 / 1.82	1.48 / 1.76	1.46 / 1.71	1.44 / 1.68
60	4.00 / 7.08	3.15 / 4.98	2.76 / 4.13	2.52 / 3.65	2.37 / 3.34	2.25 / 3.12	2.17 / 2.95	2.10 / 2.82	2.04 / 2.72	1.99 / 2.63	1.95 / 2.56	1.92 / 2.50	1.86 / 2.40	1.81 / 2.32	1.75 / 2.20	1.70 / 2.12	1.65 / 2.03	1.59 / 1.93	1.56 / 1.87	1.50 / 1.79	1.48 / 1.74	1.44 / 1.68	1.41 / 1.63	1.39 / 1.60
70	3.98 / 7.01	3.13 / 4.92	2.74 / 4.08	2.50 / 3.60	2.35 / 3.29	2.23 / 3.07	2.14 / 2.91	2.07 / 2.77	2.01 / 2.67	1.97 / 2.59	1.93 / 2.51	1.89 / 2.45	1.84 / 2.35	1.79 / 2.28	1.72 / 2.15	1.67 / 2.07	1.62 / 1.98	1.56 / 1.88	1.53 / 1.82	1.47 / 1.74	1.45 / 1.69	1.40 / 1.62	1.37 / 1.56	1.35 / 1.53
80	3.96 / 6.96	3.11 / 4.88	2.72 / 4.04	2.48 / 3.56	2.33 / 3.25	2.21 / 3.04	2.12 / 2.87	2.05 / 2.74	1.99 / 2.64	1.95 / 2.55	1.91 / 2.48	1.88 / 2.41	1.82 / 2.32	1.77 / 2.24	1.70 / 2.11	1.65 / 2.03	1.60 / 1.94	1.54 / 1.84	1.51 / 1.78	1.45 / 1.70	1.42 / 1.65	1.38 / 1.57	1.35 / 1.52	1.32 / 1.49
100	3.94 / 6.90	3.09 / 4.82	2.70 / 3.98	2.46 / 3.51	2.30 / 3.20	2.19 / 2.99	2.10 / 2.82	2.03 / 2.69	1.97 / 2.59	1.92 / 2.51	1.88 / 2.43	1.85 / 2.36	1.79 / 2.26	1.75 / 2.19	1.68 / 2.06	1.63 / 1.98	1.57 / 1.89	1.51 / 1.79	1.48 / 1.73	1.42 / 1.64	1.39 / 1.59	1.34 / 1.51	1.30 / 1.46	1.28 / 1.43
125	3.92 / 6.84	3.07 / 4.78	2.68 / 3.94	2.44 / 3.47	2.29 / 3.17	2.17 / 2.95	2.08 / 2.79	2.01 / 2.65	1.95 / 2.56	1.90 / 2.47	1.86 / 2.40	1.83 / 2.33	1.77 / 2.23	1.72 / 2.15	1.65 / 2.03	1.60 / 1.94	1.55 / 1.85	1.49 / 1.75	1.45 / 1.68	1.39 / 1.59	1.36 / 1.54	1.31 / 1.46	1.27 / 1.40	1.25 / 1.37
150	3.91 / 6.81	3.06 / 4.75	2.67 / 3.91	2.43 / 3.44	2.27 / 3.14	2.16 / 2.92	2.07 / 2.76	2.00 / 2.62	1.94 / 2.53	1.89 / 2.44	1.85 / 2.37	1.82 / 2.30	1.76 / 2.20	1.71 / 2.12	1.64 / 2.00	1.59 / 1.91	1.54 / 1.83	1.47 / 1.72	1.44 / 1.66	1.37 / 1.56	1.34 / 1.51	1.29 / 1.43	1.25 / 1.37	1.22 / 1.33
200	3.89 / 6.76	3.04 / 4.71	2.65 / 3.88	2.41 / 3.41	2.26 / 3.11	2.14 / 2.90	2.05 / 2.73	1.98 / 2.60	1.92 / 2.50	1.87 / 2.41	1.83 / 2.34	1.80 / 2.28	1.74 / 2.17	1.69 / 2.09	1.62 / 1.97	1.57 / 1.88	1.52 / 1.79	1.45 / 1.69	1.42 / 1.62	1.35 / 1.53	1.32 / 1.48	1.26 / 1.39	1.22 / 1.33	1.19 / 1.28
400	3.86 / 6.70	3.02 / 4.66	2.62 / 3.83	2.39 / 3.36	2.23 / 3.06	2.12 / 2.85	2.03 / 2.69	1.96 / 2.55	1.90 / 2.46	1.85 / 2.37	1.81 / 2.29	1.78 / 2.23	1.72 / 2.12	1.67 / 2.04	1.60 / 1.92	1.54 / 1.84	1.49 / 1.74	1.42 / 1.64	1.38 / 1.57	1.32 / 1.47	1.28 / 1.42	1.22 / 1.32	1.16 / 1.24	1.13 / 1.19
1000	3.85 / 6.66	3.00 / 4.62	2.61 / 3.80	2.38 / 3.34	2.22 / 3.04	2.10 / 2.82	2.02 / 2.66	1.95 / 2.53	1.89 / 2.43	1.84 / 2.34	1.80 / 2.26	1.76 / 2.20	1.70 / 2.09	1.65 / 2.01	1.58 / 1.89	1.53 / 1.81	1.47 / 1.71	1.41 / 1.61	1.36 / 1.54	1.30 / 1.44	1.26 / 1.38	1.19 / 1.28	1.13 / 1.19	1.08 / 1.11
∞	3.84 / 6.64	2.99 / 4.60	2.60 / 3.78	2.37 / 3.32	2.21 / 3.02	2.09 / 2.80	2.01 / 2.64	1.94 / 2.51	1.88 / 2.41	1.83 / 2.32	1.79 / 2.24	1.75 / 2.18	1.69 / 2.07	1.64 / 1.99	1.57 / 1.87	1.52 / 1.79	1.46 / 1.69	1.40 / 1.59	1.35 / 1.52	1.28 / 1.41	1.24 / 1.36	1.17 / 1.25	1.11 / 1.15	1.00 / 1.00

Source: Reprinted with permission from George W. Snedecor and William G. Cochran, *Statistical Methods*, 6th ed., © 1967, by the Iowa State University Press, Ames, Iowa.

Durbin-Watson Statistic

Significance Points of d_L and d_U : 5%

n	$k'=1$ d_L	d_U	$k'=2$ d_L	d_U	$k'=3$ d_L	d_U	$k'=4$ d_L	d_U	$k'=5$ d_L	d_U
15	1.08	1.36	0.95	1.54	0.82	1.75	0.69	1.97	0.56	2.21
16	1.10	1.37	0.98	1.54	0.86	1.73	0.74	1.93	0.62	2.15
17	1.13	1.38	1.02	1.54	0.90	1.71	0.78	1.90	0.67	2.10
18	1.16	1.39	1.05	1.53	0.93	1.69	0.82	1.87	0.71	2.06
19	1.18	1.40	1.08	1.53	0.97	1.68	0.86	1.85	0.75	2.02
20	1.20	1.41	1.10	1.54	1.00	1.68	0.90	1.83	0.79	1.99
21	1.22	1.42	1.13	1.54	1.03	1.67	0.93	1.81	0.83	1.96
22	1.24	1.43	1.15	1.54	1.05	1.66	0.96	1.80	0.86	1.94
23	1.26	1.44	1.17	1.54	1.08	1.66	0.99	1.79	0.90	1.92
24	1.27	1.45	1.19	1.55	1.10	1.66	1.01	1.78	0.93	1.90
25	1.29	1.45	1.21	1.55	1.12	1.66	1.04	1.77	0.95	1.89
26	1.30	1.46	1.22	1.55	1.14	1.65	1.06	1.76	0.98	1.88
27	1.32	1.47	1.24	1.56	1.16	1.65	1.08	1.76	1.01	1.86
28	1.33	1.48	1.26	1.56	1.18	1.65	1.10	1.75	1.03	1.85
29	1.34	1.48	1.27	1.56	1.20	1.65	1.12	1.74	1.05	1.84
30	1.35	1.49	1.28	1.57	1.21	1.65	1.14	1.74	1.07	1.83
31	1.36	1.50	1.30	1.57	1.23	1.65	1.16	1.74	1.09	1.83
32	1.37	1.50	1.31	1.57	1.24	1.65	1.18	1.73	1.11	1.82
33	1.38	1.51	1.32	1.58	1.26	1.65	1.19	1.73	1.13	1.81
34	1.39	1.51	1.33	1.58	1.27	1.65	1.21	1.73	1.15	1.81
35	1.40	1.52	1.34	1.58	1.28	1.65	1.22	1.73	1.16	1.80
40	1.44	1.54	1.39	1.60	1.34	1.66	1.29	1.72	1.23	1.79
45	1.48	1.57	1.43	1.62	1.38	1.67	1.34	1.72	1.29	1.78
50	1.50	1.59	1.46	1.63	1.42	1.67	1.38	1.72	1.34	1.77
55	1.53	1.60	1.49	1.64	1.45	1.68	1.41	1.72	1.38	1.77
60	1.55	1.62	1.51	1.65	1.48	1.69	1.44	1.73	1.41	1.77
65	1.57	1.63	1.54	1.66	1.50	1.70	1.47	1.73	1.44	1.77
70	1.58	1.64	1.55	1.67	1.52	1.70	1.49	1.74	1.46	1.77
75	1.60	1.65	1.57	1.68	1.54	1.71	1.51	1.74	1.49	1.77
80	1.61	1.66	1.59	1.69	1.56	1.72	1.53	1.74	1.51	1.77
85	1.62	1.67	1.60	1.70	1.57	1.72	1.55	1.75	1.52	1.77
90	1.63	1.68	1.61	1.70	1.59	1.73	1.57	1.75	1.54	1.78
95	1.64	1.69	1.62	1.71	1.60	1.73	1.58	1.75	1.56	1.78
100	1.65	1.69	1.63	1.72	1.61	1.74	1.59	1.76	1.57	1.78

Significance Points of d_L and d_U : 1%

n	$k'=1$ d_L	d_U	$k'=2$ d_L	d_U	$k'=3$ d_L	d_U	$k'=4$ d_L	d_U	$k'=5$ d_L	d_U
15	0.81	1.07	0.70	1.25	0.59	1.46	0.49	1.70	0.39	1.96
16	0.84	1.09	0.74	1.25	0.63	1.44	0.53	1.66	0.44	1.90
17	0.87	1.10	0.77	1.25	0.67	1.43	0.57	1.63	0.48	1.85
18	0.90	1.12	0.80	1.26	0.71	1.42	0.61	1.60	0.52	1.80
19	0.93	1.13	0.83	1.26	0.74	1.41	0.65	1.58	0.56	1.77
20	0.95	1.15	0.86	1.27	0.77	1.41	0.68	1.57	0.60	1.74
21	0.97	1.16	0.89	1.27	0.80	1.41	0.72	1.55	0.63	1.71
22	1.00	1.17	0.91	1.28	0.83	1.40	0.75	1.54	0.66	1.69
23	1.02	1.19	0.94	1.29	0.86	1.40	0.77	1.53	0.70	1.67
24	1.04	1.20	0.96	1.30	0.88	1.41	0.80	1.53	0.72	1.66
25	1.05	1.21	0.98	1.30	0.90	1.41	0.83	1.52	0.75	1.65
26	1.07	1.22	1.00	1.31	0.93	1.41	0.85	1.52	0.78	1.64
27	1.09	1.23	1.02	1.32	0.95	1.41	0.88	1.51	0.81	1.63
28	1.10	1.24	1.04	1.32	0.97	1.41	0.90	1.51	0.83	1.62
29	1.12	1.25	1.05	1.33	0.99	1.42	0.92	1.51	0.85	1.61
30	1.13	1.26	1.07	1.34	1.01	1.42	0.94	1.51	0.88	1.61
31	1.15	1.27	1.08	1.34	1.02	1.42	0.96	1.51	0.90	1.60
32	1.16	1.28	1.10	1.35	1.04	1.43	0.98	1.51	0.92	1.60
33	1.17	1.29	1.11	1.36	1.05	1.43	1.00	1.51	0.94	1.59
34	1.18	1.30	1.13	1.36	1.07	1.43	1.01	1.51	0.95	1.59
35	1.19	1.31	1.14	1.37	1.08	1.44	1.03	1.51	0.97	1.59
40	1.25	1.34	1.20	1.40	1.15	1.46	1.10	1.52	1.05	1.58
45	1.29	1.38	1.24	1.42	1.20	1.48	1.16	1.53	1.11	1.58
50	1.32	1.40	1.28	1.45	1.24	1.49	1.20	1.54	1.16	1.59
55	1.36	1.43	1.32	1.47	1.28	1.51	1.25	1.55	1.21	1.59
60	1.38	1.45	1.35	1.48	1.32	1.52	1.28	1.56	1.25	1.60
65	1.41	1.47	1.38	1.50	1.35	1.53	1.31	1.57	1.28	1.61
70	1.43	1.49	1.40	1.52	1.37	1.55	1.34	1.58	1.31	1.61
75	1.45	1.50	1.42	1.53	1.39	1.56	1.37	1.59	1.34	1.62
80	1.47	1.52	1.44	1.54	1.42	1.57	1.39	1.60	1.36	1.62
85	1.48	1.53	1.46	1.55	1.43	1.58	1.41	1.60	1.39	1.63
90	1.50	1.54	1.47	1.56	1.45	1.59	1.43	1.61	1.41	1.64
95	1.51	1.55	1.49	1.57	1.47	1.60	1.45	1.62	1.42	1.64
100	1.52	1.56	1.50	1.58	1.48	1.60	1.46	1.63	1.44	1.65

Note: k' = number of explanatory variables excluding the constant term.

Source: J. Durbin and G. S. Watson, "Testing for Serial Correlation in Least Squares Regression," *Biometrika,* **38**, 159–177 (1951). Reprinted with the permission of the author and the *Biometrika* trustees.

APPENDIX 9

Wilcoxon *W*

Wilcoxon Signed Rank Test: Left- and Right-Tail Critical Values

n	Two-Tail Test Probability: 0.2 / One-Tail Test Probability: 0.1	0.1 / 0.05	0.05 / 0.025	0.02 / 0.01	0.01 / 0.005
4	1, 9	0, 10	0, 10	0, 10	0, 10
5	3, 12	1, 14	1, 15	0, 15	0, 15
6	4, 17	3, 18	1, 20	0, 21	0, 21
7	6, 22	4, 24	3, 25	1, 27	0, 28
8	9, 27	6, 30	4, 32	2, 34	1, 35
9	11, 34	9, 36	6, 39	4, 41	2, 43
10	15, 40	11, 44	9, 46	6, 49	4, 51
11	18, 48	14, 52	11, 55	8, 58	6, 60
12	22, 56	18, 60	14, 64	10, 68	8, 70
13	27, 64	22, 69	18, 73	13, 78	10, 81
14	32, 73	26, 79	22, 83	16, 89	13, 92
15	37, 83	31, 89	26, 94	20, 100	16, 104
16	43, 93	36, 100	30, 106	24, 112	20, 116
17	49, 104	42, 111	35, 118	28, 125	24, 129
18	56, 115	48, 123	41, 130	33, 138	28, 143
19	63, 127	54, 136	47, 143	38, 152	33, 157
20	70, 140	61, 149	53, 157	44, 166	38, 172

Source: R. L. McCormack, "Extended Tables of the Wilcoxon Matched Pairs Signed Rank Statistics." *J. Am. Stat. Assoc.* **60** (1965), pp. 864–871.

For larger sample sizes, standard normal tables can be used for the test statistic

$$z = \frac{W - \dfrac{n(n+1)}{4}}{\sqrt{\dfrac{n(n+1)(2n+1)}{24}}}$$

Wilcoxon Signed Rank Test: Left- and Right-Tail Critical Values (Two Sample Test) 5% and *10%* significance levels (2.5% and *5%* for One-Tail Test, n_1 is the smaller sample)

	$n_1 = 3$	4	5	6	7	8	9	10
$n_2 = 3$	5, 16	6, 18	6, 21	7, 23	7, 26	8, 28	8, 31	9, 33
	6, 15	*7, 17*	*7, 20*	*8, 22*	*9, 24*	*9, 27*	*10, 20*	*11, 31*
4	6, 18	11, 25	12, 28	12, 32	13, 35	14, 38	15, 41	16, 44
	7, 17	*12, 24*	*13, 27*	*14, 30*	*15, 33*	*16, 36*	*17, 39*	*18, 42*
5	6, 21	12, 28	18, 37	19, 41	20, 45	21, 49	22, 53	24, 56
	7, 20	*13, 27*	*19, 36*	*20, 40*	*22, 43*	*24, 46*	*25, 50*	*26, 54*
6	7, 23	12, 32	19, 41	26, 52	28, 56	29, 61	31, 65	32, 70
	8, 22	*14, 30*	*20, 40*	*28, 50*	*30, 54*	*32, 58*	*33, 63*	*35, 67*
7	7, 26	13, 35	20, 45	28, 56	37, 68	39, 73	41, 78	43, 83
	9, 24	*15, 33*	*22, 43*	*30, 54*	*39, 66*	*41, 71*	*43, 76*	*46, 80*
8	8, 28	14, 38	21, 49	29, 61	39, 73	49, 87	51, 93	54, 98
	9, 27	*16, 36*	*24, 46*	*32, 58*	*41, 71*	*52, 84*	*54, 90*	*57, 95*
9	8, 31	15, 41	22, 53	31, 65	41, 78	51, 93	63, 108	66, 114
	10, 29	*17, 39*	*25, 50*	*33, 63*	*43, 76*	*54, 90*	*66, 105*	*69, 111*
10	9, 33	16, 44	24, 56	32, 70	43, 83	54, 98	66, 114	79, 131
	11, 31	*18, 42*	*26, 54*	*35, 67*	*46, 80*	*57, 95*	*69, 111*	*83, 127*

Source: F. Wilcoxon and R. A. Wilcox. *Some Approximate Statistical Procedures*, American Cyanamid Company, 1964.

For larger sample sizes, standard normal tables can be used for the test statistic:

$$z = \frac{W - \dfrac{n_1(n+1)}{2}}{\sqrt{\dfrac{n_1 n_2(n+1)}{12}}}$$

Kolmogorov–Smirnov Critical Values

Kolmogorov-Smirnov Critical Values for Various Significance Levels

n	0.1	0.05	0.01
1	0.950	0.975	0.995
2	0.776	0.842	0.929
3	0.642	0.708	0.828
4	0.564	0.624	0.733
5	0.510	0.565	0.669
6	0.470	0.521	0.618
7	0.438	0.486	0.577
8	0.411	0.457	0.543
9	0.388	0.432	0.514
10	0.368	0.410	0.490
11	0.352	0.391	0.468
12	0.338	0.375	0.45
13	0.325	0.361	0.433
14	0.314	0.349	0.418
15	0.304	0.338	0.404
16	0.295	0.328	0.392
17	0.286	0.318	0.381
18	0.278	0.309	0.371
19	0.272	0.301	0.363
20	0.264	0.294	0.356
25	0.24	0.27	0.32
30	0.22	0.24	0.29
35	0.21	0.23	0.27
> 35	$\dfrac{1.22}{\sqrt{n}}$	$\dfrac{1.36}{\sqrt{n}}$	$\dfrac{1.63}{\sqrt{n}}$

Source: F. J. Massey, Jr., "Kolmogorov-Smirnov Test for Goodness-of-Fit," *J. Am. Stat. Assoc.* **46** (1951), pp 68–78.

ADF Critical Values

**Augmented Dickey-Fuller (ADF) Test Left-Hand Critical Values (t test)
and Right-Hand Critical Values (F Test): 5% Level of Significance**

n	No Intercept, No Trend	Intercept, No Trend	Intercept, Trend	F Statistic
25	−2.26	−3.33	−3.95	7.24
50	−2.25	−3.22	−3.80	6.73
100	−2.24	−3.17	−3.73	6.49
250	−2.23	−3.14	−3.69	6.34
500	−2.23	−3.13	−3.68	6.30
∞	−2.23	−3.12	−3.66	6.25

Source: W. A. Fuller, *Introduction to Statistical Time Series*, Wiley, New York, 1976; D. A. Dickey and W. A. Fuller, "Likelihood Ratio Statistics for Autoregressive Time Series with a Unit Root," *Econometrica* **49** (1981), pp. 1057–1072.

Data Sources on the Web

The following are selected data sources on the Web used in this text*.

Sachs and Warner Openess Dates
http://www.nuff.ox.ac.uk/Economics/Growth/datasets/sachs/sachs.htm

World Bank Data and Current World Development Indicators
http://www.worldbank.org/data/

St. Louis Federal Reserve, Economic Time-Series Data Base
http://www.stls.frb.org/fred/

Bureau of Labor Statistics
http://www.bls.gov/

Federal Reserve Board of Governors
http://www.federalreserve.gov/releases/

Statistical Abstract of the United States
http://www.census.gov/prod/www/statistical-abstract-us.html

Economic Report of the President
http://www.gpo.ucop.edu/catalog/

Penn-World Tables
http://cansim.epas.utoronto.ca:5680/pwt/pwt.html

NASA Goddard Institute for Space Studies
http://www.giss.nasa.gov/data/update/gistemp/station_data/

New York Stock Exchange
http://www.nyse.com

Yahoo.com Stock Quotes
http://quote.yahoo.com

*Since Websites often change, we will keep an updated list on the textbook Website.

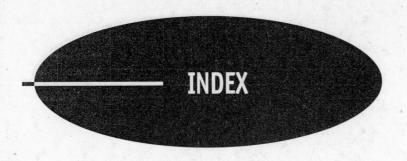

INDEX